T0309981

Linear and Convex Optimization

Linear and Convex Optimization

A Mathematical Approach

Michael H. Veatch
Gordon College

This edition first published 2021
© 2021 by John Wiley and Sons, Inc.

The right of Michael H. Veatch to be identified as the author of this work has been asserted in accordance with law.

Registered Office
John Wiley & Sons, Inc., 111 River Street, Hoboken, NJ 07030, USA

Editorial Office
111 River Street, Hoboken, NJ 07030, USA

For details of our global editorial offices, customer services, and more information about Wiley products visit us at www.wiley.com.

Wiley also publishes its books in a variety of electronic formats and by print-on-demand. Some content that appears in standard print versions of this book may not be available in other formats.

Library of Congress Cataloging-in-Publication Data

Names: Veatch, Michael H., author. | John Wiley and Sons, Inc., publisher.
Title: Linear and convex optimization : a mathematical approach / Michael Veatch, Gordon College.
Description: Hoboken, NJ : Wiley, 2021. | Includes index.
Identifiers: LCCN 2020025965 (print) | LCCN 2020025966 (ebook) | ISBN 9781119664048 (cloth) | ISBN 9781119664024 (adobe pdf) | ISBN 9781119664055 (epub)
Subjects: LCSH: Mathematical optimization. | Nonlinear programming. | Convex functions.
Classification: LCC QA402.5 .V395 2021 (print) | LCC QA402.5 (ebook) | DDC 519.6–dc23
LC record available at https://lccn.loc.gov/2020025965
LC ebook record available at https://lccn.loc.gov/2020025966

Cover Design: Wiley
Cover Image: © Hybrid_Graphics/Shutterstock

Set in 9.5/12.5pt STIXTwoText by SPi Global, Chennai, India

SKY10023218_121420

To Dad, who introduced me to operations research, and Christian and Jackie, who were always curious how things worked

Contents

Preface

This book is about optimization problems that arise in the field of operations research, including linear optimization (continuous and discrete) and convex programming. Linear programming plays a central role because these problems can be solved very efficiently; it also has useful connections with discrete and convex optimization. Convex optimization is not included in many books at this level. However, in the past three decades new algorithms and many new applications have increased interest in convex optimization. Like linear programming, large applied problems can now be solved efficiently and reliably. Conceptually, convex programming fits better with linear programming than with general nonlinear programming.

These types of optimization are also appropriate for this book because they have a clear theory and unifying mathematical principles, much of which is included. The approach taken has three emphases.

1. Modeling is covered through a broad range of applications to fields such as on-line marketing and inventory management, retail pricing, humanitarian response and rural development, public sector planning, health delivery, finance, manufacturing, service systems, and transportation. Many of these tell the story of successful applications of operations research.

2. A mathematical approach is used to communicate in a concise, unified manner. Matrix notation is introduced early and used extensively. Questions of correctness are not glossed over; the mathematical issues are described and, where the level is appropriate, proofs presented. Connections are made with some other topics in undergraduate mathematics. This approach grew out of my 30 years of teaching these topics to undergraduate mathematics students.

3. The reasoning behind algorithms is presented. Rather than introducing algorithms as arbitrary procedures, whenever possible reasons are given for why one might design such an algorithm. Enough analysis of algorithms is presented to give a basic understanding of complexity of algorithms and what makes an algorithm efficient. Algorithmic thinking is taught, not assumed.

Many introductory operations research textbooks emphasize models and algorithms without justifications and without making use of mathematical language; such books are ill-suited for mathematics students. On the other hand, texts that treat the subject more mathematically tend to be too advanced and detailed, at the expense of applications. This book seeks a middle ground.

The intended audience is junior and senior mathematics students in a course on optimization or (deterministic) operations research. The background required is a good knowledge of linear algebra and, in a few places, some calculus. These are reviewed in the appendix. The coverage and approach is intentionally kept at an undergraduate level. Material is often organized by depth, so that more advanced topics or approaches appear at the end of sections and chapters and are not needed for continuity. For example, the many ways to speed up the simplex method are saved for the last section of Chapter 9.

In keeping with this audience, the number of different problem types and algorithms is kept to a minimum, emphasizing instead unified approaches and more general problems. In particular, heuristic algorithms are only mentioned briefly. They are used for hard problems and use many different approaches, while this book focuses on problems that have efficient algorithms or at least unified approaches. The goal is to introduce students to optimization, not to be a thorough reference, and to appeal to students who are curious about other uses of mathematics. The many applications in the early chapters make the case that optimization is useful. The latter chapters connect the solution of these problems to the linear algebra and other mathematics that this audience is familiar with.

The book is also specifically written for the instructor who is mathematically trained, not for a specialist in operations research and optimization. The careful treatment of algorithmic thinking and the introduction to complexity of algorithms are intended to assist these instructors. The mathematical style throughout the book should be more accommodating to mathematics professors. It is also intended to support learning objectives more likely to be found in a mathematics department, including why the algorithms are correct and how they use theoretical results such as convexity and duality. Being able to perform an algorithm by hand is not a primary objective; it plays a supporting role to understanding the notation and reasoning of the algorithm. Calculations that are well-understood by mathematics students, such as solving a linear system or performing row operations, are not elaborated on. The somewhat more advanced material at the end of sections or chapters is also intended to support instructors who are not specialists, allowing them to extend their knowledge and explore the literature.

Chapters 1–4 are devoted to introducing optimization and optimization modeling. Convex models appear later with the other material on convex optimization. In my experience teaching mathematics students, they find modeling challenging. These chapters assume technology is available to solve problems, so that the focus can stay on formulation, as well as interpreting solutions. They build steadily in sophistication, starting with numerical instances but soon moving to algebraic formulations to make clear the distinction between model structure and data. The features of the models also build, particularly when using logical variables. In contrast with the business case study approach, each model has a limited number of features and focuses on some novel feature. I have found that mathematics students relate better to a succession of simpler models, from which they learn different modeling principles, than to a long case study.

Chapters 5–8 discuss iterative algorithms, giving some easily explained examples from discrete optimization, and the theoretical background for linear programming. This includes a little computational complexity, convexity and the study of polyhedra, optimality conditions for linear programming, and duality theory for linear programming. It is unorthodox to cover all of these before introducing the simplex method. However, conceptually these topics fit together and do not depend on the simplex method; putting them together emphasizes this fact. Chapter 8 on duality is independent of Chapter 9 on the simplex method, so that they can be covered them in either order. I typically skip the topics in Sections 5.3, 7.5, 8.4, and 8.5 to arrive at the simplex method about the middle of the semester.

Chapters 9–11 present the simplex method, including sensitivity analysis, and other algorithms for linear programming. A developmental approach is taken to presenting the simplex method. Starting with geometric reasoning about why it works, it is presented in the "naive" form first, where the so-called inverse basis matrix is computed from scratch each iteration. While this form is computationally inefficient, it is very easy to explain to a mathematics student, both for computing and justifying that the algorithm works. When working examples, technology can be used to invert and multiply matrices. After completing the picture of why the method works (degeneracy, two-phase simplex), Section 9.4 takes up the issue of making the simplex method more efficient, including the tableau form and revised simplex method. Instructors who wish to start with the tableau can use the material found here. Chapter 10 on sensitivity analysis, which depends Chapter 8, can be skipped without loss of continuity; however, the interpretation of dual variables as shadow prices and partial derivatives is enriching, even in an age when sensitivity analysis can be done quickly by solving modified linear programs. An interpretation of strong duality in terms of average costs is also given in Section 10.3. Chapter 11 presents three more algorithms for linear programming, all of which rely on duality: the dual simplex, transportation simplex, and a primal-dual, or path following, interior point method. The transportation simplex method is presented first as a minimum cost flow problem, then specialized to transportation problems.

Chapters 12 and 13 present integer programming algorithms. These relate to the earlier material because of the importance of linear programming to establish bounds when solving an integer program. Integer programming also has greater modeling power, as demonstrated by the many applications in Chapter 4. Chapters 14 and 15 introduce convex programming, including some motivating applications. The earlier chapters are designed to prepare the reader to understand convex programming more readily. The KKT optimality conditions and duality theorems are a generalization of Lagrangian duality (Section 8.4). Necessary and sufficient conditions for a global optimum follow from convexity theory, already applied to linear programs in Chapter 6. Chapter 15 culminates in the primal-dual interior point method, which was presented for linear programs in Section 11.3. Quadratic programming is also introduced and the connection between the primal-dual interior point method and sequential quadratic programming is made.

Supplemental material will be available at the web site www.gordon.edu/michaelveatch/optimization for the book. A full solution manual will be made available to instructors.

The book contains the amount of material covered in a typical two-semester sequence of undergraduate classes. A semester course focusing on linear programming could cover Chapters 1, 2, Sections 3.1–3.2, 5, 6, Sections 7.1–7.4 and 8.1–8.3, 9, 10 plus some other topics from these chapters and Chapter 11. A course on linear and integer programming could cover Chapters 1, 2, Sections 3.1–3.2, 4, 5, 6, Sections 7.1–7.4 and 8.1–8.3, 9, 12, and 13. A somewhat more advanced course on linear and convex programming could cover Chapters 1–3, 5–7.4, 8, 9, Sections 11.1–11.3, 14, and 15.

Several more advanced or specialized topics have been included at the end of chapters or sections that are optional and can be easily skipped. Section 3.3 shows that a dynamic program can be solved as a linear program, an approach that relates to machine learning. Section 5.3 on computational complexity, while not difficult, is only occasional mentioned in the later chapters. Section 7.5 extends the optimality conditions needed to solve linear programs to general polyhedra. Section 8.4 introduces Lagrangian duality for linear programs and shows that it is equivalent to the usual dual; it is only needed if convex programming (Chapter 14) is being covered. Farkas' lemma is presented in Section 8.5, providing another approach to duality theorems. The computational strategies in Section 9.4 are important to the simplex method but are not used in the sequel. The interior point algorithm in Section 11.4 is computationally more involved. It is closely related to Section 15.5.

I want to express my deep appreciation to the many people who helped make this book possible. First, I want to thank David Rader, Larry Leemis, and Susan Martonosi for encouraging me to undertake the project. I am grateful to my former and current students Isaac Bleecker, Mackenzie Hinds, Joe Iriana, Stephen Rizzo, and Michael Yee for reviewing portions of the draft. I also thank my friend

John Sanderson for drawing the figures, my colleague Jonathan Senning for his technical advice, and students Isaac Bleecker, Jessica Guan, Seth McKinney, and Yi Zhou for their help with the exercises and figures.

Most of all, I am grateful for my wife Cindy's confidence in me and acceptance of my working odd hours on the project. Now we are both authors.

Wenham, MA   Michael H. Veatch
March, 2020

About the Companion Website

This book is accompanied by a companion website:

www.wiley.com/go/veatch/convexandlinearoptimization

The website includes the instructor solutions manual.

1

Introduction to Optimization Modeling

1.1 Who Uses Optimization?

Data-driven decision-making is on the rise. Many businesses, governmental organizations, and nonprofits collect large amounts of data and seek to use them to inform the decisions they make. In addition to data and statistical analysis, a mathematical model of the system is often needed to find the best or even a viable option. Examples include planning the flow of goods through a supply chain, scheduling personnel shifts, or choosing an investment portfolio. The approach of optimization is to develop a model describing which potential decisions could be taken, in light of physical, logical, financial, or other limitations, and assess them using a single objective. This objective could be profit or loss in a business setting, expected return on an investment, a physical measure such as minimizing time to complete a task, a health measure such as expected lives saved, or in the government and nonprofit sector, a measure of social welfare. The mathematical model is an optimization problem; the study of them is called optimization.

Once a system has been modeled, an algorithm is required to find the best, or nearly best, decision. Fortunately, there are general algorithms that can be used to solve broad classes of problems. Algorithms for several classes of problems will be the main topic of this book.

Optimization is a very broad concept. There are minimization principles in physics, such as Snell's law of diffraction, or surface tension minimizing the area of a surface. In statistics, the least squares principle is an easily-solved minimization problem while new techniques involve challenging optimization problems to fit models to data. Control engineers use minimization principles to develop feedback controls for systems such as autopilots and robots. Many techniques in machine learning can also be described as optimization problems.

Linear and Convex Optimization: A Mathematical Approach, First Edition. Michael H. Veatch.
© 2021 John Wiley & Sons, Inc. Published 2021 by John Wiley & Sons, Inc.
Companion website: www.wiley.com/go/veatch/convexandlinearoptimization

This book focuses on optimization to support decisions in businesses, government, and organizations. The models used in these applications tend to have many interrelated variables, require significant amounts of data, and often have a complex structure unique to the application. The field that addresses these problems is called *operations research*. The field had its origins during World War II, when British military leaders asked scientists to work on military problems, such as the deployment of radars and the management of antisubmarine operations. These efforts were called military operations research; after the war similar techniques were applied to industry. Operations research employs mathematical models, not all of which use optimization. The same approach and mathematical methods are used in engineering problems, but we focus on decision-making applications. Today, operations research is a profession, an approach to applied problems, and a collection of mathematical techniques. It is sometimes called "the science of better" as it seeks to increase value and efficiency through better decision-making.

Operations research is sometimes also called management science or operational research. Similar techniques are used in industrial engineering and operations engineering, with a somewhat narrower view of applications. There is a strong overlap with business analytics, which refers to any use of data to improve business processes, whether statistical methods or optimization.

While the use of optimization models for decision-making has become common, most of the decisions to which they are applied are not fully automated; rather, the models provide guidelines or insights to managers for making decisions. For some problems, models can be fairly precise and identify good decisions. For many others, the model is enough of an approximation that the goal of modeling is insights, not numbers. We only briefly cover the modeling process, at the beginning of Chapter 2. However, the examples we present give some appreciation for how models can be useful for decisions.

A prime example of the use of optimization models is the airline industry (Barnhart et al., 2003). Starting in the late 1960s with American Airlines and United Airlines, flight schedules, routing, and crew assignment in the airline industry were based on large-scale, discrete optimization models. By about 1990, airlines started using revenue management models to dynamically adjust prices and use overbooking, generating significant additional revenues. Optimization models have also been used more recently for air traffic flow management.

Optimization has also been of great value for e-business and the sharing economy. For example, the Motivate bikeshare systems in New York City, Chicago, and San Francisco use optimization to reallocate the docking stations throughout the city (Freund et al., 2019). First they estimate the user dissatisfaction function, due to not having a bike or docking station available for potential users, as a function of the docking capacity and initial inventory at each location. Then a large optimization problem is solved to find the allocation of docking capacity that minimizes this user dissatisfaction. Optimization was also used in the design of an incentive scheme to crowdsource rebalancing, where users are given incentives to take trips that move bikes to locations where they are needed. The next section presents a question arising in disaster response to introduce the basic concepts of optimization modeling.

1.2   Sending Aid to a Disaster

International responses to rapid-onset disasters are frequent and expensive. After a major disaster, such as the 2014 earthquake in Nepal, food, shelter, and other items are needed quickly in large quantities. Local supplies may be insufficient, in which case airlift to a nearby airport is an important part of a timely response. The number of flights in the first few days may be limited by cost and other factors, so the aid organizations seek to send priority items first.

To illustrate the decisions involved, consider an organization that is sending tents and nonperishable food from their stockpile to people whose homes were destroyed in a hurricane. Because they are delivering to a small airport, they are using a medium-sized cargo aircraft, the C-130. It has space for six pallets and a payload capacity of 40 000 lb for this length of flight. In the next week or so, the responding organizations will try to fly in enough supplies to meet all the high-priority needs. But the first flights will be loaded before the needs assessment is even complete, just knowing that a large number of people need food and shelter. The organization has estimated the criticality of each item on a 10-point scale (10 being the most critical). They are based on the deprivation likely to be experienced by recipients if the item is not delivered until later. However, for many reasons, not all aid meets high-priority needs, so the organization also considers the cost (market value) of the aid, and defines expected benefit to be the average of criticality and cost. Their objective is to load the aircraft to have the greatest expected benefit. Table 1.1 lists the weight, cost, criticality, and expected benefit for both items. They have time to re-pack one pallet with a mixture of tents and food, so the number of pallets of each item does not have to be an integer. However, only five pallets

Table 1.1   Data for sending aid.

	Tents	Food
Weight (1000 lbs)	7.5	5
Cost ($1000/pallet)	11	4
Criticality	5	8
Expected benefit	8	6

of food are immediately available – they do not stockpile large quantities of food because it eventually expires.

The first idea one might have is to load six pallets of tents, the item with the largest expected benefit. However, this load is not allowed because it exceeds the weight limit. Further, since the number of pallets does not have to be an integer, trying other loads may not find the best one. Instead, we formulate the problem mathematically. Let

$$x = \text{pallets of tents loaded,}$$
$$y = \text{pallets of food loaded.}$$

Then the expected value of aid loaded is

$$f(x, y) = 8x + 6y \text{ (expected value of aid)}$$

and we want to maximize f over a domain that contains the possible values of (x, y). First, we have the logical restrictions $x, y \geq 0$. The weight limit requires that

$$7.5x + 5y \leq 40 \text{ (weight).}$$

In this expression, 7.5 is the weight per pallet of tents, so $7.5x$ is the weight of tents. Similarly, $5y$ is the weight of food. The left-hand side, then, is the total weight of the load, which must be less than or equal to the payload capacity of 40 (these quantities are in 1000s of lbs). The space limit requires that

$$x + y \leq 6 \text{ (pallets).}$$

The left-hand side is the total number of pallets. This total does not have to be an integer; a total of 5.4 pallets would mean that one pallet is only loaded 40% full. Finally, only five pallets of food are ready, so

$$y \leq 5 \text{ (food available).}$$

These inequalities define the domain of $f(x, y)$. We will call them *constraints* and the function f to be maximized the *objective function*. Optimizing a function whose domain is defined by constraints is a *constrained optimization* problem. The complete problem is

$$
\begin{aligned}
\max \quad & 8x + 6y \\
\text{s.t.} \quad & 7.5x + 5y \leq 40 \\
& x + y \leq 6 \\
& y \leq 5 \\
& x, y \geq 0.
\end{aligned}
\tag{1.1}
$$

We have abbreviated "subject to" as "s.t."

Constrained optimization problems have three components:

Components of an Optimization Problem

1. The decision variables.
2. The objective function to be maximized or minimized, as a function of the decision variables.
3. The constraints that the decision variables are subject to.

The formulation of this problem consists of the Eqs. (1.1) and the definitions of the variables. It is essential to define the variables and also good practice to label or describe the objective function and each constraint. This problem is an example of a *linear program* because the variables are continuous and the objective function and all constraints are linear.

Now that we have formulated the loading decisions as a linear program, how do we solve it? For x and y to satisfy the constraints, they must lie in the intersection of the half planes defined by these inequalities, shown in Figure 1.1. Most linear inequalities can be conveniently graphed by finding the x and y intercept of the corresponding equation, drawing a line between them, and checking a point not on the line, such as $(0, 0)$, to see if it satisfies the inequality. If the point satisfies the inequality, then the half-plane is on the same side of the line as that point; if the point does not satisfy the inequality, the half-plane is on the other side of the line as that point. For the first constraint (weight), the x intercept is $5\frac{1}{3}$, the y

Figure 1.1 Region satisfying constraints for sending aid.

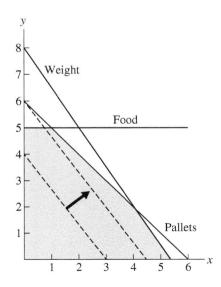

intercept is 8, and (0,0) is on the correct side of the line. Other constraints, such as $y \leq 5$, have horizontal or vertical boundary lines. Once all of the constraints have been graphed, we can identify the region (or possibly a line or a point) satisfying all the constraints. We are seeking the point in this region that has the largest value of f. One way to find this point graphically is to plot contour lines $f(x, y) = k$ for one or two values of k. For example, in Figure 1.1 the contours

$$8x + 6y = 24,$$
$$8x + 6y = 36$$

are graphed as dashed lines. They both intersect the shaded region, so there are points satisfying the constraints with objective function values of 24 and 36. Since all contour lines are parallel, we can visualize sweeping the line up and to the right without rotating it to find larger objective function values. The farthest contour line that touches the shaded region is shown in Figure 1.2. The point where it touches the region has the largest objective function value. This point lies on the constraint lines for weight and pallets, so we can find it by solving these two constraints with equality in place of inequality:

$$7.5x + 5y = 40,$$
$$x + y = 6,$$

with solution $(x, y) = (4, 2)$ and objective function value 44. This agrees with Figure 1.2, where the contour line drawn is $f(x, y) = 44$. Thus, the optimal load is four pallets of tents and two pallets of food, with an expected value of 44.

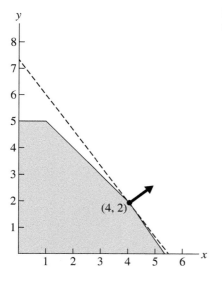

Figure 1.2 Optimal point and contour for sending aid.

To summarize, to solve a linear program with two variables graphically:

Solving a Linear Program Graphically

1. Graph the region that satisfies all of the constraints.
2. Draw a contour line of the objective function and determine which direction improves the objective function. A second contour line can be used to determine which direction is improving.
3. Sweep the contour line in the improving direction, generating parallel lines, until it is about to leave the region. The last line intersecting the region has the optimal objective function value. The point(s) where this line intersects the region is optimal.
4. Find the coordinates of an optimal point by solving a system of two linear equations from the constraints whose lines pass through this point.

If the set of points satisfying the constraints is nonempty and bounded, then an optimal solution exists and the method earlier will find it. The other cases are discussed in Section 1.3.

Viewing the problem graphically brings up several questions about linear programs that will be answered in later chapters. First, the optimal value occurred at a corner point of the region, where two (or more) constraint lines intersect. Will this always be the case? Second, the idea of sweeping the contour line until it leaves the region assumes that contour lines $f(x, y) = k$ intersect points satisfying the constraints for k in a single interval. Thus, an optimal objective function value is one where lines with slightly larger k do not touch the region. Is it possible for the contour line to leave the region and reenter it? That is, for $k^* < k_1 < k_2$ if the line for k^* touches the region and the line for k_1 does not, it possible for the line for k_2 to touch the region? If so, we would have to do more checking to find the optimal k^*. Related to this is the question of local maxima. The point $(4, 2)$ is called a *local maximum* as it has the largest objective function value of all points in a small circle around $(4, 2)$. It is easier to check that it is a local maximum than examining all points that satisfy the constraints to check that it is a global maximum. Is it always true that a local maximum is also a global maximum? Finally, the region in Figure 1.2 is defined by the constraints, but the optimal point depends on the objective function. For example, if instead we wish to maximize

$$C(x, y) = 11x + 4y \text{ (cost of aid, \$1000s)}, \tag{1.2}$$

the optimal load is $x = 5\frac{1}{3}$ pallets of tents and $y = 0$ pallets of food, with a cost of \$58 667. However, for small changes in the coefficients of f, the point $(4, 2)$ is still optimal. How much can an objective function coefficient be changed without changing the optimal point?

The fractional solution $(5\frac{1}{3}, 0)$ makes sense in this problem: one pallet is re-packed with $\frac{1}{3}$ as much food. In other problems, fractional solutions may not make sense. They can either be interpreted as approximate solutions that must be converted to integers to be implemented, or a different optimization problem can be studied, where the variables are discrete, only taking on integer values.

1.2.1 The Modeling Process

As we move from this small linear program to more general and complex models, some guidelines on modeling will be helpful. When formulating a mathematical program, it is not always obvious what the decision variables should be. There are often multiple approaches to defining the variables and there can be pros and cons to different approaches. The objective is intended to reflect the priorities of the decision-maker. To pose the situation as a mathematical program, an objective must be chosen. This may be very clear, as when maximizing profit, or much more subtle, as in the example previously. The constraints may include physical, logical, financial, or other limitations. They may also reflect policies that the decision-maker applies. Finally, a constraint may simply be an equation that defines an auxiliary variable that is used to write the problem more concisely.

Another theme that will emerge is the difference between specific and general models. We will often use *algebraic notation* for (some of) the constants in a model, rather than numbers. This allows us to separate the data that goes into a model from its structure. It is this structure that defines the general model. There are multiple levels of generality to a model used in an application. One way to generalize is to allow any size of the problem: instead of two items being loaded, there could be n items. To make a model more specific, we often add constraints with a special structure to describe a specific situation.

The examples in this book take a verbal description of an optimization problem and translate it into mathematics. There is much more involved in using models to guide decisions. The overall modeling process may involve:

1. *Problem definition.* The first step is to identify the problem that the decision-maker faces. This includes specifying the objective and the system to be studied.
2. *Data collection.* The next step is to collect data that is relevant to the model. This may entail observing the system or obtaining data that is already recorded.
3. *Model formulation.* In this step the analyst formulates a mathematical model of the decision problem. It involves translation into mathematical language, but also abstraction or simplification of a complex operation into a manageable problem.

4. *Model solution.* For optimization problems, this step finds an optimal or near-optimal solution. Generally this is done numerically using an algorithm, rather than algebraically.
5. *Verification and validation.* This step tries to determine if the model is an adequate representation of the decision problem. Verification involves checking for internal consistency in that the mathematical model agrees with its specification or assumptions. Validation seeks to build confidence that the model is sufficiently accurate. For optimization problems, accuracy may be gauged by whether the optimal solution produced by the model is reasonable or whether the objective function is accurate over a variety of feasible solutions. Validation methods that have been developed for statistical or simulation models might also be used.
6. *Implementation.* Once an optimal solution is obtained from the model, it may need to be modified. For example, fractional values may need to be converted to integers. The result may be one or more near-optimal solutions that are presented to the decision-maker. In other situations, the model may provide insights that help the decision-maker with their decisions, or a system might be implemented so that the model is run repeatedly with new data to support future decisions.

These steps are not just done in this order. The verification and validation process, feedback obtained about the model, and changing situations often lead to multiple cycles of revising the model. When applications are presented in this book, we focus on Step 3, model formulation, and mostly on the translation process, not the simplifying assumptions. Whether a model is useful, however, depends on the other steps. It is important to ask "Where does the data come from?" as in Albright and Winston (2005). But it is also necessary to ask "Where does the model come from?," that is, what are the model assumptions and why were they made?

1.3 Optimization Terminology

Now we introduce more terminology for constrained optimization problems, as well as summarizing the terms introduced in the last section. These problems are called *mathematical programs*. The variables that we optimize over are the *decision variables*. They may be *continuous variables*, meaning that they can take on values in the real numbers, including fractions, or they may *discrete variables* that can only take on integer values. The function being optimized is the *objective function*. The domain over which the function is optimized is the set of points that satisfy the constraints, which can be equations or inequalities. We distinguish two kinds of constraints. *Functional constraints* can have any form, while *variable bounds* restrict the allowable values of one variable. Many problems have nonnegativity constraints as their variable bounds. In (1.1), the food constraint $y \leq 5$ is listed with the functional constraints, making three functional constraints plus two nonnegativity constraints (variable bounds), but it is also a bound, so the problem can also be said to have two functional constraints and three bounds.

Let $\mathbf{x} = (x_1, \ldots, x_n)$ be the decision variables and $S = \{\mathbf{x} \in \mathbb{R}^n : \mathbf{x}$ satisfies the constraints$\}$ be the solution set of the constraints. Departing somewhat from the usual meaning of a "solution" in mathematics, we make the following definitions

A **solution** to a mathematical program is any value $\mathbf{x}$ of the variables.
A **feasible solution** is a solution that satisfies all constraints, i.e. $\mathbf{x} \in S$.
The **feasible region** is the set of feasible solutions, i.e. S.
The **value** of a solution is its objective function value.
An **optimal solution** is a feasible solution that has a value at least as large (if maximizing) as any other feasible solution.

To solve a mathematical program $\max_{x \in S} f(\mathbf{x})$ is to find an optimal solution $\mathbf{x}^*$ and the optimal value $f(\mathbf{x}^*)$, or determine that none exists. We can classify each mathematical program into one of three categories.

Existence of Optimal Solutions   When solving a mathematical program:

The problem has one or more **optimal solutions** (we include in this category having solutions whose value is arbitrarily close to optimal, but this distinction is rarely needed), or
The problem is an **unbounded problem**, meaning that, for a maximization, there is a feasible solution $\mathbf{x}$ with value $f(\mathbf{x}) > k$ for every k, or
The problem is **infeasible**: there are no feasible solutions, i.e. $S = \emptyset$.

An unbounded problem, then, is one whose objective function is unbounded on S (unbounded above for maximization and below for minimization). One can easily specify constraints for a decision problem that contradict each other and have no feasible solution, so infeasible problems are commonplace in applications. Any realistic problem could have bounds placed on the variables, leading to a bounded feasible region and a bounded problem. In practice, these bounds are often omitted, leaving the feasible region unbounded. A problem with an unbounded feasible region will still have an optimal solution for certain objective functions. Unbounded *problems*, on the other hand, indicate an error in the formulation. It should not be possible to make an infinite profit, for example.

Figure 1.3 Problem has optimal solution for dashed objective but is unbounded for dotted objective.

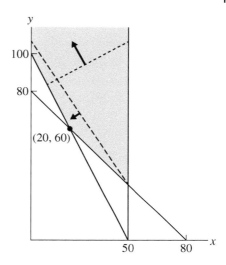

Example 1.1 (Unbounded region). Consider the linear program

$$
\begin{aligned}
\min \quad & 3x + 2y \\
\text{s.t.} \quad & 2x + y \geq 100 \\
& x + y \geq 80 \\
& x \quad\quad \leq 50 \\
& x, y \geq 0.
\end{aligned}
\tag{1.3}
$$

The feasible region is shown in Figure 1.3; the dashed line has an objective function value of 210. Sweeping this line down, the minimum value occurs at the corner point $(20, 60)$ with an optimal value of 180. The feasible region is unbounded. Now suppose the objective function is changed to $x - 2y$. The dotted line shows the contour line $x - 2y = -160$; sweeping the line upward decreases the objective without bound, showing that the problem is unbounded.

1.4 Classes of Mathematical Programs

In Section 1.2 we introduced the concept of a linear program. This section introduces algebraic and matrix notation for linear programs. It then defines the three main classes of mathematical programs: linear programs, integer programs, and nonlinear programs. Each of these will be studied in later chapters, though for nonlinear programs we will focus on the more tractable subclass of convex programs. A *mathematical program* with variables $x_1, \dots, x_n$ and objective function

$f(x_1, \ldots, x_n)$ can be stated abstractly in terms of a feasible region $S \subset \mathbb{R}^n$ as

$$\max/\min \ f(x_1, \ldots, x_n)$$
$$\text{s.t.} \ (x_1, \ldots, x_n) \in S.$$

However, we will always describe the feasible region using constraint equations or inequalities. The notation for the constraints is introduced in the following text.

1.4.1 Linear Programs

We have already seen two examples of linear programs.

General Form of a Linear Program

$$\max/\min \ \sum_{j=1}^{n} c_j x_j$$

$$\text{s.t.} \ \sum_{j=1}^{n} a_{ij} x_j \begin{pmatrix} \leq \\ \geq \\ = \end{pmatrix} b_i, \quad i = 1, \ldots, m$$

$$x_j \begin{pmatrix} \geq 0 \\ \leq 0 \\ \text{u.r.s.} \end{pmatrix}, \quad j = 1, \ldots, n.$$

There are n decision variables, $x_1, \ldots, x_n$ and m functional constraints. The constraints can use a mixture of "$\leq$", "$\geq$", and "$=$". Each variable may have the bound $x_j \geq 0$, $x_j \leq 0$, or no bound, which we call unrestricted in sign (u.r.s.). The distinguishing characteristics of a linear program are (i) the objective function and all constraints are linear functions and (ii) the variables are continuous, i.e. fractional values are allowed. They are often useful as approximate models even when these assumptions do not fully hold.

We will use matrix notation for linear programs whenever possible. Let $\mathbf{b} = (b_1, \ldots, b_m)$, $\mathbf{c} = (c_1, \ldots, c_n)$, $\mathbf{x} = (x_1, \ldots, x_n)$, and

$$\mathbf{A} = \begin{bmatrix} a_{11} & \cdots & a_{1n} \\ \vdots & & \vdots \\ a_{m1} & \cdots & a_{mn} \end{bmatrix}.$$

Here $\mathbf{b}$, $\mathbf{c}$, and $\mathbf{x}$ are column vectors. If all the constraints are equalities, they can be written $\mathbf{Ax} = \mathbf{b}$. Similarly, "$\geq$" constraints can be written $\mathbf{Ax} \geq \mathbf{b}$.

Example 1.2. Consider the linear program

$$\begin{array}{rl}
\min & 9x_1 + 12x_2 + 6x_3 \\
\text{s.t.} & 3x_1 + x_2 - 2x_3 \geq 4 \\
& x_1 + 2x_2 + 4x_3 \geq 6 \\
& x_j \geq 0.
\end{array} \tag{1.4}$$

Converting the objective function and constraints to matrix form, we have

$$\min \quad \begin{bmatrix} 9 & 12 & 6 \end{bmatrix} \begin{bmatrix} x_1 \\ x_2 \\ x_3 \end{bmatrix}$$

$$\text{s.t.} \quad \begin{bmatrix} 3 & 1 & -2 \\ 1 & 2 & 4 \end{bmatrix} \begin{bmatrix} x_1 \\ x_2 \\ x_3 \end{bmatrix} \geq \begin{bmatrix} 4 \\ 6 \end{bmatrix}$$

$$x_j \geq 0.$$

If we let

$$\mathbf{A} = \begin{bmatrix} 3 & 1 & -2 \\ 1 & 2 & 4 \end{bmatrix}, \quad \mathbf{b} = \begin{bmatrix} 4 \\ 6 \end{bmatrix}, \quad \mathbf{c} = \begin{bmatrix} 9 \\ 12 \\ 6 \end{bmatrix}, \quad \mathbf{x} = \begin{bmatrix} x_1 \\ x_2 \\ x_3 \end{bmatrix}$$

then this linear program can be written

$$\begin{array}{rl}
\min & \mathbf{c}^T \mathbf{x} \\
\text{s.t.} & \mathbf{A}\mathbf{x} \geq \mathbf{b} \\
& \mathbf{x} \geq \mathbf{0}.
\end{array}$$

It is important to distinguish between the structure of an optimization problem and the data that provides a specific numerical example. The structure of (1.4) is a minimization linear program with "≥" constraints and nonnegative variables; the data are the values in **A**, **b**, and **c**.

To write a mixture of "≤" and "≥" constraints, it is convenient to use submatrices

$$\mathbf{A} = \begin{bmatrix} \mathbf{A}_1 \\ \mathbf{A}_2 \end{bmatrix}, \quad \mathbf{b} = \begin{bmatrix} \mathbf{b}_1 \\ \mathbf{b}_2 \end{bmatrix}$$

and write, e.g.

$$\mathbf{A}_1 \mathbf{x} \leq \mathbf{b}_1,$$
$$\mathbf{A}_2 \mathbf{x} \geq \mathbf{b}_2.$$

1.4.2 Integer Programs

Many practical optimization problems involve making discrete quantitative choices, such as how many fire trucks of each type a fire department should purchase, or logical choices, such as whether or not each drug being developed by a pharmaceutical company should be chosen for a clinical trial. Both types of situations can be modeled by integer variables.

Consider again the linear program (1.1) with the alternative objective function (1.2). The variables represent the number of pallets of tents and food. If we restrict the variables to be integers, i.e. we can only load whole pallets, then the problem becomes an *integer program* and can be stated

$$
\begin{aligned}
\max \quad & 11x + 4y \\
\text{s.t.} \quad & 7.5x + 5y \leq 40 \\
& x + y \leq 6 \\
& y \leq 5 \\
& x, y \geq 0, \text{integer.}
\end{aligned}
\tag{1.5}
$$

Without the integer restriction, it is a linear program with optimal solution $(5\frac{1}{3}, 0)$ and optimal value $58\frac{2}{3}$. However, because this solution is not integer (i.e. not all of its variables have integer values), it is not feasible for the integer program. Only solutions that have integer values and satisfy the constraints are feasible, as shown in Figure 1.4. Once we have determined the feasible points, an optimal solution can be found graphically by sweeping the contour line $11x + 4y = k$ in

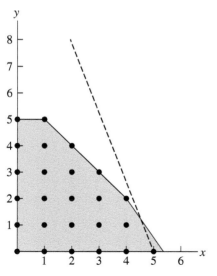

Figure 1.4 Feasible integer solutions for (1.5).

the improving direction (upward). The last point touched by the line is optimal. In this case the last point is $(5, 0)$ and the optimal value is $k = 55$.

Although the graphical method was fairly simple for this example, it is quite different than when solving linear programs. Note that:

- There is not necessarily an optimal solution at the intersection of two constraints. In our example, $(5, 0)$ lies only on the constraint line $y = 0$. In other examples, the optimal solution may not lie on any constraint.
- The optimal solution is not necessarily obtained by rounding the linear program optimal solution. In fact, there is no limit to how far the optimal solution could be from the linear programming solution.
- The integer program can be infeasible when the linear program is feasible.

These difficulties suggest that integer programs are harder to solve than linear programs, which we will see is true. Even solving a two-variable integer program graphically can be tedious. However, we do not necessarily need to generate all integer feasible solutions. For example, if we start by generating the feasible point $(4, 2)$, then we draw the contour line through it and check for integer points in the linear program's feasible region and above the contour. Checking points to the right of $x = 4$ (because this contour does not touch the feasible region for $x < 4$), $(5, 1)$ is infeasible but $(5, 0)$ qualifies, so it is optimal. The idea of using a feasible integer solution to eliminate other integer points from consideration will be used in Chapter 13.

The general form of an integer program is the same as a linear program with the added constraint that the variables are integers. If only some of the variables are restricted to be integer, it is called a *mixed integer program*. When variables represent logical choices, they are usually defined so that $x_j = 1$ if true (decide to take the action) and $x_j = 0$ if false (decide not to take the action). A variable whose possible values are $0, 1$ is called a *binary variable*. An integer program with all binary variables is called a *binary program*.

General Form of an Integer Program

$$\max/\min \ \sum_{j=1}^{n} c_j x_j$$

$$\text{s.t.} \ \sum_{j=1}^{n} a_{ij} x_j \begin{pmatrix} \leq \\ \geq \\ = \end{pmatrix} b_i, \quad i = 1, \ldots, m$$

$$x_j \begin{pmatrix} \geq 0 \\ \leq 0 \\ \text{u.r.s.} \end{pmatrix}, \text{ integer}, \quad j = 1, \ldots, n.$$

1.4.3 Nonlinear Programs

An optimization problem where the objective function and constraints may be nonlinear is called a *nonlinear program*. The variables are assumed to be continuous, as in a linear program. Thus, nonlinear programs are more general than linear programs, but do not include integer programs. While linear programs can be described in matrix notation, nonlinear programs are described in terms of functions.

General Form of a Nonlinear Program

$$\max/\min \ f(x_1, \dots, x_n)$$

$$\text{s.t. } g_i(x_1, \dots, x_n) \begin{pmatrix} \leq \\ \geq \\ = \end{pmatrix} b_i, \quad i = 1, \dots, m.$$

We have not stated the nonnegativity constraints separately. However, a nonnegativity constraint can be included in the functional constraints if needed.

The ability to solve a nonlinear program depends on the type of objective function f and constraints g_i. A class of more easily solved nonlinear programs called *convex programs* is discussed in Chapters 14 and 15. Also, Chapter 3 shows how certain piece-wise linear convex programs can be converted into linear programs, which is useful because convex programs are somewhat harder to solve than linear programs.

Problems

For Exercises 1–6, solve the linear program graphically.

1
$$\begin{aligned}
\max \ & 2x + y \\
\text{s.t.} \ & x + y \leq 3 \\
& 4x + y \leq 9 \\
& x, y \geq 0.
\end{aligned}$$

2
$$\begin{aligned}
\max \ & 0.5x + y \\
\text{s.t.} \ & 3x + y \leq 7 \\
& 2x - y \leq 4 \\
& x, y \geq 0.
\end{aligned}$$

3
$$\max\ 30x + 15y$$
$$\text{s.t.}\quad x + 2y \le 200$$
$$2x + y \le 200$$
$$4x + y \le 380$$
$$x, y \ge 0.$$

4 a) Solve as stated. b) Change "min" to "max" and solve.
$$\min\ 2x + y$$
$$\text{s.t.}\quad x - 0.5y \ge 0$$
$$0.75x + y \ge 275$$
$$0.5x - y \le 0$$
$$x, y \ge 0.$$

5
$$\max\ x + y$$
$$\text{s.t.}\quad x + 3y \le 9$$
$$-2x + y \le -5$$
$$x + 2y \ge 10$$
$$x, y \ge 0.$$

For Exercises 6–8, solve the integer program graphically.

6
$$\max\ 3x + 4y$$
$$\text{s.t.}\quad -x + y \le 1$$
$$3x + 2y \le 14$$
$$x, y \ge 0, \text{integer.}$$

7
$$\max\ 3x + y$$
$$\text{s.t.}\quad 4x - y \le 0$$
$$2x + y \le 9$$
$$x + y \le 8$$
$$x, y \ge 0, \text{integer.}$$

8
$$\max\ 5x + 4y$$
$$\text{s.t.}\quad -4x + 4y \le 12$$
$$6x + 2y \le 13$$
$$x, y \ge 0, \text{integer.}$$

9 For the linear program (1.1)
(a) Suppose the objective is to *minimize* the cost of aid given in (1.2). What is the optimal solution? Explain why minimizing cost is not a reasonable objective for this problem.
(b) Find an objective function for which $x = 1, y = 5$ is optimal. Show graphically that this point is optimal.

2

Linear Programming Models

A great variety of real-world problems can be modeled as linear programs. Many businesses and other organizations have successfully used linear programming or other optimization models to improve their operations and decision-making. This chapter presents some of the linear programming models that have been useful. We will cite specific successful applications, but will generally present smaller examples that illustrate the approach.

2.1   Resource Allocation

Many problems involve deciding how much of each product to produce or sell when the products compete for resources. Here is an example of such a *product mix* problem.

Example 2.1 (Kan Jam).   Wild Sports produces the Kan Jam yard game, which consists of a flying disc and two plastic cans used as goals. In addition to the Ultimate Disc Game, they have begun producing a Splash Pool Disc Game. They also sell the disc individually. The primary materials used are plastic to make the goals, cardboard packaging, and the discs, which are manufactured at a separate location. Because of uncertainty in when these materials will be resupplied, they would like to make a one month production plan using the on-hand materials. The material requirements and availability are shown in Table 2.1, along with the retail price on Amazon and the profit for each product. Because the pool version is smaller, they have set its price slightly lower but it still has a larger profit margin. However, the demand for this new product is limited. The forecasted demand at this price is 2000 for the month. How many of each product should be produced to maximize profit for the month?

Linear and Convex Optimization: A Mathematical Approach, First Edition. Michael H. Veatch.
© 2021 John Wiley & Sons, Inc. Published 2021 by John Wiley & Sons, Inc.
Companion website: www.wiley.com/go/veatch/convexandlinearoptimization

Table 2.1 Data for Kan Jam production.

Material	Ultimate	Pool	Disc	Available
Plastic (oz)	48	40	0	240 000
Flying discs	1	1	1	8000
Packaging (ft^2)	10	10	2	75 000
Selling price ($)	40	38	9	—
Profit margin ($)	10	12	2	—

To formulate as a linear program, let U, P, and D denote the number of Ultimate, Pool, and separate discs produced this month, respectively. We can express profit as

$$10U + 12P + 2D.$$

Each material imposes a constraint on the production quantities. For example, the total amount of plastic used must be less than or equal to the amount of plastic available, so

$$48U + 40P \leq 240\,000.$$

Because the objective only considers this month's sales, there is no reason to produce more of a product than can be sold this month. This creates one more constraint, $P \leq 2000$. Finally, production quantities must be nonnegative. Putting these together, the linear program is

$$
\begin{array}{llll}
\max & 10U + 12P + 2D \\
\text{s.t.} & 48U + 40P & \leq 240\,000 & \text{(Plastic)} \\
& U + P + D & \leq 8000 & \text{(Flying discs)} \\
& 10U + 10P + 2D & \leq 75\,000 & \text{(Packaging)} \\
& P & \leq 2000 & \text{(Demand for Pool)} \\
& U, P, D \geq & 0.
\end{array}
$$

The optimal solution is to produce $3333\frac{1}{3}$ Ultimate games, 2000 Pool games, and $2666\frac{2}{3}$ separate discs, for a profit of $62 666.67. When we round the solution to 3333 and 2667, the profit is $62 664 so the change is negligible.

To illustrate another kind of constraint, suppose that, because of the uncertainty of demand quantities, Wild Sports preferred to describe demand for the Pool game as a percentage of sales. They might wish to limit Pool game production to 50% of the production of Ultimate games. This constraint can be stated as $P \leq 0.5U$ or

$$-0.5U + P \leq 0.$$

Replacing $P \le 2000$ with this constraint, the new linear program has an optimal solution (rounded) of 3529 Ultimate games, 1765 Pool games, and 2706 separate discs, for a profit of $61 882.

Careful cost accounting is needed. The profit margin should include all variable costs, including the cost of materials listed in the table, other material, and labor. Unused material may have a holding cost assigned when it is carried into the next month, or may lose value. Labor and equipment may be a fixed or variable cost depending on the time frame and other considerations. Here is a product mix problem where the profit margin must be calculated.

Example 2.2 (Auto parts). A division of an auto parts manufacturer makes four parts, each of which requires time on each of three machines. The division is charged for the time that it uses on the machines. The time required, cost, time available each month, and selling price for batches of 100 parts are shown in Table 2.2. Ten batches of each part must be produced each month to fulfill a contract with a car manufacturer. How many of each product should be produced each month to maximize revenue for the month?

First we compute the profit margins:

Profit per batch of Part $1 = 570 - [2(160) + 4(10) + 6(15)] = 120.$

Similar computations give profits of 40, 80, and 60 for Parts 2, 3, and 4, respectively. Let x_j be the batches of Part j produced, $j = 1, 2, 3, 4$. The linear program to maximize profit is

$$
\begin{aligned}
\max \quad & 120x_1 + 40x_2 + 80x_3 + 60x_4 \\
\text{s.t.} \quad & 2x_1 + x_2 + 3x_3 + 2x_4 \le 200 \text{ (drilling)} \\
& 4x_1 + 2x_2 + x_3 + 2x_4 \le 300 \text{ (polishing)} \\
& 6x_1 + 2x_2 + x_3 + 2x_4 \le 500 \text{ (finishing)} \\
& x_j \ge 10.
\end{aligned}
$$

The optimal solution is $\mathbf{x} = (61, 10, 16, 10)$, i.e. produce 61 batches of Part 1, etc. with a profit of $9600.

Table 2.2 Data for auto parts production.

Machine	Part 1	Part 2	Part 3	Part 4	Hours	Cost ($/h)
Drilling	2	1	3	2	200	160
Polishing	4	2	1	2	300	10
Finishing	6	2	1	2	500	15
Price ($/100)	570	250	585	430	—	—

Any linear program in the form

$$\max \sum_{j=1}^{n} c_j x_j$$

$$\text{s.t.} \sum_{j=1}^{n} a_{ij} x_j \leq b_i, \quad i = 1, \ldots, m \tag{2.1}$$

$$x_j \geq 0, \quad j = 1, \ldots, n$$

can be considered a *resource allocation* problem that chooses levels of *activities* to maximize profit. We interpret x_j as the level of activity j, b_i as the amount of resource i available, and a_{ij} as the amount of resource i consumed by one unit of activity j. Some examples of resource allocation are:

- Investment portfolios. Activities are the different types of investment and resources can include funds available, a risk budget, limitations on exposure to certain markets, etc.
- Glass pane production and cutting at Libbey–Owens–Ford, where various tints of glass are produced and then cut into different sizes (Martin et al., 1993). Each size/tint combination is an activity. They develop a large linear programming model with many additional constraints to plan their production.
- Choosing tenants at a shopping mall (Bean et al., 1988). Each type of store (e.g. women's clothing) is an activity. The resources include floor space and cost of building the stores.

2.1.1 Production Process Models

In the examples so far, activities are outputs that generate profit. In production process models, the output from one activity can be used as an input to another. For example, four hours of labor may be needed to produce a unit of product 1, then two hours of labor and 0.5 units of product 1 might be needed to produce a unit of product 2. In addition to the variables x_1 and x_2 for the amount of product 1 and 2 produced, let y_1 be the units of product 1 sold. Then the constraint

$$y_1 = x_1 - 0.5x_2$$

holds. Profit is calculated using y_1 (sales), not x_1 (production). In general, production process models have constraints of the form

$$\text{(units of product } j \text{ sold)} = \text{(units of product } j \text{ produced)}$$
$$- \text{(units of product } j \text{ used in other products)},$$

which can be written in matrix form as

$$\mathbf{y} = \mathbf{x} - \mathbf{A}\mathbf{x} = (\mathbf{I} - \mathbf{A})\mathbf{x}, \tag{2.2}$$

$$\mathbf{y} \geq \mathbf{0},$$

where $\mathbf{A} = [a_{ij}]$ is a square matrix and a_{ij} is the amount of product i consumed to produce one unit of product j.

A production process model typically has resources, with constraints in the form (2.1), and products, with the process constraints (2.2). However, one can collect the resources and products into a single set of activities. In this view a resource is just an activity that is not sold, so it does not appear in the objective function. Then all of the constraints can be written in the form (2.2), using a larger matrix $\mathbf{A}$ to include the resources, plus bounds $\mathbf{x} \leq \mathbf{b}$ to limit the amount of a resource used. In this form, (2.2) is known in economics as the *Leontief input–output model*, where the activities are thought of as sectors of the economy instead of activities of a single firm. In the economics context, x_j is the total production of sector j and y_j is the portion of that production that goes to consumers – not toward other production. For example, in the agriculture sector, x_j might be the tons of grain produced nationally in a year. Then y_j would be the tons of grain used for food, with the rest being used for animal feed or seed. The technology matrix $\mathbf{A}$ measures how efficient the economy is, with small values more efficient. Similarly, in production process models, it measures the firm's efficiency. For the example earlier, a more efficient firm might only use 0.3 units of product 1 to produce a unit of product 2.

2.2 Purchasing and Blending

This section presents examples where the decisions variables represent purchasing different items and the objective is to minimize cost.

Example 2.3 (Online ads). Custom Tees is a clothing retailer that sells high-quality T-shirts with personalized logos. They are planning an online advertising campaign with two web companies. The first sells banner and popup ad space on their web pages; Custom Tees can select the parts of web pages where their ad appears. The second is a social media company that sells ads that appear on their visitor's newsfeed; Custom Tees can select which categories of visitors will see their ads, based on various characteristics. These choices allow them to show their ads to visitors who are more likely to be interested in clicking on them and making a purchase. Both companies charge when a visitor to the site clicks on an ad (called engagement) rather than when they visit a page containing the ad (called an impression). Customers in the 18–25 age range are particularly valuable to them, so they have set goals of 500 000 clicks by visitors in this age range and 600 000 clicks from older visitors. After making their placement choices and running test ads, Custom Tees was able to estimate the percent of clicks in the 18–25 age range for each type of ad.

Another concern is that with this large of an ad campaign, repeat visitors to the web sites are likely to see many Custom Tees ads. Although they only pay when the visitor clicks, some of these clicks will be by people who have already clicked and been directed to the Custom Tees site. These repeats are not as valuable as new visitors, so they have also set goals for unique visitors at 50% of the total click goals in each age group. Forecasting unique visitors as a function of total visitors is more challenging. They recognize that this relationship is not linear, but in the absence of more detailed data they have estimated it using a linear function that assumes a certain percent of clicks are unique visitors. They have specified a maximum number of clicks for each type of ad, partly because of the concern about unique visitor clicks. Table 2.3 lists this data and the prices the companies charge per click.

To formulate these decisions as a linear program, let x_i be the number of clicks (in 1000s) on type i ads, for $i = 1$ (banner), 2 (popup), 3 (newsfeed). The total clicks from age 18–25 visitors must be at least 500 000, giving the constraint

$$0.4x_1 + 0.3x_2 + 0.7x_3 \geq 500.$$

Apply the unique click percentages, the unique clicks constraint for age 18–25 visitors is

$$0.4(0.4x_1) + 0.75(0.3x_2) + 0.9(0.7x_3) \geq 0.5(500).$$

For age 26 and over visitors, the total clicks constraint is

$$0.6x_1 + 0.7x_2 + 0.3x_3 \geq 600$$

Table 2.3   Data for Custom Tees ads.

	Banner	Popup	Newsfeed
Cost ($ per 1000 clicks)	75	100	120
Clicks from age 18–25	40%	30%	70%
Clicks from age 26 and over	60%	70%	30%
Unique clicks	40%	75%	90%
Maximum clicks (1000s)	600	300	300

and unique clicks constraint is

$$0.4(0.6x_1) + 0.75(0.7x_2) + 0.9(0.3x_3) \geq 0.5(600).$$

The additional constraints are limits on the number of clicks and nonnegativity. The linear program minimizes total cost:

$$
\begin{array}{llll}
\min & 75x_1 + & 100x_2 + & 120x_3 \\
\text{s.t.} & 0.4x_1 + & 0.3x_2 + & 0.7x_3 \geq 500 & \text{(18--25 total clicks)} \\
& 0.16x_1 + & 0.225x_2 + & 0.63x_3 \geq 250 & \text{(18--25 unique clicks)} \\
& 0.6x_1 + & 0.7x_2 + & 0.3x_3 \geq 600 & \text{(over 26 total clicks)} \\
& 0.24x_1 + & 0.525x_2 + & 0.27x_3 \geq 300 & \text{(over 26 unique clicks)} \\
& & & x_1 \leq 600 & \text{(max clicks-banner)} \\
& & & x_2 \leq 300 & \text{(max clicks-popup)} \\
& & & x_3 \leq 300 & \text{(max clicks-newsfeed)} \\
& & & x_j \geq 0.
\end{array}
$$

The optimal solution $\mathbf{x} = (600, 225, 275)$ purchases 600 000 clicks through banner ads, 225 000 through popup, and 275 000 through newsfeed for a total cost of $100 500.

Another successful application of optimization is placing television ads on different shows to reach more of a target audience. The traditional segmentation of viewers by demographics (age bracket and gender) has recently been refined by Turner Broadcasting System, Inc. (Carbajal et al., 2019). They use data on viewer characteristics and integer programs to optimize ad placement, increasing the impressions delivered to target audiences by 27%.

2.2.1 Online Advertising Markets

Online advertising is an area where the large amounts of data available have led to growing use of optimization. In Chickering and Heckerman (2003), a targeted approach to advertising is described using linear programming. They consider a web site publisher that contracts with advertisers to deliver a quota of their advertisements on its web site. Regardless of whether they charge for impressions (page views) or clicks, the publisher has an incentive to improve clicks. Thus, their problem is to decide which ad to show each visitor to maximize the number of clicks for a given number of impressions. Visitors are divided into clusters to estimate how likely they are to click on each ad. The data are (i) p_{ij}, the fraction of visitors in cluster $j = 1, \ldots, n$ that will click on ad $i = 1, \ldots, m$ when it is delivered to them, (ii) c_j, the number of visitors in cluster j, and (iii) q_i, the required number of impressions (the quota) for ad i. Let x_{ij} be the number of impressions of ad i shown to customers in cluster j. The linear program is

$$\max \quad \sum_{i=1}^{m} \sum_{j=1}^{n} p_{ij} x_{ij}$$

$$\text{s.t.} \quad \sum_{j=1}^{n} x_{ij} \geq q_i, \quad i = 1, \ldots, m \ \text{(quota for ad)}$$

$$\sum_{i=1}^{m} x_{ij} \leq c_j, \quad j = 1, \ldots, n \ \text{(visits in cluster)}$$

$$x_{ij} \geq 0.$$

The data used in the model are readily available. In Chickering and Heckerman (2003), one day of visitor data on the msn.com news website is adequate. The clusters are the types of news stories; the site had about 20 types at the time. Combined with about 500 advertisements, this gives $500 \times 20 = 10\,000$ variables. A shortcoming of the model is that most variables are zero in the optimal solution (we will see why in Section 11.2 where we call this type of model a transportation problem), which leads to visitors seeing the same advertisements repeatedly. They resolve this issue using modified linear programs to smooth the solution. These linear programs were easily solved.

More recent approaches use more granular targeting, based on a visitor's history on the site or other information about the visitor, to classify visitors into many more types (larger n). The number of ads is also much larger, due to more targeting and ad exchanges, where advertisers can place ads with many different publishers. As large m and n were needed, the linear programming approach was used for a time, with up to one million variables, but eventually became impractical. The marketplace for ads also evolved, with advertisers bidding for ad locations. As described in Chen et al. (2011), a more recent approach uses a combination of "offline" linear programming and "online" real-time policies that provide more granularity and adapt to market changes.

2.2.2 Blending Problems

One of the early uses of linear programs was to minimize cost when blending ingredients in refineries to make gasoline. Similar problems can be posed when choosing foods for diets, making metals out of alloys, or fertilizers out of chemicals. The following example is representative.

Example 2.4 (Producing steel). A company makes two types of steel out of three alloys. The steel must meet the carbon and nickel requirements in Table 2.4. The percent of carbon and aluminum in the alloys, prices, and amount of alloy

Table 2.4 Data for producing steel.

	Alloy 1	Alloy 2	Alloy 3	Steel 1	Steel 2
Percent carbon	3	4	3.5	≥ 3.6	≥ 3.4
Percent nickel	1	1.5	1.8	≤ 1.5	≤ 1.7
Price ($/ton)	380	400	440	650	600
Available (tons)	40	50	80	—	—

available are shown. Also, all steel must be at most 40% alloy 1. How much of each alloy should be used in each type of steel and how much steel should be produced to maximize profit?

The decisions x_{ij} are the tons of alloy $i = 1, 2, 3$ to use in producing steel $j = 1, 2$. Once these decisions are made, the total amount of steel is determined. However, it is convenient to define additional variables z_j for the amount steel j. The amount of a steel produced is the weight of the alloys in it, so

$$z_1 = x_{11} + x_{21} + x_{31}, \tag{2.3}$$
$$z_2 = x_{12} + x_{22} + x_{32}.$$

The profit is the revenue from the steel minus the cost of the alloys used:

$$\text{profit} = 650z_1 + 600z_2$$
$$- [380(x_{11} + x_{12}) + 400(x_{21} + x_{22}) + 440(x_{31} + x_{32})].$$

The amount of each alloy used is constrained by the amount available:

$$x_{11} + x_{12} \leq 40,$$
$$x_{21} + x_{22} \leq 50,$$
$$x_{31} + x_{32} \leq 80.$$

Now we must write the percentage constraints. The fraction of alloy 1 in steel 1 is

$$\frac{x_{11}}{x_{11} + x_{21} + x_{31}} = \frac{\text{tons of alloy 1 in steel 1}}{\text{tons of steel 1}}.$$

For steel 1 to be at most 40% alloy 1, the constraint is

$$\frac{x_{11}}{x_{11} + x_{21} + x_{31}} \leq 0.40.$$

This ratio constraint is not linear, but is easily converted to the equivalent linear constraint

$$x_{11} \leq 0.40(x_{11} + x_{21} + x_{31}).$$

Also, steel 2 must be at most 40% alloy 1:

$$x_{12} \leq 0.40(x_{12} + x_{22} + x_{32}).$$

The constraints for the percent of carbon and nickel are also ratio constraints. Assuming the percentages are by weight, the fraction of steel 1 that is carbon is

$$\frac{0.03x_{11} + 0.04x_{21} + 0.035x_{31}}{x_{11} + x_{21} + x_{31}} = \frac{\text{tons of carbon in steel 1}}{\text{tons of steel 1}},$$

so the carbon constraint for steel 1 is

$$0.03x_{11} + 0.04x_{21} + 0.035x_{31} \geq 0.036(x_{11} + x_{21} + x_{31}).$$

Adding the three remaining ratio constraints and nonnegativity, the linear program to maximize profit is

$$
\begin{aligned}
\max \quad & 650(x_{11} + x_{21} + x_{31}) + 600(x_{12} + x_{22} + x_{32}) \\
& -[380(x_{11} + x_{12}) + 400(x_{21} + x_{22}) + 440(x_{31} + x_{32})] \\
\text{s.t.} \quad & x_{11} \leq 0.40(x_{11} + x_{21} + x_{31}) \\
& x_{12} \leq 0.40(x_{12} + x_{22} + x_{32}) \\
& 0.03x_{11} + 0.04x_{21} + 0.035x_{31} \geq 0.036(x_{11} + x_{21} + x_{31}) \\
& 0.03x_{12} + 0.04x_{22} + 0.035x_{32} \geq 0.034(x_{12} + x_{22} + x_{32}) \\
& 0.01x_{11} + 0.015x_{21} + 0.018x_{31} \leq 0.015(x_{11} + x_{21} + x_{31}) \\
& 0.01x_{12} + 0.015x_{22} + 0.018x_{32} \leq 0.017(x_{11} + x_{21} + x_{31}) \\
& x_{11} + x_{12} \leq 40 \\
& x_{21} + x_{22} \leq 50 \\
& x_{31} + x_{32} \leq 80 \\
& x_{ij} \geq 0.
\end{aligned}
$$

The first two constraints are for the percent of alloy 1, the next two for carbon, then two for nickel in the two types of steel. The last three constraints are for alloy availability. The optimal solution, shown in Table 2.5, has a profit of $37 100.

Note that the linear program only contains the variables x_{ij}. If we add the variables z_j, we must also add the constraints (2.3), which define z_j as the total amount

Table 2.5 Solution for producing steel.

	Alloy 1	Alloy 2	Alloy 3	Total
Steel 1	28	50	32	110
Steel 2	12	0	48	60

of steel j. In general, *when we add auxiliary variables we must add their defining equality constraints*; however, when we substitute new variables to eliminate original variables, we do not add constraints.

A mathematical program can give unrealistic solutions if we neglect important constraints when formulating it. It is very easy to do this in blending problems. For different data, it could have been optimal to produce only one type of steel. If that is not realistic, minimum production constraints need to be added to the model. A classic example is the *diet problem*, one of the first linear programs solved by computer in the late 1940s by Dantzig (1990). The linear program finds the combination of foods that meet daily nutritional requirements at minimum cost. Without other constraints, the solutions include absurd amounts, such as two pounds of bran per day. Even when upper bounds are added, the diets are totally unappealing. However, institutions do plan menus using integer program versions of the diet problem with more constraints; see Lancaster (1992).

2.3 Workforce Scheduling

Linear programs have been used to plan work schedules in situations where many employees do the same type of work.

Example 2.5 (Police shifts). During each four-hour period, a town's police force requires the number of police on duty shown in Table 2.6. Each police officer works for two consecutive four-hour periods. The pay per shift varies slightly. For example, the shift starting at midnight costs the town $500. How many officers should be assigned to each shift to minimize the total cost?

There is a decision variable for each shift (not each time period). Let $x_{0,8}$ be the number of officers working the shift from hour 0 to 8, i.e. midnight to 8:00 a.m.,

Table 2.6 Requirements and costs for police shifts.

	Officers needed	Cost of shift starting
Midnight to 4:00 a.m.	15	$500
4:00 a.m. to 8:00 a.m.	13	$480
8:00 a.m. to noon	11	$450
Noon to 4:00 p.m.	11	$460
4:00 p.m. to 8:00 p.m.	9	$470
8:00 p.m. to midnight	7	$490

and so forth, with $x_{20,4}$ for the last shift, hour 20 to 4, i.e. 8:00 p.m. to 4:00 a.m. Two shifts cover each time period. For example, the number of officers working from 4:00 a.m. to 8:00 a.m. is

$$x_{0,8} + x_{4,12}.$$

The linear program to minimize cost is

$$\min \; 500x_{0,8} + 480x_{4,12} + 450x_{8,16}$$
$$+ \; 460x_{12,20} + 470x_{16,24} + 490x_{20,4}$$

$$
\begin{array}{llll}
\text{s.t.} & x_{20,4} + x_{0,8} & \geq 15 & \text{(midnight to 4:00 a.m.)} \\
& x_{0,8} + x_{4,12} & \geq 13 & \text{(4:00 a.m. to 8:00 a.m.)} \\
& x_{4,12} + x_{8,16} & \geq 11 & \text{(8:00 a.m. to noon)} \\
& x_{8,16} + x_{12,20} & \geq 11 & \text{(noon to 4:00 p.m.)} \\
& x_{12,20} + x_{16,24} & \geq 9 & \text{(4:00 p.m. to 8:00 p.m.)} \\
& x_{16,24} + x_{20,4} & \geq 7 & \text{(8:00 p.m. to midnight)} \\
& x_{i,j} & \geq 0. &
\end{array}
$$

The optimal solution is $x_{0,8} = 13$, $x_{4,12} = 0$, $x_{8,16} = 11$, $x_{12,20} = 4$, $x_{16,24} = 5$, $x_{20,4} = 2$ for a daily cost of \$16 620. Fractional solutions would not make sense for this problem; fortunately, the optimal solution is integral. In fact, linear programs with this particular form of constraint and integer right-hand sides always have integer optimal solutions.

The model can be extended to shifts of different lengths and assigning employees shifts for a week or longer, not just a single day. The constraints coefficient matrix has $a_{ij} = 1$ if shift j covers time period i and 0 otherwise. An important application of shift scheduling is inbound telephone call centers, where agents handle incoming calls for a business. A linear programming approach for call centers with multiple skills is described in Bhulai et al. (2008). It extends the model above by considering multiple groups of agents, each with a different set of skills. Incoming calls are assigned to agents that have the needed skill. Skills could be the language, e.g. English or Spanish, or whether the call is about billing, technical support, etc.

2.4 Multiperiod Problems

Work scheduling models consider time, but in a rather static way. Now we consider dynamic models where there are flows between time periods. An important example is multiperiod inventory.

Table 2.7 Demand and labor available for gift baskets

	Demand	Regular labor (h)	Overtime labor (h)
Week 1	700	450	40
Week 2	1500	550	200
Week 3	2800	600	320
Week 4	1800	600	160

Example 2.6 (Gift baskets). A fine foods company makes gift baskets for the holiday season and supplies them to retailers. The season only lasts four weeks. They have sufficient capacity to produce the foods, but assembling the baskets is labor intensive. The demand, hours of regular labor available, and hours of over-time labor available each week are shown in Table 2.7. The demand must be met. Regular labor costs \$30/h and overtime \$45/h. Only the labor used is charged to this product. The food and other material in a basket costs \$25. Baskets are sold for \$65. It take 0.4 hours of labor to assemble a basket. Each week, baskets are assem-bled, then used to meet the demand. Each basket that is not sold in a week incurs a \$4 holding cost. At the beginning of the four weeks there is no inventory. Unsold baskets at the end of week 4 have a salvage value of \$30 and do not incur a hold-ing cost. How much regular and overtime labor should be planned for assembling baskets each week to maximize profit?

It will be easier to write equations if the decision variables are production quan-tities, not labor hours. Let

P_t = number of baskets assembled with regular labor, week $t = 1, 2, 3, 4,$

Q_t = number of baskets assembled with overtime labor, week $t = 1, 2, 3, 4.$

They are constrained by the available labor, e.g. $P_1 \leq \frac{450}{0.4} = 1125$. From these we can compute the inventory at the end of each week. Let

I_t = number of baskets on hand at the end of week $t = 1, 2, 3, 4.$

Inventory satisfies a conservation constraint each week:

inventory this week = inventory last week

+ production this week − demand this week.

To write this constraint algebraically, also define d_t as the demand in week $t = 1, 2, 3, 4$. Then

$$I_t = I_{t-1} + P_t + Q_t - d_t. \tag{2.4}$$

Actually, the equation is different for week 1, since there is no "inventory last week." However, if we define the constant I_0 to be the initial inventory (at the beginning of week 1), then $I_0 = 0$ and (2.4) also holds for $t = 1$. Because demand must be met, inventory must be nonnegative. In some settings, demand that cannot be met from inventory is backordered and counted as negative inventory; in this case (2.4) still applies.

To compute profit, note that the quantity sold is the quantity produced minus the final inventory, for which the salvage value is received. Thus, the net revenue after subtracting the cost of material is

$$\text{net revenue} = (65 - 25)(P_1 + Q_1 + P_2 + Q_2 + P_3 + Q_3 + P_4 + Q_4) - 35I_4.$$

The other costs are regular labor, overtime labor, and holding costs:

$$\text{other cost} = 0.4(30)(P_1 + P_2 + P_3 + P_4) + 0.4(45)(Q_1 + Q_2 + Q_3 + Q_4)$$
$$+ 4(I_1 + I_2 + I_3).$$

The difference between these two is profit. The linear program to maximize profit is

$$\max \quad 28(P_1 + P_2 + P_3 + P_4) + 22(Q_1 + Q_2 + Q_3 + Q_4)$$
$$- 4(I_1 + I_2 + I_3) - 35I_4$$
$$\text{s.t.} \quad I_1 = P_1 + Q_1 - 700$$
$$I_2 = I_1 + P_1 + Q_1 - 1500$$
$$I_3 = I_2 + P_2 + Q_2 - 2800$$
$$I_4 = I_3 + P_3 + Q_3 - 1800$$
$$P_1 \le 1125, \ P_2 \le 1375, \ P_3 \le 1500, \ P_4 \le 1500$$
$$Q_1 \le 100, \ Q_2 \le 500, \ Q_3 \le 800, \ Q_4 \le 400$$
$$P_t, Q_t, I_t \ge 0.$$

The optimal solution is to assemble 1125 baskets in week 1, 1575 in week 2, 2300 in week 3, and 1800 in week 4, with a profit of $178 900. This results in an inventory of 425 in week 1, 500 in week 2, and none in weeks 3 and 4. All of the regular labor is used each week. The overtime labor used is 80 hours in week 2, 320 in week 3, and 120 in week 4.

The production schedule uses regular labor in week 1 instead of overtime labor in week 2 to meet week 2 demand; this makes sense because the savings in labor ($6 per basket) is more than the holding cost ($4 per basket). It also uses overtime labor in week 2 to meet week 3 demand; the less expensive categories of labor (all week 3 labor and week 2 regular labor) are fully used.

2.4.1 Multiperiod Financial Models

Multiperiod optimization models are widely used in finance. Most applications use probability models to deal with market uncertainty; however, one area where linear programming is used is cash flow planning. Given a portfolio of possible investments with different duration, return on investment, and interest payments during the life of the investment, how much should be invested in each at each time to maximize the final cash position? There might be other planned cash receipts and disbursements over time. The cash position at each time must be nonnegative – this is a liquidity constraint. There may be other constraints that limit the amount placed in various investments, such as a limit on total risk. For a model with three monthly investment periods where the only investments are bonds of each duration, let

r_{st} = return on bond purchased in month $s = 1, \ldots, t - 1$

   that matures in month $t = 1, \ldots, 4$,

R_t = cash inflow (receipts) in month $t = 1, 2, 3$,

D_t = cash outflow (disbursements) in month $t = 1, 2, 3$.

Define the variables

x_{st} = investment in bonds in month $s = 1, \ldots, t - 1$

   that matures in month $t = 1, \ldots, 4$,

y_t = cash position in month $t = 1, \ldots, 4$.

There are six investments: $x_{12}, x_{13}, x_{14}, x_{23}, x_{24}, x_{34}$. Investments, inflow, and outflow occur at the beginning of the month. Bonds also mature at the beginning of the month. Cash position is calculated after these transactions. The cash position in month 1 is

$$y_1 = y_0 - x_{12} - x_{13} - x_{14}, \tag{2.5}$$

where y_0 is the initial cash position, a constant. Then in month 2,

$$y_2 = y_1 + R_1 - D_1 - x_{23} - x_{24} + (1 + r_{12})x_{12}. \tag{2.6}$$

The last term is the principal and interest on the bond purchased in month 1 that matures in month 2. Similarly,

$$y_3 = y_2 + R_2 - D_2 - x_{34} + (1 + r_{13})x_{13} + (1 + r_{23})x_{23}, \tag{2.7}$$

$$y_4 = y_3 + R_3 - D_3 + (1 + r_{14})x_{14} + (1 + r_{24})x_{24} + (1 + r_{34})x_{34}. \tag{2.8}$$

The linear program to maximize the final cash position is

$$\max \ y_4$$

$$\text{s.t. } (2.5)–(2.8)$$

$$x_{st}, y_t \geq 0.$$

The constraints have the same form as inventory constraints.

2.5 Modeling Constraints

The mathematical programming framework assumes that constraints must be satisfied and that there is a single objective function to optimize. In practice, many constraints don't necessarily have to hold. Additional labor could be hired, labor could be reassigned, more material purchased, or policy guidelines modified. There is often more than a single objective. In some situations, an organization's goal of efficiency can be stated in two or more ways by swapping some constraints with the objective function. We provide several examples of how constraints might be modified. Modifying constraints will be studied in Chapter 10.

Example 2.7 (Cross-trained labor). For Example 2.2, suppose that the available polishing and finishing hours can be shared. This might be the case if the limit is on the machine operator labor hours, not the machines themselves, while the drilling hours are for the machine. Cross training the polishing and finishing operators allows them to operate either machine. For this combined department, the time required by each part is the sum of its polishing and finishing times, and the time available is the sum of the polishing and finishing time available, so we add the two constraints. The new constraint

$$10x_1 + 4x_2 + 2x_3 + 4x_4 \leq 800$$

replaces the polishing and finishing constraints. The new optimal solution is $\mathbf{x} = (70, 10, 10, 10)$ with a profit of $10\,200.

Pooling the polishing and finishing labor increased the optimal profit from $9600 to $10\,200. We know before solving that the optimal value can only improve because any feasible solution to Example 2.2, where the two departments work separately, is also feasible for the new problem.

Next we give an example of why a constraint might not have to be met. These are sometimes called *soft constraints.*

Example 2.8 (Additional labor). For Example 2.2 (without the changes in Example 2.7), suppose that additional polishing time can be purchased at $15/h and additional finishing time at $22.50/h. Up to 80 hours can be purchased in each department. Let y_2 and y_3 be the hours purchased in each department. They are added to the available hours and their cost is subtracted from the objective. The new linear program is

$$
\begin{array}{lrrrrrrl}
\max & 120x_1 + 40x_2 + 80x_3 + 60x_4 - 15y_2 - 22.5y_3 & & \\
\text{s.t.} & 2x_1 + x_2 + 3x_3 + 2x_4 & \leq 200 & \text{(drilling)} \\
& 4x_1 + 2x_2 + x_3 + 2x_4 - y_2 \leq 300 & & \text{(polishing)} \\
& 6x_1 + 2x_2 + x_3 + 2x_4 - y_3 \leq 500 & & \text{(finishing)} \\
& x_j \geq 10 & & \\
\end{array}
$$

$$0 \leq y_2 \leq 80, \ 0 \leq y_3 \leq 80.$$

The optimal solution is $\mathbf{x} = (70, 10, 10, 10)$ with a profit of $9750. It uses 30 hours of overtime in polishing ($y_2 = 30$).

The final example illustrates why the objective and a constraint might be swapped.

Example 2.9 (Online ads revisited). For Example 2.3, another way to design an efficient advertising campaign is to maximize the number of unique clicks given an advertising budget of $105 000. The required number of clicks and unique clicks in each age category must still be met and the maximum number of clicks from each type of ad still applies. What was the objective is now the left-hand side of the budget constraint. Using the unique click percentages from Table 2.3, the number of unique clicks (1000s) is

$$0.4x_1 + 0.75x_2 + 0.9x_3.$$

The linear program to maximize unique clicks is

$$
\begin{array}{lrlll}
\max & 0.4x_1 + 0.75x_2 + 0.9x_3 & & \\
\text{s.t.} & 75x_1 + 100x_2 + 120x_3 & \leq 105\,000 & \text{(budget)} \\
& 0.4x_1 + 0.3x_2 + 0.7x_3 & \geq 500 & \text{(18–25 total clicks)} \\
& 0.16x_1 + 0.225x_2 + 0.63x_3 & \geq 250 & \text{(18–25 unique clicks)} \\
& 0.6x_1 + 0.7x_2 + 0.3x_3 & \geq 600 & \text{(over 26 total clicks)} \\
& 0.24x_1 + 0.525x_2 + 0.27x_3 & \geq 300 & \text{(over 26 unique clicks)} \\
& x_1 & \leq 600 & \text{(max clicks-banner)} \\
& x_2 & \leq 300 & \text{(max clicks-popup)} \\
& x_3 & \leq 300 & \text{(max clicks-newsfeed)} \\
& x_j & \geq 0. & \\
\end{array}
$$

The optimal solution $\mathbf{x} = (520, \ 300, \ 300)$ purchases 520 000 clicks through banner ads, 300 000 through popup, and 300 000 through newsfeed, with 703 000 unique clicks.

Because we kept all the constraints from Example 2.3, which has an optimal cost of $100 500, this linear program is infeasible for budgets of less than that. Because the budget we used is only slightly larger, the feasible region is fairly small and the two linear programs give similar solutions.

For comparison, consider the calculus problem of maximizing the area of a rectangle with a fixed perimeter. Interchanging the objective and the constraint, one can minimize the perimeter of a rectangle with a fixed area. The solution to both problems is a square, so, for corresponding values of the constants, the two problems have the same solution. In the advertising problem, there is a not an exact equivalence because there are multiple constraints.

2.6 Network Flow

An important class of linear programs are called *minimum cost network flow problems*. They have a network structure and material flows through the network to meet requirements. The flow could be physical goods, moving through a supply chain, electricity in a distribution grid, a financial transaction, or an information flow. Such models are widely used. There are very efficient algorithms for solving this type of linear program; one is presented in Section 11.2. First we consider a model called the transportation problem, which has a specific network structure.

2.6.1 Transportation Problems

In a transportation problem, material moves directly from its current locations, called supply points, to destinations, called demand points. The following example is based on Aarvik and Randolph (1975).

Example 2.10 (Electricity transmission). A small electric utility has three electric power plants that provide power to two cities. The capacity of each plant and the peak demand of each city, which must be met at the same time, are listed in Table 2.8. Any plant can send electricity to any city, as shown in Figure 2.1, but the transmission costs per million kwh differ depending on the distance. How much should be sent to each city to minimize transmission cost?

To formulate as a linear program, define the decision variables

$$x_{ij} = \text{electricity (million kwh) sent from plant } i = 1, 2, 3 \text{ to city } j = 1, 2.$$

Figure 2.1 Electricity transmission network.

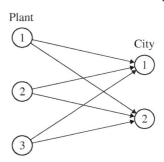

Table 2.8 Transmission costs, supply, and demand.

	Transmission costs ($/million kwh)		Supply (M kwh)
	City 1	City 2	
Plant 1	14	22	30
Plant 2	18	12	25
Plant 3	10	16	45
Demand (M kwh)	40	60	—

The total transmission cost is expressed as

$$14x_{11} + 22x_{12} + 18x_{21} + 12x_{22} + 10x_{31} + 16x_{32}.$$

There are two types of constraints. First, the electricity supplied by each plant cannot exceed the capacity. For example, the power supplied by Plant 1 is

$$\sum_{j=1}^{2} x_{1j} = x_{11} + x_{12}$$

which cannot exceed 30. The three supply constraints are

$$x_{11} + x_{12} \leq 30,$$
$$x_{21} + x_{22} \leq 25,$$
$$x_{31} + x_{32} \leq 45.$$

Second, the electricity received by each city must meet its peak demand. The power received by City 1 is

$$\sum_{i=1}^{3} x_{i1} = x_{11} + x_{21} + x_{31}.$$

The two demand constraints are

$$x_{11} + x_{21} + x_{31} \geq 40,$$
$$x_{12} + x_{22} + x_{32} \geq 60.$$

We assume that electricity cannot be sent from a city to a plant, so $x_{ij} \geq 0$ for $i = 1, 2, 3$ and $j = 1, 2$. Putting these together, the linear program is

$$\min \ 14x_{11} + 22x_{12} + 18x_{21} + 12x_{22} + 10x_{31} + 16x_{32}$$

$$\text{s.t.} \qquad\qquad\qquad\qquad x_{11} + x_{12} \leq 30 \ \ (\text{Plant 1})$$
$$x_{21} + x_{22} \leq 25 \ \ (\text{Plant 2})$$
$$x_{31} + x_{32} \leq 45 \ \ (\text{Plant 3})$$
$$x_{11} + x_{21} + x_{31} \geq 40 \ \ (\text{City 1})$$
$$x_{12} + x_{22} + x_{32} \geq 60 \ \ (\text{City 2})$$
$$x_{ij} \geq 0.$$

The optimal solution, found in Section 11.2, is $x_{11} = 30$, $x_{22} = 25$, $x_{31} = 10$, and $x_{32} = 35$ with a cost of \$1380 (all other variables are zero).

To generalize, we use the following notation:

---

m = number of supply points
s_i = amount that can be supplied by supply point $i = 1, \dots, m$
n = number of demand points
d_j = demand at demand point $j = 1, \dots, n$
c_{ij} = cost of shipping one unit from supply point i to demand point j.

Let x_{ij} be the amount of flow from supply point i to demand point j.

Transportation Problem

$$\min \ \sum_{i=1}^{m} \sum_{j=1}^{n} c_{ij} x_{ij}$$

$$\text{s.t.} \ \sum_{j=1}^{n} x_{ij} \leq s_i, \quad i = 1, \dots, m \ (\text{supply}) \qquad\qquad (2.9)$$

$$\sum_{i=1}^{m} x_{ij} \geq d_j, \quad j = 1, \dots, n \ (\text{demand})$$

$$x_{ij} \geq 0.$$

We call a solution **x** a *flow*. In Example 2.10, the supply and demand constraints are active at the optimal solution because all of the available electricity is needed to meet demand. Thus, they could be written as equality constraints. This occurs when total supply equals total demand, $\sum_{i=1}^{m} s_i = \sum_{j=1}^{n} d_j$. Transportation problems with this property are called *balanced*. If total supply is less than total demand the problem is infeasible. If there is excess supply, a dummy demand point with zero costs can be added to create a balanced transportation problem.

A transportation problem with supplies and demands equal to one is called an *assignment problem*. For example, a class of medical students might be assigned to residency positions at hospitals by having each student give their preference for each position. The students are the supplies, the positions are the demands, and c_{ij} is the value, or utility, that student i has for position j. Total utility is maximized (or the negative of utility can be minimized). Although it appears that a constraint is needed that the variables be binary (0 or 1), it turns out that every balanced assignment problem has an optimal solution with every $x_{ij} = 0$ or 1, so it can be solved as a linear program without the integer constraint.

Many-to-one assignment problems can be handled in a similar way. For example, suppose we are assigning 10 jobs to 7 machines. Each machine has a maximum number of jobs that can be assigned to it. The machines are modeled as the supply points and the number of jobs they can handle is the supply. The jobs are the demand points, each with a demand of 1, so that one machine is assigned to each job. Each job has a processing time on each machine. The objective is to minimize the total processing time.

Many variations of the transportation problem are useful. Shipments from certain supplies to certain demands may not be allowed; these variables can be eliminated from the model or their costs set to a large number so that they are not used. The variables may have upper bounds, representing the capacity of that route, or lower bounds, representing the minimum quantity that must be sent on that route. When there are capacity limits, different commodities being shipped are competing for the same capacity, so a multi-commodity version of the problem may be needed.

Like any linear program, (2.9) assumes that the cost from supply point i to demand point j is proportional to the flow x_{ij} and that these costs are added to obtain the total cost. When the flow on a route represents the number of truckloads, the cost per item of a partial truckload will be higher than for a full truckload. If supplies and demands are integers, then partial truckloads will not be needed. Even if some are fractional, the proportionality assumption may be close enough to obtain accurate results. Other cost functions can be modeled with integer variables; see Chapter 4.

2.6.2 Minimum Cost Network Flow

An important generalization of the transportation problem is to allow other network structures. There could be addition locations that are neither supply points nor demand points that material passes through. For example, if supply points are factories and demand points are stores, products may be sent from factories to distribution centers and then to stores. In many applications, this is known as a transshipment problem. If we further generalize by allowing flows between other pairs of locations, it is called the *minimum cost network flow problem*.

To describe this problem, it will be convenient to use terminology from graph theory. We could have used graphs to describe the transportation problem as well, but that would be less convenient. A *directed graph* is a set of *nodes*, or vertices, $\mathcal{V}$ and a set of *arcs*, $\mathcal{A}$. Each arc is an ordered pair of nodes:

$$\mathcal{A} = \{(i,j) : i,j \in \mathcal{V} \text{ and there is an arc from } i \text{ to } j\}.$$

Let c_{ij} be the cost of sending one unit of flow from node i to node j and x_{ij} be the amount of flow from node i to node j for $(i,j) \in \mathcal{A}$. Note that a reverse flow, from node j to node i, uses a different arc (j,i); this allows the costs to differ, $c_{ij} \neq c_{ji}$.

Each node has an external supply b_i, which is positive at supply nodes, negative at demand nodes, and zero at other nodes. We assume the problem is balanced, meaning that $\sum_{i \in \mathcal{V}} b_i = 0$. Any feasible solution to such a problem has *flow balance* at each node:

$$(\text{flow out of node } i) - (\text{flow into node } i) = b_i. \tag{2.10}$$

For example, if i is a supply node and there are no arcs into i, then

$$\text{flow out of node } i = \text{supply at node } i.$$

Similarly, if i is a demand node with no arcs out of i, then $b_i < 0$ and

$$\text{flow into node } i = \text{demand at node } i = -b_i.$$

We can write the minimum cost network flow problem as

Minimum Cost Network Flow Problem

$$\min \sum_{(i,j) \in \mathcal{A}} c_{ij} x_{ij}$$

$$\text{s.t.} \sum_{j:(i,j) \in \mathcal{A}} x_{ij} - \sum_{k:(k,i) \in \mathcal{A}} x_{ki} = b_i, \ i \in \mathcal{V} \tag{2.11}$$

$$x_{ij} \geq 0.$$

The first summation in the constraint is over all arcs out of node i; the second is over all arcs into node i.

Example 2.11 (Soybeans). US soybean exports are transported by truck to terminals, by river or rail to US ports, then by ocean to foreign ports. The United States is able to remain competitive against lower production costs, particularly in Brazil, because transport costs are lower. Because ocean freight costs fluctuate, suppliers need to reevaluate supply routes. Suppose the routes in Figure 2.2 currently have the costs listed in Table 2.9. A supplier currently has the quantities of soybeans listed at each terminal and will ship the quantities listed to Shanghai, China and Tokyo, Japan. These amounts represent about 1% of their annual soybean imports from the United States. Although Portland, OR is closer to both destinations, other factors have increased its ocean freight costs. What quantity of soybeans should they ship on each route to minimize transportation costs?

Define the decision variables

$$x_{ij} = \text{soybeans (1000 metric tons) sent from node } i \text{ to node } j,$$

where (i,j) is an arc in the network. As in the last example, the constraints for the terminals (supply points) and destinations are

$$x_{\text{MN,PTL}} + x_{\text{MN,NO}} \leq 150 \text{ (Minneapolis)},$$
$$x_{\text{FAR,PTL}} + x_{\text{FAR,NO}} \leq 100 \text{ (Fargo)},$$
$$x_{\text{SFA,PTL}} + x_{\text{SFA,NO}} \leq 100 \text{ (Sioux Falls)},$$
$$x_{\text{PTL,SHA}} + x_{\text{NO,SHA}} \geq 320 \text{ (Shanghai)},$$
$$x_{\text{PTL,TOK}} + x_{\text{NO,TOK}} \geq 30 \text{ (Tokyo)}.$$

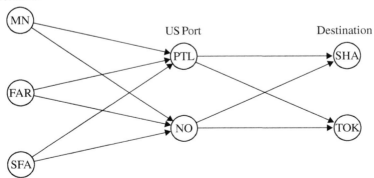

Figure 2.2 Transportation network for soybeans.

Table 2.9 Soybean shipping costs, supply, and demand.

From \ To	Shipping costs ($/metric ton)				Supply
	Portland	New Orleans	Shanghai	Tokyo	(10^3 metric tons)
Minneapolis, MN	36	30	—	—	150
Fargo, ND	24	40	—	—	100
Sioux Falls, SD	38	32	—	—	100
Portland, OR	—	—	66	60	—
New Orleans, LA	—	—	70	62	—
Demand	—	—	320	30	—

The US ports are transshipment nodes and have the flow balance constraints

$$x_{MN,PTL} + x_{FAR,PTL} + x_{SFA,PTL} = x_{PTL,SHA} + x_{PTL,TOK} \quad \text{(Portland)},$$
$$x_{MN,NO} + x_{FAR,NO} + x_{SFA,NO} = x_{NO,SHA} + x_{NO,TOK} \quad \text{(New Orleans)}.$$

The linear program minimizes shipping cost subject to these constraints and non-negativity constraints. Keeping the units as in Table 2.9, the objective function units are $/(metric ton) × (1000 metric tons) = $1000:

$$\begin{aligned}
\min \quad & 36x_{MN,PTL} + 30x_{MN,NO} + 24x_{FAR,PTL} + 40x_{FAR,NO} \\
& + 38x_{SFA,PTL} + 32x_{SFA,NO} + 66x_{PTL,SHA} + 60x_{PTL,TOK} \\
& + 70x_{NO,SHA} + 62x_{NO,TOK} \\
\text{s.t.} \quad & x_{MN,PTL} + x_{MN,NO} \leq 150 \\
& x_{FAR,PTL} + x_{FAR,NO} \leq 100 \\
& x_{SFA,PTL} + x_{SFA,NO} \leq 100 \\
& x_{PTL,SHA} + x_{NO,SHA} \geq 320 \\
& x_{PTL,TOK} + x_{NO,TOK} \geq 30 \\
& x_{MN,PTL} + x_{FAR,PTL} + x_{SFA,PTL} = x_{PTL,SHA} + x_{PTL,TOK} \\
& x_{MN,NO} + x_{FAR,NO} + x_{SFA,NO} = x_{NO,SHA} + x_{NO,TOK} \\
& x_{ij} \geq 0.
\end{aligned}$$

The optimal solution ships 150 000 metric tons from Minneapolis to New Orleans, 100 000 from Fargo to Portland, and 100 000 from Sioux Falls to New Orleans. All 100 000 tons in Portland are shipped to Shanghai, 220 000 from New Orleans to Shanghai, and 30 000 from New Orleans to Tokyo with a cost of $33.96 million.

2.6.3 Shortest Path Problems

One does not have to formulate the entire model to solve a transshipment problem. The route used between each terminal and destination is the lowest cost route. For example, from Minneapolis to Shanghai, the route through New Orleans is used, with a cost of $100 per metric ton. After these lowest cost routes are found, all that remains is a transportation problem from the terminals to the destinations, with only six arcs. This situation illustrates a general principle about models: if a problem can be decomposed into two or more smaller models, that is usually preferable. Here the transshipment problem can be decomposed into finding least expensive routes (called the *shortest path problem*) and a transportation problem. However, if capacities are added to the arcs, then the transshipment problem is needed.

The shortest path problem on a network is to find the path, or sequence of arcs, from an origin node to a destination node with minimum total cost. To formulate it as a minimum cost network flow, the origin is given a supply of 1 and the destination a demand of 1. We can also find the shortest path to the destination from every other node by giving each of these nodes a supply of 1. If the network has n nodes, the destination is given a demand of $n - 1$. For example, to find the shortest path from the three terminals to Tokyo in the soybean exporting network, we can use a demand of 3 at Tokyo and a supply of 1 at each terminal. The arcs into Shanghai are not needed. Eliminating these arcs, the shortest path problem is

$$
\begin{aligned}
\min \quad & 36x_{MN,PTL} + 30x_{MN,NO} + 24x_{FAR,PTL} + 40x_{FAR,NO} \\
& + 38x_{SFA,PTL} + 32x_{SFA,NO} + 60x_{PTL,TOK} + 62x_{NO,TOK} \\
\text{s.t.} \quad & x_{MN,PTL} + x_{MN,NO} \leq 1 \\
& x_{FAR,PTL} + x_{FAR,NO} \leq 1 \\
& x_{SFA,PTL} + x_{SFA,NO} \leq 1 \\
& x_{PTL,TOK} + x_{NO,TOK} \geq 3 \\
& x_{MN,PTL} + x_{FAR,PTL} + x_{SFA,PTL} = x_{PTL,TOK} \\
& x_{MN,NO} + x_{FAR,NO} + x_{SFA,NO} = x_{NO,TOK} \\
& x_{ij} \geq 0.
\end{aligned}
$$

In the optimal solution $x_{MN,NO} = x_{FAR,PTL} = x_{SFA,NO} = 1$, $x_{PTL,TOK} = 1$, and $x_{NO,TOK} = 2$, the values are the number of origin nodes whose shortest path contains the arc. We can easily infer that the shortest paths are MN→NO→TOK, FAR→PTL→TOK, and SFA→NO→TOK.

Some shortest path problems do not involve movement of material. For example, choosing when to replace or maintain equipment to minimize maintenance and replacement costs can be formulated as a shortest path problem.

There also algorithms to solve shortest path and other network problems that do not require formulating them as minimum cost network flow problems. One approach is dynamic programming, described in Section 3.3. For a thorough treatment of network problems including their applications, see Ahuja et al. (1993).

Problems

1 For Example 2.1, suppose that production of Ultimate, Pool, and Disc must be at least 3500, 1000, and 2800, respectively, and that the upper bound for Pool production is replaced by requiring Pool to be no more than 25% of the games produced. What constraints should be added to the linear program? What constraint should be removed?

2 For Example 2.2, suppose that the available drilling and polishing hours are for the machine operators, while the finishing hours are for the machine. If the drilling and polishing operators are cross-trained so that they can operate either machine, what constraints should be removed and added to the linear program? Do you think optimal profit will increase, decrease, or stay the same?

3 A company produces two products that require labor for assembly and packaging. The labor requirements, selling price, cost of raw materials, and labor available each day are shown as follows.

	Labor (h)		Material	Selling
	Assembly	Packaging	cost ($)	price ($)
Product A	1.5	0.75	7.20	54
Product B	2	2	5.40	48
Available	900	800	—	—

(a) Formulate and solve a linear program to maximize the daily profit of the company.
(b) Suppose up to 300 hours of additional assembly labor can be scheduled at a cost of $15/h. Formulate a linear program to incorporate this change.

4 (Based on Robbins and Tuntiwongpiboom (1986)) A hospital schedules four general types of surgery, each with different requirements for operating room time, length of patient stay, and nursing time used, shown in the following table. The profit for each type and the resources available each week are also shown. The hospital must schedule at least 10 type 1, 10 type 2, and 20 type 3 surgeries each week. Also, the demand for type 1 surgeries is limited to 25 per week. There are enough surgeons affiliated with the hospital to meet the demand for each specialty. Formulate and solve a linear program to find the hospital's optimal mix of surgery types.

Surgery type	OR time (h)	Stay (d)	Nurse time (h)	Profit ($)
1	7	5	30	14 000
2	4	2	10	10 500
3	2	1	5	3500
4	1	0	1	2100
Available	456	800	40 000	—

5 Wood Built Bookshelves (WBB) is a small wood shop that produces three types of bookshelves: models A, B, and C. Each bookshelf requires a certain amount of time for cutting each component, assembling them, and then staining them. WBB also sells unstained versions of each model. The times, in hours, for each phase of construction and the profit margins for each model, as well as the amount of time available in each department over the next two weeks, are given as follows.

Model	Labor (h)			Profit margin ($)	
	Cutting	Assembling	Staining	Stained	Unstained
A	1	4	7	60	30
B	0.5	3	5	40	20
C	2	6	8	75	40
Labor available	200	700	550	—	—

Since this is the holiday season, WBB can sell every unit that it makes. Also, due to the previous year's demand, at least 20 model B bookshelves (stained or unstained) must be produced. Finally, to increase sales of the stained bookshelves, at most 50 unstained models are to be produced. Formulate and solve a linear program to determine what WBB should produce in order to maximize its holiday profits. Source: (Rader, 2010, Exercise 2.4).

6 (Based on Bean et al. (1988)) A developer is building a new shopping mall and is trying to allocate retail space among various stores. The plaza will include women's clothing stores, children's clothing stores, shoe stores, electronic stores, jewelry stores, and sporting goods stores. To simplify its communications, all stores in each class are owned by a single company; for example, all jewelry stores are owned by one large corporation, even though the stores are named differently. Each of the six store classes have minimum and maximum space requirements (in square feet) and have provided estimates on the per square-foot annual profit margins for the developer. The developer has also estimated the per square-foot cost of building each type of store, since they have different physical requirements. These values are provided in the table as follows.

Store	Min space	Max space	Cost ($/ft^2)	Profit ($/ft^2/yr)
Women's	1500	6500	120	50
Children's	750	5000	100	30
Shoe	800	3000	80	70
Electronic	1000	4000	150	50
Jewelry	1000	5000	75	90
Sporting	1000	4000	90	60

The shopping plaza will encompass 15 000 square feet of retail space. In addition, the developer would like to ensure that at least 45% of the space is given to clothing stores, while at most 15% is provided to sporting goods stores. If the developer has a construction budget of $25 million, formulate and solve a linear program that tells them how to allocate space to the various store classes in order to satisfy the various requirements and maximize their profit margins. Source: (Rader, 2010, Exercise 2.16).

7 John Gilligan plans to open a microbrewery along the beer mile in Beverly, MA. His business plan for the first two years is to focus on two beers, an IPA and a stout. He must decide how much to spend on brewing and advertising each product each year. After rent and other fixed expenses, he has $20 000 available to spend on brewing and advertising. He has decided to save revenue from year 1 to use on brewing and advertising in year 2 and not use year 2 revenue. The brewing costs per quart sold (not all product is sold) are $1.40 for IPA and $1.60 for stout in year 1; in year 2 they are projected to be $1.70 and $1.50. The selling prices are $7 and $8 in both years. Without any spending on advertising, he anticipates he can sell 1500 quarts of IPA and 500 quarts of stout each year. He estimates each dollar spent on advertising IPA in year 1 will increase demand that year by 0.4 quarts; year 2 spending will increase year 2 demand by 0.5 quarts. For stout, year 1 advertising will increase demand by 0.25 quarts and year 2 by 0.3 quarts. He wants to produce between 40% and 70% IPA.

(a) Formulate and solve a linear program to maximize Gilligan's profit over the two years.

(b) The assumption that advertising one product does not increase demand for the other is questionable. Formulate a modified problem where any advertising has the same impact on sales as if it was divided evenly between IPA and stout.

8 A company is planning a television ad campaign and needs to decide how many 30-second ads to purchase on each program. Their target audience includes males and females ages 18–55. Programs have different audiences, broken down into gender and age category by a rating service. For the programs they are considering, the number of viewers in each category and the cost are shown in the following table. The company has specified the number of viewers it wants to reach in each category; views are called *impressions*, which is more than the number of unique individuals reached because a person may view more than one ad.

	This is us	Grey's anatomy	NFL SN night	The bachelor	Required
Cost ($1000)	140	80	120	50	impress.
	Number of viewers (millions)				(million)
Men 18–35	8	0.5	6	2	30
Men 36–55	3	3	5	1	30
Women 18–35	10	3	0.8	4	40
Women 36–55	4	6	0.5	3	50

(a) Formulate and solve a linear program to minimize their advertising cost.

(b) Another reasonable objective for the company is to maximize the number of impressions, given an advertising budget of $1 million, if they must achieve the specified number of impressions in each category. Formulate and solve this linear program. The budget should be converted to 1000, since the ad costs are in $1000s.

9   For Example 2.3, another reasonable objective is the number of unique clicks of all ages. Formulate and solve a linear program to maximize this objective, subject to a budget of $105 000. Assume that the constraints from Example 2.3 still apply.

10   For Example 2.4, suppose steel 2 must be 40% to 60% of total production (by weight). Write the additional constraints, using the variables x_{ij} defined there.

11   Three different processes can be used to grow above-ground diamonds. Two inputs that they use are diamond seeds, originating from mined diamonds, and high temperature and pressure chambers. The processes result in different quality distributions of the resulting diamonds, which can be summarized by the amount of high and low quality diamond. Shown in the following table are the inputs and outputs for one batch of each process, and the costs per batch (including the cost of the inputs). The manager of the facility has 400 diamond chips and 3600 hours of chamber time available for the next month. The selling price is $1800 per carat for high quality diamond and $1200 for low. Formulate and solve a linear program to maximize their profit.

	Inputs/outputs per batch		
	Process 1	**Process 2**	**Process 3**
Diamond chips	3	1	5
Chamber time (h)	50	10	30
High-quality diamond (carat)	4	1	3
Low-quality diamond (carat)	3	1	4
Cost ($1000/batch)	5	2	4

12   A semiconductor wafer fabrication facility has 20 000 silicon wafers to process into three products. The yield of their processes is not 100%; some wafers will be lost during processing. The first stage of coating the wafers yields 65% product 1, 30% product 2, and 5% waste. Product 1 can then be further enhanced in the second stage of production, yielding 30% product 2, 65% product 3, and 5% waste. An alternative process takes uncoated wafers and yields 20% product 1, 30% product 2, 38% product 3, and 12% waste. They must sell 3000 wafers of product 1, 4000 of product 2, and 5000 of product 3. Formulate and solve a linear program that minimizes the number of wafers wasted.

13 Delilah is planning her evening at the fair with friends. To accommodate everyone's interests, they need to eat three times, have six chances at prizes, and see animals six times. The activities she can choose from, the tickets required, and the requirements they meet are shown in the following text. She has given each activity an excitement rating and wants the total excitement score to be at least 20. Activities can be done multiple times, but Delilah can only go on the Zipper twice, or she might get sick. They can only eat at the grill once and the truck show only happens twice.

	Zipper	Ferris wheel	Bumper cars	Truck show	Racing pigs	Shoot	Frogs	Grill	Ice cream	Barn
Tickets	4	3	4	6	3	3	3	4	3	4
Food	0	0	0	1	0	0	0	1	1	1
Excite	4	1	2	3	1	2	1	0	0	0
Prizes	0	0	1	0	1	2	1	0	0	0
Animals	0	0	0	0	2	0	1	0	0	2

(a) Formulate and solve a linear program that minimizes the tickets she needs. You may ignore the requirement that the number of times she does an activity must be an integer.

(b) Add the requirement that at least one-third of the tickets must be spent on rides (the first three activities) and resolve.

14 Three airports in New York City run 24 hours a day, 7 days a week. In a given day, there are requirements for the total number of air traffic controllers that must be at the airports. These are given as follows.

Hours	Controllers needed
12 a.m. to 4 a.m.	8
4 a.m. to 8 a.m.	10
8 a.m. to 12 p.m.	16
12 p.m. to 4 p.m.	21
4 p.m. to 8 p.m.	18
8 p.m. to 12 a.m.	12

Air traffic controllers can either work 8-hour or 12-hour shifts, starting at the times stated earlier (12-hour shifts can start only at 12 a.m./p.m. or

8 a.m./p.m.). Those working 8-hour shifts cost \$40/h in salary and benefits, and those working 12-hour shifts cost \$35/h.

(a) Formulate and solve a linear program to minimize the dispatcher labor costs.

(b) Suppose at most one-third of its controllers can work 12-hour shifts. Repeat (a). You may ignore the requirement that the number of employees must be an integer. Source: (Rader, 2010, Exercise 2.6).

15 For Example 2.5, some of the officers are sergeants and the rest are lieutenants. Of the officers needed, at least two must be sergeants from midnight to 8:00 a.m., five from 8:00 a.m. to 8:00 p.m., and three from 8:00 p.m. to midnight. The costs in Example 2.5 are for lieutenants; sergeants cost 50% more. Formulate and solve a linear program to minimize the total cost.

16 A seaport unloads shipping containers off of rail cars, holds them in its yard on a transport system, and then loads them onto ships. The monthly demand for containers by ships, the cost per container in the following table. The demand must be met each month. There is enough capacity to load the containers onto the ships. There is a holding cost of \$20 for each container held in the yard until the next month. The holding cost is applied at the end of months 1, 2, and 3 after all the unloading and loading is done for the month. Because of space constraints, no more than 500 containers may be in the yard at the end of a month. At the beginning of month 1 there are 200 containers in the yard. No containers should be in the yard after month 4. Formulate and solve a linear program to minimize the loading and holding costs.

	Month			
	1	2	3	4
Demand	450	700	500	750
Unloading cost (\$/container)	75	100	105	130
Unloading capacity	800	500	450	700

17 For Exercise 2.16, now suppose that in addition to containers departing by ship (outbound) there are containers arriving by ship (inbound) that are held in the yard and then loaded onto rail cars. All inbound containers must be taken each month. There is enough capacity to load and unload the containers onto the ships. The cost of loading or unloading a container on a rail car is the same. The maximum number of containers that can be loaded or unloaded onto rail cars is shown below as the capacity. The holding cost of

$20 applies to all containers held in the yard at the end of the month, no more than 500 containers total may be in the yard at the end of the month, and no containers should be in the yard after month 4. At the beginning of month 1 there are 200 outbound containers and 100 inbound containers in the yard. Formulate and solve a linear program to minimize the loading, unloading, and holding costs.

	Month			
	1	2	3	4
Demand (outbound)	450	700	500	750
Arrivals (inbound)	250	400	500	350
(Un)loading cost ($/container)	75	100	105	130
Capacity	1000	900	700	1200

18 For Example 2.10, suppose that electricity can be sent between the two cities with a transmission cost of $5 per million kwh. Formulate and solve a linear program to minimize the transmission costs.

19 Firefighters are being deployed from two base camps to three wildfire locations. There are 20 teams at camp 1 and 15 at camp 2. Each fire needs 15 teams to be adequately staffed. The costs of transporting teams to the fires are shown in the following table. It is estimated that for every team they are short of this staffing level, there will be additional losses of $18 000, $16 000, and $22 000 at fire 1, 2, and 3, respectively. Formulate and solve a balanced transportation problem to minimize the sum of transportation costs and additional losses.

	Fire 1	Fire 2	Fire 3
Camp 1	$2000	$6000	$4000
Camp 2	$1000	$9000	$7000

20 The US government sells various public goods, such as oil rights to public land, in sealed bid auctions. Suppose m companies bid on each of n sites. Let c_{ij} be the bid ($/acre) of company i for site j and d_j be the acres at site j being sold. Suppose the government will not sell more than 30% of the land being auctioned to one company. Formulate a linear program to maximize the government's revenue.

21 For Example 2.11, modify the linear program for each of the following.
(a) At most 210 000 metric tons can go through New Orleans.
(b) At most 150 000 metric tons can be shipped from New Orleans to Tokyo at the given rate. Additional amounts on this route cost $68 per metric ton.

22 Bottled water is produced by Purewater. Currently, it has three production plants in town, one that can produce 1000 cases per week, another that produces 750 cases per week, while the third produces only 500 cases per week. Purewater uses two distributors to deliver its water bottles to the three stores that sell it. The per week demand for the three stores is 700, 600, and 800, respectively. In addition, cost ($/case) to transport the water between locations is given in the table as follows. Formulate and solve a minimum cost network flow model to find the optimal routing of the water bottles from the production plants to the stores. Source: (Rader, 2010, Exercise 2.36).

	Distributor 1	Distributor 2	Store 1	Store 2	Store 3
Plant 1	8	14	—	—	—
Plant 2	12	10	—	—	—
Plant 3	16	12	—	—	—
Distributor 1	—	—	10	8	12
Distributor 2	—	—	6	15	9

23 The U.N. World Food Program transports grain from its humanitarian response depots in Brindisi (Italy) and Dubai (UAE) to relief programs in Yemen and South Sudan. For Yemen, the possible ports of entry are (1) Aden, (2) Hudayduh, and (3) Saleef, all in Yemen. For South Sudan, they are (4) Mombasa (Kenya), (5) Djibouti, and (6) Port Sudan (Sudan). Grain is sent by ship to the port, then trucked to the program. Supply, demand, and costs are shown as follows.

	Shipping cost ($/ton)						Supply
	Port of entry						
Depot	1	2	3	4	5	6	(ton)
Brindisi	350	275	270	410	350	375	850
Dubai	260	260	240	360	390	300	450

	Trucking cost ($/ton)						Demand
	Port of Entry						
Program	1	2	3	4	5	6	(ton)
Yemen	115	100	150	—	—	—	600
S. Sudan	—	—	—	100	130	140	700

(a) Formulate algebraically as a minimum cost network flow problem. Draw the network, labeling depots with their supply, programs with their demand, and edges with their cost.

(b) Notice that once we know how much is being sent to each port of entry, we know how much must be sent from ports of entry to programs. Find a smaller formulation of the problem that eliminates the variables for the amounts shipped from ports of entry to programs. *Hint*: Add the trucking cost to the shipping cost, giving the total cost per ton when a given source and port of entry are used. Add a constraint for each program's demand. This formulation is not a transportation problem because of these constraints.

(c) Find the best port of entry to use for each depot and program. Find the total cost per ton from each depot to each program when this port is used. Draw the network for the transportation problem from depots to programs using these costs, labeling depots with their supply, programs with their demand, and edges with their cost. Do not include ports of entry as nodes in this network.

(d) Solve the transportation problem in (c). Combining this with the best ports of entry, state the optimal shipping and trucking plan.

24 A school district is revising its assignment of students to its four elementary schools. There are 20 neighborhoods, with S_{ik} grade $k = 1, 2, 3, 4, 5, 6$ students in neighborhood i. The distance from neighborhood i to school j is d_{ij} and the capacity for grade k students at school j is C_{jk}.

(a) If students in different grades can be assigned to different schools, formulate a linear program for assigning grade 6 students to schools to minimize their total distance traveled.

(b) Now assume that the same proportion of students in each grade within a neighborhood must be assigned to a school. For example, suppose neighborhood 1 has 30 fifth graders and 45 sixth graders. If 10 of the fifth graders are assigned to school 1, then 15 of the sixth graders must be also. Formulate a linear program for assigning students to schools to minimize their total distance traveled.

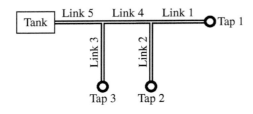

Figure 2.3 Water pipe network for Exercise 2.25.

25 (Based on Babonneau et al. (2019)) The nonprofit organization Agua Para La Vida designs gravity-driven water systems for rural communities in Nicaragua. A system has PVC pipes of various diameters connecting a water tank to various taps (delivery points). The pipes form a tree structure, that is, there is only one route from the source to each tap. Pipe sizes must be large enough so that the total headloss due to friction in the pipes is less that the drop in elevation between the source and each tap. A system with three taps and five segments of pipe, or links, is shown in Figure 2.3. There are four diameters of pipe to choose from. A link may be a mixture of more than one diameter. Let c_i be the cost (Nicaraguan córdoba per meter) of a pipe with diameter $i = 1, 2, 3, 4$, a_i be the headloss of pipe with diameter i, L_j be the length (meters) of link $j = 1, 2, 3, 4, 5$, h_0 be the elevation (meters) of the tank, and h_k be the elevation (meters) of tap $k = 1, 2, 3$. For example, the headloss for 10 m of diameter 4 pipe is $10a_4$ meters. Formulate a linear program to minimize the cost of the pipes.

3

Linear Programming Formulations

Linear programs are a remarkably broad and flexible class of optimization problems. This chapter focuses on the mathematical form of linear programs, and their relationship to some nonlinear problems. We will see that there are several forms of a linear program and that we can convert between forms. Changing forms (Section 3.1) will be convenient when we consider algorithms to solve them, since the algorithm will only need to solve one form of linear program. Section 3.2 describes how certain piecewise linear problems can be converted to linear programs. These include minimax problems and goal programming. Section 3.3 introduces dynamic programming, an important class of optimization problems that can be solved as linear programs.

3.1   Changing Form

The change that we will most often need to make to a linear program is to convert inequality constraints to equality constraints. The algorithms for linear programming that we will present apply to problems with equality constraints $\mathbf{Ax} = \mathbf{b}$ plus nonnegativity constraints $\mathbf{x} \geq \mathbf{0}$. This may seem surprising, since the feasible regions that we encountered in Chapter 1 are described by inequalities, not equations. However, equality constraints are easier to work with.

3.1.1   Slack Variables

We begin with a definition.

> The **slack** in a constraint at a solution is the right-hand side of the constraint minus the left-hand side.

Linear and Convex Optimization: A Mathematical Approach, First Edition. Michael H. Veatch.
© 2021 John Wiley & Sons, Inc. Published 2021 by John Wiley & Sons, Inc.
Companion website: www.wiley.com/go/veatch/convexandlinearoptimization

For the constraint

$$\sum_{j=1}^{n} a_{ij}x_j \leq b_i, \tag{3.1}$$

the slack is

$$s_i = b_i - \sum_{j=1}^{n} a_{ij}x_j.$$

Now, $\mathbf{x}$ satisfies (3.1) if and only if $s_i \geq 0$. Thus, (3.1) is equivalent to

$$\sum_{j=1}^{n} a_{ij}x_j + s_i = b_i,$$

$$s_i \geq 0.$$

We call s_i a *slack variable*. Any "$\leq$" constraint can be converted to an "$=$" constraint by adding a nonnegative slack variable. This variable must not appear in any other constraint; to convert multiple constraints, one slack variable is needed for each constraint.

For the constraint

$$\sum_{j=1}^{n} a_{ij}x_j \geq b_i,$$

$\mathbf{x}$ is feasible if the slack is nonpositive. We could add a nonpositive slack variable, but it is more consistent to subtract a nonnegative *surplus variable*, so that the constraint is replaced by

$$\sum_{j=1}^{n} a_{ij}x_j - s_i = b_i,$$

$$s_i \geq 0.$$

Example 3.1. Consider the constraints

$$
\begin{aligned}
5x_1 + 7x_2 &\leq 14, \\
4x_1 + 2x_2 &\leq 20, \\
3x_1 + 4x_2 &= 36, \\
-x_1 + 2x_2 &\geq 4, \\
x_j &\geq 0.
\end{aligned}
$$

To convert them to equality constraints, we add slack variables to constraints 1 and 2 and subtract a surplus variable from constraint 4:

$$
\begin{aligned}
5x_1 + 7x_2 + s_1 &= 14, \\
4x_1 + 2x_2 + s_2 &= 20, \\
3x_1 + 4x_2 &= 36, \\
-x_1 + 2x_2 - s_4 &= 4, \\
x_j, \; s_i &\geq 0.
\end{aligned}
$$

Another important conversion is to change from unrestricted in sign variables to nonnegative variables. Given an unrestricted variable x_j in any linear program, we can replace it by $x^+ - x^-$, adding the nonnegativity constraints $x^+ \geq 0$ and $x^- \geq 0$. This substitution works because any real number can be written as the difference of two nonnegative numbers. For example, $7 = 7 - 0 = 9 - 2$ and $-4 = 0 - 4$. Each solution in the original corresponds to multiple solutions in the nonnegative form, but the two forms are equivalent: they have the same set of achievable objective function values, so they have the same optimal value.

Example 3.2. Consider the problem

$$
\begin{array}{lrcl}
\max & 2x + 11y + 4z \\
\text{s.t.} & 7x \quad\quad + z & \leq & 35 \\
& 9x + 6y - 3z & \leq & 27 \\
& -4x + y + 2z & \leq & 0 \\
& y \geq 0, \quad x, z \text{ u.r.s.}
\end{array}
$$

The equivalent problem with nonnegative variables is

$$
\begin{array}{lrcl}
\max & 2x^+ - 2x^- + 11y + 4z^+ - 4z^- \\
\text{s.t.} & 7x^+ - 7x^- \quad\quad + z^+ - z^- & \leq & 35 \\
& 9x^+ - 9x^- + 6y - 3z^+ + 3z^- & \leq & 27 \\
& -4x^+ + 4x^- + y + 2z^+ - 2z^- & \leq & 0 \\
& x^+, x^-, y, z^+, z^- \geq 0.
\end{array}
$$

We can also easily make the following conversions.

Equivalent Forms of Linear Programs

- Convert equality to inequality constraints. The constraints $\mathbf{Ax} = \mathbf{b}$ are equivalent to the constraints $\mathbf{Ax} \leq \mathbf{b}$ and $\mathbf{Ax} \geq \mathbf{b}$.
- Convert nonpositive variables to nonnegative variables. The variable $x_j \leq 0$ can be replaced by $x'_j = -x_j$, where $x'_j \geq 0$.
- Convert nonzero lower or upper bounds to nonnegative variables. For example, if $x_j \geq 8$ we make the substitution $x_j = x'_j + 8$, where $x'_j \geq 0$.
- Convert a minimization to a maximization. The objective $\min \mathbf{c}^T\mathbf{x}$ is equivalent to $\max -\mathbf{c}^T\mathbf{x}$ in the sense that both have the same optimal solutions and the optimal value is the negative of the original.

3.2 Linearization of Piecewise Linear Functions

The importance of linear programs lies not only in their ability to model many situations but also in their tractability. Certain nonlinear programs can be converted to linear programs by adding variables or constraints. That may sound unappealing. However, linear programming algorithms are so much faster than algorithms for nonlinear programs that this benefit can far outweigh the increased number of variables or constraints. Here is an example of a piecewise linear objective.

Example 3.3 (Outsourced labor). Consider the production problem in Example 2.2. Suppose drilling is being outsourced so that the manufacturer must pay for the labor. This labor costs \$20/h up to 400 hours, after which it costs \$30/h due to overtime. As before, x_j is the number of batches of part j produced. The drilling labor used (h) is

$$w = 2x_1 + x_2 + 3x_3 + 2x_4$$

and the profit contribution (the negative of the cost) is

$$p(w) = \begin{cases} -20w, & w \le 400 \\ -8000 - 30(w - 400), & w > 400. \end{cases}$$

Including this cost, the profit to be maximized is

$$120x_1 + 40x_2 + 80x_3 + 60x_4 + p(w).$$

Note that $p(w)$ consists of two linear pieces and that it is concave. Thus, we can write it as the minimum of two linear functions:

$$p(w) = \min\{-20w, 4000 - 30w\}$$

as shown in Figure 3.1. The linear program is

$$
\begin{aligned}
\max \quad & 120x_1 + 40x_2 + 80x_3 + 60x_4 + \lambda \\
\text{s.t.} \quad & \lambda \le -20w \\
& \lambda \le 4000 - 30w \\
& w = 2x_1 + x_2 + 3x_3 + 2x_4 \\
& w, x_j \ge 0
\end{aligned}
$$

plus the polishing and finishing constraints of Example 2.2.

Notice that converting p to a linear function was done by adding one variable and two constraints. The linear formulation is equivalent to $p(w)$ because, for any value of w, λ is feasible if $\lambda \le p(w)$ and the value that achieves the maximum is $\lambda = p(w)$.

A very similar method applies to piecewise linear constraints.

Example 3.4 (Auto parts with profit constraint). Reconsider Example 3.3 with a constraint that profit must be at least \$5000, rather than a profit objective.

Figure 3.1 Profit contribution (the negative of cost) for labor.

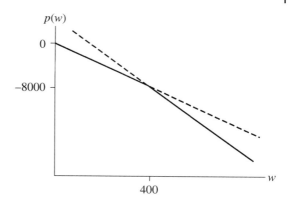

The constraint is

$$120x_1 + 40x_2 + 80x_3 + 60x_4 + p(w) \geq 5000,$$

which is equivalent to the linear constraints

$$120x_1 + 40x_2 + 80x_3 + 60x_4 - 20w \geq 5000,$$
$$120x_1 + 40x_2 + 80x_3 + 60x_4 - 30w \geq 1000.$$

It should be clear how to extend this method to functions with more than two linear pieces. For example, if labor beyond 500 hours costs \$40/h, $p(w)$ is the minimum of three linear functions and one more constraint must be added to the linear program. The method can also be used for minimization problems where the piecewise linear function is a maximum of linear functions. Next we describe the general method.

Minimax Problems A problem of the form $\min\limits_{x} \max\{g_1(x_1, \ldots, x_n), \ldots g_k(x_1, \ldots, x_n)\}$ is called a *minimax* problem. It is equivalent to

$$\min \lambda \tag{3.2}$$
$$\text{s.t. } \lambda \geq g_i(x_1, \ldots, x_n), \quad i = 1, \ldots, k.$$

The *maximin* problem $\max\limits_{x} \min\{g_1(x_1, \ldots, x_n), \ldots g_k(x_1, \ldots, x_n)\}$ is equivalent to

$$\max \lambda \tag{3.3}$$
$$\text{s.t. } \lambda \leq g_i(x_1, \ldots, x_n), \quad i = 1, \ldots, k.$$

As Example 3.3 suggests, the equivalence still holds when there are additional constraints and objective function terms.

We are interested in the case where the functions g_i are linear. It is important to note that not all piecewise linear functions can be written as a minimax or maximin. The ones that can be are those that are convex (for minimax). Convex and concave functions of more than one variable are discussed in Chapter 6.

Example 3.5 (Inventory surplus and shortage). In multiperiod inventory problems, the production of a product is scheduled to manage the inventory level at the end of each time period. Focusing on one period, let x be the inventory level at the end of the period, with positive values representing inventory on hand and negative values representing customer orders that have not been filled. A typical cost function is

$$c(x) = \begin{cases} -bx, & x < 0, \\ hx, & x \geq 0. \end{cases}$$

The backorder cost $b > 0$ is usually much larger than the holding cost $h > 0$. To minimize cost, we note that $c(x) = \max\{-bx, hx\}$ and apply the minimax method to obtain

$$\min \lambda$$
$$\text{s.t. } \lambda \geq -bx$$
$$\lambda \geq hx.$$

With no other constraints, the optimal solution is simply $\lambda = x = 0$. When other constraints and objective function terms are added, it might not be feasible or optimal to have an inventory level of 0.

3.2.1 Linearization without Adding Constraints

The minimax method replaces a nonlinear constraint with a set of constraints, one for every linear piece, and adds one variable. One common method for solving linear programs, the simplex method of Chapter 9, is slowed down more by constraints than by variables. Thus, if we wish to use this algorithm and are concerned about the amount of computation, avoiding adding constraints becomes important. There is an alternative method that only adds variables. We illustrate this method on Example 3.5. The positive and negative part of a number are defined as

$$x^+ = \max\{0, x\},$$
$$x^- = \max\{0, -x\}.$$

These parts combine as $x = x^+ - x^-$. We make this substitution wherever x appears. The problem min $c(x)$ becomes

$$\min \ bx^- + hx^+,$$

$$\text{s.t.} \ \ x^+, x^- \geq 0.$$

Although there are feasible solutions where the variables x^+ and x^- are both positive, recalling that $b, h > 0$, those values are not optimal. At any optimal solution, at most one is positive, and they represent the positive and negative part of some x.

Example 3.6. Consider the nonlinear optimization problem

$$\begin{aligned} \min \quad & 8|x| + 10|y| + 4z \\ \text{s.t.} \quad & x - 5y + 2z = 2 \\ & z \geq 0. \end{aligned}$$

Since $|x| = x^+ + x^-$, this problem can be linearized as

$$\begin{aligned} \min \quad & 8x^+ + 8x^- + 10y^+ + 10y^- + 4z \\ \text{s.t.} \quad & x^+ - x^- - 5y^+ + 5y^- + 2z = 2 \\ & x^+, x^-, y^+, y^-, z \geq 0. \end{aligned}$$

3.2.2 Goal Programming

Another area in which piecewise linear objective functions are used is Decision-making with multiple objectives. An investment portfolio decision, for example, may seek to maximize return and minimize risk. There are several ways to model this decision as an optimization problem:

- Maximize expected return subject to the risk being less than some limit.
- Minimize risk subject to the expected return being greater than some goal.
- Maximize a weighted sum of expected return and the negative of risk.

The first two have the disadvantage that goals for risk and return are modeled as "hard" constraints when in fact violating the goals might be desirable. The third has the disadvantages that we must choose weights that represent the decision-maker's priorities and that goals are not incorporated at all. Approaches to these decision problems are called multi-criteria decision-making (MCDM) or multi-attribute decision-making. Much of the theory of MCDM involves how the decision-maker interacts with the model, eliciting her preferences or providing her with options.

A popular method for MCDM that addresses the disadvantages earlier is *goal programming*. Each objective is assigned a goal and a weight. The optimization problem minimizes the weighted sum of the amounts by which the goals are not met. For example, suppose there are three objectives, maximizing $f_1(\mathbf{x}), f_2(\mathbf{x})$, and $f_3(\mathbf{x})$ with goals of 100, 200, and 300 and weights of 8, 4, and 2, respectively. The goal programming objective is

$$\min 8[100 - f_1(\mathbf{x})]^+ + 4[200 - f_2(\mathbf{x})]^+ + 8[300 - f_3(\mathbf{x})]^+.$$

An equivalent form is

$$\begin{aligned}
\min \quad & 8s_1 + 4s_2 + 2s_3 \\
\text{s.t.} \quad & f_1(\mathbf{x}) + s_1 \geq 100 \\
& f_2(\mathbf{x}) + s_2 \geq 200 \\
& f_3(\mathbf{x}) + s_3 \geq 300 \\
& s_i \geq \quad 0,
\end{aligned}$$

where s_i is the shortfall for goal i.

Another approach to MCDM eliminates decisions that are dominated by another decision that is as good for all objectives and better for one objective. The remaining decisions are called effective and collectively are called the *efficient frontier*.

3.3 Dynamic Programming

Many optimization problems can be stated as minimax problems with linear functions. In this section we introduce a powerful optimization approach called *dynamic programming*. The key idea is that decisions are made in stages and that the problem can be solved by starting with the last stage and working backwards. First we describe a recursive approach called Bellman's equation, then we show how it can be stated as a linear program. Although solving as a linear program is not usually the fastest method, it illustrates the generality of linear programming. We begin with an example.

Example 3.7 (Shortest path with stages). Starting at the corner of 5th Avenue and 20th Street in New York City, a delivery person must walk to the corner of 10th Avenue and 25th Street to deliver a package. The streets between these points form a 5×5 grid as shown in Figure 3.2. During the morning rush hour, pedestrian traffic makes her walking pace slightly different on each block. What path should she take to deliver the package as soon as possible? Each path includes five blocks walking north and five walking west. We can think of the problem in 10 stages. At each stage she decides whether to walk north or west. Number the stages $n = 0, \ldots, 9$, where n is the number of blocks she has walked.

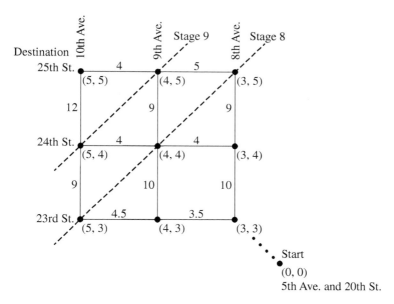

Figure 3.2 Street grid. Each block is labeled with its travel time and each intersection with its coordinates.

Call her position (W, N) if she is W blocks west of her starting point, on $(5 + W)$th Avenue, and N blocks north, on $(20 + N)$th Street. At stage $n = W + N$, she is at an intersection somewhere on a diagonal line running northeast.

We will make the decision to walk north or west for each possible location, working backwards through the stages, then combine them to find the optimal path. When the decision is made for a location, we will have found $v(W, N)$, the optimal time from that location to the destination. Starting at the last stage, where she has one block left, we compute the optimal time to walk from each intersection that she might be at to the destination. She might be at $(4, 5)$ or $(5, 4)$, and their times are the time to walk the last block, so $v(4, 5) = 4$ and $v(5, 4) = 12$. Next, for stage 8 she might be at $(3, 5)$, $(4, 4)$, or $(5, 3)$. The optimal time from $(4, 4)$ is the minimum over her two choices:

$$v(4, 4) = \min\{9 + v(4, 5), \ 4 + v(5, 4)\}, \tag{3.4}$$
$$= \min\{9 + 4, \ 4 + 12\} = 13.$$

The first term in the minimum is for walking north, incurring that block's time plus the time from $(5, 4)$ to the destination; the second term is if she walks west, incurring that block's time plus the time from $(4, 5)$. The optimal decision is to walk north, since it achieves the minimum. From $(3, 5)$ she must walk west; the time for the block is added to the time from $(4, 5)$:

$$v(3, 5) = 5 + v(4, 5) = 5 + 4 = 9.$$

Similarly, at $(5, 3)$ she must walk north and

$$v(5, 3) = 9 + v(5, 4) = 9 + 12 = 21.$$

Continuing backward by stages, using the times shown in Figure 3.2 we can also compute

$$
\begin{aligned}
v(3, 4) &= \min\{9 + v(3, 5), \ 4 + v(4, 4)\} \\
&= \min\{9 + 9, \ 4 + 13\} = 17, \\
v(4, 3) &= \min\{10 + v(4, 4), \ 4.5 + v(5, 3)\} \\
&= \min\{10 + 13, \ 4 + 21\} = 23, \\
v(3, 3) &= \min\{10 + v(3, 4), \ 3.5 + v(4, 3)\} \\
&= \min\{10 + 17, \ 3.5 + 23\} = 26.5.
\end{aligned}
$$

So far, the optimal path from $(3, 3)$ proceeds west, north, north, and then west.

Example 3.7 is a shortest path problem with a special structure. Because she was walking through a grid of streets, heading north or west, all possible paths followed a stage structure, with a stage for each of the 10 blocks traveled. Shortest path problems on graphs without this structure require somewhat different algorithms, developed by Bellman (1958), Ford (1956), and Dijkstra (1959). A comprehensive treatment is given in Ahuja et al. (1993).

In the language of dynamic programming, we found the optimal cost (travel time) from a *state* (location) to a final *stage* (the destination) as the minimum of the cost for the current stage (one block) plus the cost for the remaining stages, given the action taken in the state (travel north or west). This approach finds the optimal decision at each state in a stage, not all of which are part of the optimal solution. Equation (3.4) is an example of *Bellman's equation*.

To generalize, a dynamic program has stages $n = 0, \ldots, T$, possible states $x \in S_n$ at stage n, and possible decisions (or actions) $a \in A(x)$ in state x. The cost in the current stage n of taking action a in state x is $c_n(x, a)$, and this action leads to a state $f_n(x, a)$ at the next stage. There is a terminal cost $v_T(x)$ at the last stage. The objective is to minimize the sum of the costs for all stages, given an initial state x_0 at stage 0. For our shortest path problem, the stage n is the number of blocks that have been traveled. The states in stage n are the intersections $\mathbf{x} = (W, N)$ located at n blocks from the starting point. The possible actions at (W, N) are to walk north, $a = \mathcal{N}$ (unless $N = 5$), or west, $a = \mathcal{W}$ (unless $W = 5$). The cost in the current stage, $c(\mathbf{x}, a)$, is the time to walk a block in direction a from location $\mathbf{x}$. The state at the next stage is the next intersection, which can be written

$$f(\mathbf{x}, a) = \begin{cases} \mathbf{x} + \mathbf{e}_1, & a = \mathcal{W}, \\ \mathbf{x} + \mathbf{e}_2, & a = \mathcal{N}. \end{cases}$$

Given this description of a dynamic program, Bellman's equation can be used to compute the optimal action and the optimal "cost-to-go" $v_n(x)$ in each state at each stage. This is the sum of the costs from the current stage to the last stage. For our shortest path problem, we omitted the stage subscript because a state $\mathbf{x}$ only appears in one stage. Bellman's equation for this problem is

$$v(\mathbf{x}) = \min\{c(\mathbf{x}, \mathcal{W}) + v(\mathbf{x} + \mathbf{e}_1),\ c(\mathbf{x}, \mathcal{N}) + v(\mathbf{x} + \mathbf{e}_2)\} \tag{3.5}$$

for all states $\mathbf{x}$. The computation is recursive: first we set $v(5,5) = 0$ at the destination ($n = 10$), then proceed backwards through stages $n = 9, \dots, 0$, as described earlier. If W or $N = 5$, that term is omitted from the minimum (because moving in that direction is not allowed). An optimal action in state $\mathbf{x}$ is one that achieves the minimum.

Notice that Bellman's equation is a minimum of two linear functions of the variables $v(\mathbf{x})$. As in the minimax problem, that allows us to rewrite it as a linear program:

$$\max v(0, 0)$$

$$\text{s.t. } v(\mathbf{x}) \le c(\mathbf{x}, \mathcal{W}) + v(\mathbf{x} + \mathbf{e}_1),\quad W = 0, 1, 2, 3, 4,\ N = 0, 1, 2, 3, 4, 5 \tag{3.6}$$

$$v(\mathbf{x}) \le c(\mathbf{x}, \mathcal{N}) + v(\mathbf{x} + e\mathbf{x}_2),\quad W = 0, 1, 2, 3, 4, 5,\ N = 0, 1, 2, 3, 4,$$

where $v(5,5) = 0$. There is a constraint for every combination of state and possible action, for a total of 60 constraints. A solution v^* to (3.5) is a feasible solution to (3.6), with an active constraint at every $\mathbf{x} \ne (5,5)$. Because of the active constraints, it can be shown that v^* is in fact an optimal solution to (3.6), so that solving the linear program solves Bellman's equation. Intuitively, the maximization increases v at each stage, in backwards order, to remove slack from a constraint at each $\mathbf{x}$, and removing the slack gives a solution to (3.5).

For the general dynamic program described earlier, Bellman's equation is

$$v_n(x) = \min_{a \in A(x)} \{c_n(x, a) + v_{n+1}(f_n(x, a))\},\quad x \in S_n \tag{3.7}$$

for $n = 0, \dots, T - 1$. Now we have a minimum of several linear functions of the variables $v_n(x)$. If there is only one initial state $x_0 \in S_0$, the equivalent linear program is

$$\max v(x_0)$$

$$\text{s.t. } v_n(x) \le c_n(x, a) + v_{n+1}(f_n(x, a)),\quad x \in S_n,\ a \in A(x)$$

for $n = 0, \dots, T - 1$. There is a variable $v_n(x)$ for every combination of state and stage and a constraint for every combination of state, stage, and possible action.

Although solving the linear program requires somewhat more computation than solving (3.7) directly, it has the appeal of using readily available linear programming solvers.

Two generalizations are worth noting. First, if there are multiple possible initial states, we can find the optimal solution for all of them in one linear program by changing the objective to

$$\max \sum_{x \in S_0} v(x).$$

Second, each equation in (3.7) only has one variable on the right side, representing the next state visited. The linear programming approach still works for any linear function of the variables v_{n+1}. If the dynamic program has a *random* next state, the appropriate linear function is $\sum_{y \in S_{n+1}} p_a(x, y) v_{n+1}(y)$, where $p_a(x, y)$ is now the probability of the next state being y given that the current state is x and action a is taken. Making these changes, the linear program is

$$\max \sum_{x \in S_0} v(x)$$

$$\text{s.t. } v_n(x) \le c_n(x, a) + \sum_{y \in S_{n+1}} p_a(x, y) v_{n+1}(y), \tag{3.8}$$

$$x \in S_n, \ a \in A(x)$$

for $n = 0, \dots, T - 1$. This random dynamic program is called a *Markov Decision Process*. The deterministic Eq. (3.7) is a special case where $p_a(x, y) = 1$ for some y for each x and a; this y is the next state visited (with probability 1). Dynamic programming is a vast topic; a standard reference is Bertsekas (2007). See Puterman (2014) and Bellman and Dreyfus (2015) for many examples of Markov Decision Processes.

In many applications, Markov Decision Processes have a prohibitively large number of states, making the linear program (3.8) too large to solve. In this situation, rather than try to create an approximate model with fewer states, a fruitful approach has been to approximate the functions v_n in (3.8), creating a linear program with a small number of variables but a very large number of constraints. This approach, called approximate linear programming, was first proposed in Schweitzer and Seidmann (1985) and further developed in De Farias and Van Roy (2003), De Farias and Van Roy (2004), and Veatch (2013).

Problems

1 Suppose the problem $\max_{x \in S} f(x)$ has the optimal solution x^*.
   (a) Is x^* optimal for the problem $\max_{x \in S} f(x) + d$? If so, what is its optimal value?

(b) For $k > 0$, is x^* optimal for $\max_{x \in S} kf(x)$? If so, what is its optimal value?
(c) For $k < 0$, is x^* optimal for $\max_{x \in S} kf(x)$? If so, what is its optimal value?
(d) Is x^* optimal for $\min_{x \in S}(-f(x))$? If so, what is its optimal value?

For Exercises 3.2–3.5, rewrite using equality constraints and nonnegative variables.

2

$$
\begin{aligned}
\max \quad & 3x + 5y + 2z \\
\text{s.t.} \quad & x + 2y + z \leq 20 \\
& x + y + 2z \geq 8 \\
& x, y, z \geq 0.
\end{aligned}
$$

3

$$
\begin{aligned}
\min \quad & 6x + 4y + z \\
\text{s.t.} \quad & 2x + 3y + z = 12 \\
& 3x + y + z \geq 6 \\
& y, z \geq 0, \quad x \text{ u.r.s.}
\end{aligned}
$$

4

$$
\begin{aligned}
\max \quad & 2x + 4y + z \\
\text{s.t.} \quad & 2x + 3y = 8 \\
& 3y + z = 6 \\
& x, z \geq 0, \quad y \text{ u.r.s.}
\end{aligned}
$$

5

$$
\begin{aligned}
\min \quad & 18x + 11y \\
\text{s.t.} \quad & 5x + 4y \geq 30 \\
& 3x + 4y \geq 20 \\
& x \geq 0, \, y \geq 4.
\end{aligned}
$$

6 Consider the optimization problem

$$
\begin{aligned}
\max \quad & 3x - 5|y| \\
\text{s.t.} \quad & x + 2y \leq 8 \\
& 3x - y \leq 6 \\
& x \geq 0, \, y \text{ u.r.s.}
\end{aligned}
$$

Convert this problem to a linear program without adding constraints.

7 Consider the optimization problem

$$
\begin{aligned}
\min \quad & 2|x - 4| + 3|y - 1| + |x - 6| \\
\text{s.t.} \quad & x + y = 10 \\
& x, y \text{ u.r.s.}
\end{aligned}
$$

Rewrite $2|x - 4| + |x - 6|$ as the maximum of three linear functions and $|y - 1|$ as the maximum of two linear functions. Then use the minimax approach to convert the problem to a linear program.

8 Explain why the problem

$$\max \quad |4 - 2x + y|$$
$$\text{s.t.} \quad x + y \leq 10$$
$$\quad x, y \geq 0$$

cannot be converted to a linear program.

9 An electric car manufacturer must deliver d_t cars at the end of month t. Cars held in inventory until the next month incur a cost of h dollars per car per month. Although demand fluctuates, changing production level is also costly. If they produce x_t cars in month t and x_{t-1} cars the previous month, the cost incurred is $p|x_t - x_{t-1}|$. At the beginning of month 1, their production capacity is $x_0 = 2000$, so in month 1 the cost is $p|x_1 - 2000|$. There is no inventory at the beginning of month 1 and should be no inventory at the end of the last month.

(a) Formulate a linear program to minimize holding and change in production costs over a three month period. (A car manufacturer would actually use a 12 month period.)

(b) Now assume that increasing production costs $p^+(x_t - x_{t-1})$ (hiring costs) while decreasing production costs $p^-(x_{t-1} - x_t)$ (layoff costs). Revise the linear program in (a) to handle this.

10 Four tasks must be assigned to one of five work teams. The time (in weeks) for each team to complete each task is shown in the following text.

(a) Assume each team can be assigned at most one task. Formulate and solve a linear program to minimize the total time to complete the tasks.

(b) Now assume each task consists of processing thousands of applications, so that the tasks can be subdivided between the teams. Formulate and solve a linear program to minimize the time until the last task is finished. *Hint*: Use the fraction of a task assigned to a team as the decision variables. The time when the last task is finished is the same as the time when the last team is finished. *Note*: Minimizing the time until the last task is finished when tasks cannot be subdivided is not a linear program; it will be solved in Chapter 4.

	Task 1	Task 2	Task 3	Task 4
Team 1	4	5	8	9
Team 2	8	6	6	5
Team 3	5	9	10	7
Team 4	6	7	5	11
Team 5	10	8	6	8

Figure 3.3 Travel times for Exercise 3.12.

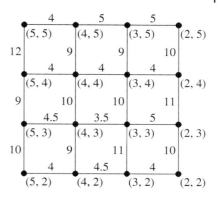

11 A large technology company has acquired a startup company and is planning a compensation package for the acquired employees. They will allocate $20 million between three forms of compensation:

- x = salaries ($ millions)
- y = stock options in the parent company ($ millions)
- z = profit sharing in the startup ($ millions).

They must limit stock options to $8 million and profit sharing to $5 million. They are considering three scenarios, where the cost of providing the compensation ($ million) are

$$
\begin{aligned}
\text{Scenario 1 Cost:} &\quad x + 0.8y + 0.7z \\
\text{Scenario 2 Cost:} &\quad x + 1.2y + 0.5z \\
\text{Scenario 3 Cost:} &\quad 1.05x + 0.75y + 0.75z
\end{aligned}
$$

They wish to minimize the worst-case cost of the scenarios.

(a) Formulate as a minimax problem.

(b) Convert to a linear program and solve.

12 For Example 3.7, some additional block travel times are shown in Figure 3.3.

(a) List the states at stage 4.

(b) Use dynamic programming to find $v(W, N)$ and the optimal action for states with $W = 2$ and $N = 2$. On a graph like Figure 3.3, label each intersection with its optimal travel time and shade the arcs that are optimal actions. For example, shade the arc from (4,4) to (4,5), showing that the optimal action at (4,4) is north.

(c) What is the optimal path from (2,2) to (5,5)?

13 A project has three stages. Stages 1 and 2 each have multiple options, which affect costs of future stages. Figure 3.4 shows the costs in $1000s. For example,

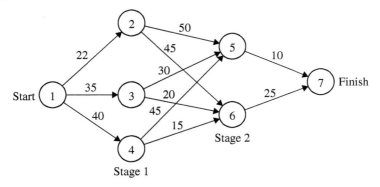

Figure 3.4 Project costs for Exercise 3.13.

if Option 2 is chosen at Stage 1 and Option 5 at Stage 2, the cost of Stage 2 is $50 000. Which options minimize the total cost of the project?

(a) Solve using dynamic programming.
(b) Formulate as a linear program.
(c) Suppose that a decision has already been made about Stage 1, but we don't know yet what the decision was. Modify the linear program in (b) to minimize the cost of the remaining stages.
(d) Now suppose that we cannot fully control the option chosen at Stage 1. If we recommend an option, it is chosen with probability 1/2, while the other two options each have probability 1/4. Modify the constraints involving Stage 1 in (b) and find the new optimal cost.

4

Integer Programming Models

Many practical optimization problems involve discrete decisions. A natural framework for modeling discrete decisions is integer programming. As noted in Section 1.4, integer programs can have a mixture of continuous, integer, and binary (0 or 1) variables. Integer programming is a much richer modeling framework than linear programming. However, constructing an integer programming model is often more challenging and more of an art than constructing a linear program. This chapter presents a variety of modeling techniques for formulating discrete optimization problems as integer programs. Applications will be presented, but we cannot survey the breadth of applications in nearly the detail of Chapter 2.

Integer optimization is more demanding computationally than linear optimization. This often affects how models are formulated. Although there are many possible linear programming formulations of a problem, the choice of formulation rarely affects our ability to solve the problem. Integer optimization is dramatically different. Without careful formulation, many practical problems cannot be solved because far too much computation would be required. The amount of computation depends critically not only on the number of integer variables but also on the choice of formulation. Section 12.2 defines the concept of a stronger (better) formulation of an integer program. Informally, if two formulations with different linear constraints have the same integer feasible solutions, but the continuous solutions of the first formulation (ignoring the integrality constraints) are a subset of those from the second formulation, then the first formulation is stronger. Thus, a stronger formulation has a smaller region satisfying the linear constraints.

We begin in Section 4.1 by considering discrete quantitative variables, then use logical variables to model fixed costs. Section 4.2 presents set covering constraints, such as when every skill needed on a team must be possessed by some member of the team. Several types of logical constraints are introduced in Section 4.3 and more applications are presented in Section 4.4. Two important discrete optimization problems, the traveling salesperson problem (TSP) and the cutting stock problem, are modeled as integer programs in Section 4.5.

Linear and Convex Optimization: A Mathematical Approach, First Edition. Michael H. Veatch.
© 2021 John Wiley & Sons, Inc. Published 2021 by John Wiley & Sons, Inc.
Companion website: www.wiley.com/go/veatch/convexandlinearoptimization

4.1 Quantitative Variables and Fixed Costs

Some problems can be formulated with linear constraints and quantitative variables, but the variables can only take on integer values. In the aircraft loading problem of Section 1.2, continuous variables were appropriate if they had time to repack one pallet with a mixture of tents and food; if not, both variables must be integers as in (1.5). Whether a variable needs to be an integer may depend on the accuracy required.

Example 4.1. (Seaport). In Exercise 2.16, a seaport unloads shipping containers off of rail cars, holds them in its yard on a transport system, and then loads them onto ships. The demand over four months totals 2400 containers, with an initial inventory of 200. Let x_t be the number of containers unloaded in month $t = 1, \ldots, 4$. Now suppose that loading onto ships requires renting cranes each month. Up to four cranes can be rented; each has a capacity of 200 containers/month. Let y_t be the number of cranes rented in month t. In this model we should add the constraint that y_t is an integer, namely, $y_t \in \{0, 1, 2, 3, 4\}$. However, the constraint that x_t is an integer is probably not needed because feasible solutions have x_t on the order of 500, so roundoff error is very small.

4.1.1 Fixed Costs

Many activities involve a fixed cost as well as a variable cost. Such cost functions can be modeled by a binary variable. Suppose we wish to model the cost

$$c(x) = \begin{cases} a + bx, & \text{if } 0 < x \le M \\ 0, & \text{if } x = 0, \end{cases}$$

where x is the level of the activity, a is the fixed cost, and b is the unit cost of the activity. We introduce the binary variable $y = 1$ if the activity occurs and $y = 0$ otherwise. Adding the constraints

$$x \le My,$$

$$x \ge 0, \ y \text{ binary}$$

forces $x = 0$ if $y = 0$; if $y = 1$ then $0 \le x \le M$. Then we can use the linear function $ay + bx$ to represent $c(x)$. The constraints imply that $ay + bx \ge c(x)$; if we minimize $ay + bx$, equality will hold.

Example 4.2. (Setup times). Consider the auto parts manufacturing problem in Example 2.2. Suppose it takes 8 hours of labor to set up the drilling machine for Part 2, 4 hours for Part 3, and 12 hours to set up for both Part 1 and Part

4, which use the same configuration. Also, suppose there are no minimum production quantities. How many of each product should be produced each month to maximize revenue for the month?

As before, let x_j be the batches of Part $j = 1, 2, 3, 4$ produced. Let

$$y_j = \begin{cases} 1, & \text{if set up for Part } j, \\ 0, & \text{otherwise,} \end{cases}$$

for $j = 1, 2, 3$. Because of the available hours, production cannot be more than 150 for Part 2, 100 for Part 3, and 100 combined for Parts 1 and 4 (the bound for Part 3 is actually 200/3, but any larger bound is equivalent). We add the constraints

$$x_1 + x_4 \leq 100y_1,$$
$$x_2 \leq 150y_2,$$
$$x_3 \leq 100y_3,$$
$$x_j \geq 0, \ y_j \ \text{binary}.$$

The second constraint, for example, forces $x_2 = 0$ if $y_2 = 0$; otherwise it does not restrict x_2 (because the available hours constraints make $x_2 \leq 150$). We also subtract the cost of the setup labor, at \$160/h, from the objective and add the set up time to the drilling hours used. The mixed integer program to maximize profit is

$$
\begin{array}{llll}
\max & 120x_1 + & 40x_2 + & 80x_3 + 60x_4 \\
& -12(160)y_1 - & 8(160)y_2 - & 4(160)y_3 \\
\text{s.t.} & 2x_1 + & x_2 + & 3x_3 + 2x_4 \\
& +12y_1 + & 8y_2 + & 4y_3 & \leq 200 & \text{(drilling)} \\
& 4x_1 + & 2x_2 + & x_3 + 2x_4 \leq 300 & & \text{(polishing)} \\
& 6x_1 + & 2x_2 + & x_3 + 2x_4 \leq 500 & & \text{(finishing)} \\
& x_1 & & + \ x_4 \leq 100y_1 & & \text{(set up 1, 4)} \\
& & x_2 & \leq 150y_2 & & \text{(set up 2)} \\
& & & x_3 \qquad \leq 100y_3 & & \text{(set up 3)} \\
& & & x_j \geq 0, \ y_j \ \text{binary}.
\end{array}
$$

The optimal solution is $\mathbf{x} = (71.6, 0, 13.6, 0)$ and $\mathbf{y} = (1, 0, 1)$, i.e. produce 71.6 batches of Part 1, etc. with a profit of \$7120. Set up is done for Parts 1, 3, and 4.

In other applications, fixed costs are for capital expenses, such as opening a warehouse. When placing an order, there may be a fixed cost of the order plus a variable cost for each item ordered.

4.2 Set Covering

A common problem involves selecting items from a set to satisfy multiple criteria. If each criterion must be satisfied by at least one selected item, it is called a *set covering* problem.

Example 4.3. (Choosing translators). A team of translators is being assembled for the summer Olympics to translate from English into the eight other languages spoken by most of the athletes: (i) French, (ii) Arabic, (iii) German, (iv) Russian, (v) Spanish, (vi) Chinese, (vii) Italian, and (viii) Portuguese. The pool consists of six translators. The languages they can translate into and the cost of sending each translator is listed in Table 4.1. Which translators should be sent to minimize the cost of being able to translate into all eight languages?

Let $x_j = 1$ if translator j is sent and 0 otherwise, $j = 1, \dots, 6$. The binary integer program to minimize cost is

$$
\begin{aligned}
\min \quad & 12x_1 + 16x_2 + 13x_3 + 15x_4 + 9x_5 + 7x_6 \\
& x_1 + x_4 + x_5 \geq 1 \ (\text{FR}) \\
& x_1 + x_3 \geq 1 \ (\text{AR}) \\
& x_2 + x_5 + x_6 \geq 1 \ (\text{GE}) \\
& x_2 + x_5 \geq 1 \ (\text{RU}) \\
& x_4 + x_5 + x_6 \geq 1 \ (\text{SP}) \\
& x_2 + x_3 \geq 1 \ (\text{CH}) \\
& x_1 + x_4 \geq 1 \ (\text{IT}) \\
& x_4 + x_6 \geq 1 \ (\text{PO}) \\
& x_j \ \text{binary.}
\end{aligned}
$$

The optimal solution sends translators 1, 2, and 6 for an optimal cost of $35\,000.

In this example, at least one person selected must have each skill, called a covering constraint. In other situations, at most one element of a set may be selected

Table 4.1 Languages and costs for translators.

Translator	Cost ($1000)	Languages
1	12	FR, AR, IT
2	16	GE, RU, CH
3	13	AR, CH
4	15	FR, SP, IT, PO
5	9	FR, GE, RU, SP
6	7	GE, SP, PO

with each characteristic; this is called a packing constraint. If exactly one element must be selected with each characteristic, it is called a partitioning constraint. Let C be the set of indices of elements with a characteristic, $x_j = 1$ if element j is chosen, and $x_j = 0$ otherwise. In the last example, for French $C = \{1, 4, 5\}$, denoting the people who can translate into French. Then

---

- A **set covering** constraint has the form $\sum_{j \in C} x_j \geq 1$.
- A **set packing** constraint has the form $\sum_{j \in C} x_j \leq 1$.
- A **set partitioning** constraint has the form $\sum_{j \in C} x_j = 1$.

Closely related to the set covering problem is the *facility location* problem. This problem chooses which facilities to open from a list of potential facilities at different locations to meet customer demands at minimum cost. Let $y_j = 1$ if facility j is selected and 0 otherwise, for $j = 1, \ldots, n$, and f_j be the fixed cost of selecting facility j. If the facilities are emergency service locations, such as fire stations, then the requirement might be that at least one facility is close enough to each of m districts, or neighborhoods, to serve it effectively. These requirements are set covering constraints, so this version of the problem is a set covering problem. As in the previous example, we can describe the requirements as subsets. Let C_i be the set of facilities that could serve district $i = 1, \ldots, m$. This facility location problem can be stated as

$$\min \sum_{j=1}^{n} f_j y_j$$
$$\text{s.t.} \sum_{j \in C_i} y_j \geq 1 \quad i = 1, \ldots, m \tag{4.1}$$
$$y_j \text{ binary.}$$

It is worth noting that, like linear programs, integer programs may be stated in matrix form. Define

$$a_{ij} = \begin{cases} 1, & \text{if district } i \text{ can be served by facility } j \\ 0, & \text{otherwise,} \end{cases}$$

$\mathbf{f} = (f_1, \ldots, f_n)$, $\mathbf{y} = (y_1, \ldots, y_n)$, and $\mathbf{A} = [a_{ij}]$. Then (4.1) can be written

$$\min \mathbf{f}^T \mathbf{y}$$
$$\text{s.t. } \mathbf{A}\mathbf{y} \geq \mathbf{1}$$
$$\mathbf{y} \text{ binary.}$$

Problem (4.1) is uncapacitated, in that any number of districts can be served by the same facility, and only has facility costs. There may be different costs of serving a district from different facilities, e.g. different transportation costs. When we add service costs, it takes on aspects of the assignment problem from Section 2.6. We will say that we are assigning customers to facilities, though a customer could represent a district or other aggregation of demand. Let d_i be the customer i demand, s_j be the service capacity of facility j, and c_{ij} be the cost of serving customer i from facility j. If customer i cannot be served by facility j, set $c_{ij} = \infty$. For example, c_{ij} might be d_i multiplied by a transportation cost. Introduce the variables $x_{ij} = 1$ if customer i is served by facility j, and 0 otherwise. The capacitated facility location problem with service costs is

$$\min \sum_{j=1}^{n} f_j y_j + \sum_{i=1}^{m} \sum_{j=1}^{n} c_{ij} x_{ij}$$

$$\text{s.t. } \sum_{j=1}^{n} x_{ij} = 1, \quad i = 1, \dots, m$$

$$\sum_{i=1}^{m} d_i x_{ij} \le s_j y_j, \quad j = 1, \dots, n$$

$$x_{ij}, \ y_j \text{ binary.}$$

The first constraint requires each customer to be served by one facility. The second requires the total demand assigned to a facility to be no more than its capacity and prevents assigning any customers to facilities unless they are opened.

Example 4.4. (Hiring consultants). A company has five projects to be done by consultants. They have a choice of three consultants. Each consultant may be assigned up to three projects. There is a fixed cost f_j of hiring consultant $j = 1, 2, 3$ and an additional cost c_{ij} of assigning project i to consultant j.

To formulate, define the variables $x_{ij} = 1$ if project i is assigned to consultant j, and 0 otherwise, and $y_j = 1$ if consultant j is used for any project and 0 otherwise. The integer program to minimize cost is

$$\min \sum_{j=1}^{3} f_j y_j + \sum_{i=1}^{5} \sum_{j=1}^{3} c_{ij} x_{ij}$$

$$\text{s.t. } \sum_{j=1}^{3} x_{ij} = 1, \quad i = 1, \dots, 5$$

$$\sum_{i=1}^{5} x_{ij} \le 3 y_j, \quad j = 1, \dots, 3$$

$$x_{ij}, \ y_j \text{ binary.}$$

4.3   Logical Constraints and Piecewise Linear Functions

In Section 4.1 we used a binary variable to represent the logical expression if $y = 0$, then $x = 0$ with the constraint $x \leq My$. This section describes how to model other logical expressions and piecewise linear functions. Consider binary variables $x_j = 1$ if item j is selected and 0 otherwise. Here are some basic relationships:

- If item 2 is selected, then so is item 1: $x_1 \geq x_2$.
- If item 2 is selected, then item 1 is not selected: $x_1 + x_2 \leq 1$.
- Either item 1 or item 2 or both must be selected: $x_1 + x_2 \geq 1$.
- At least three of items 1–5 must be selected: $x_1 + x_2 + x_3 + x_4 + x_5 \geq 3$.

A more general either-or, or disjunctive, constraint is the following. Given two constraints $\mathbf{A}_1\mathbf{x} \geq b_1$ and $\mathbf{A}_2\mathbf{x} \geq b_2$, we would like to capture the condition that at least one of them must hold. Assume that $\mathbf{A}_i \geq \mathbf{0}$ for $i = 1, 2$ and $\mathbf{x} \geq \mathbf{0}$, so that $\mathbf{A}_i\mathbf{x} \geq \mathbf{0}$. Then the condition can be expressed as

$$\mathbf{A}_1\mathbf{x} \geq b_1 y,$$
$$\mathbf{A}_2\mathbf{x} \geq b_2(1 - y),$$
$$y \text{ binary.}$$

If $y = 1$, these reduce to the first constraint; if $y = 0$, they reduce to the second.

Example 4.5. (Network intruder).   To stop the spread of an intruder on a computer network, the security software must either isolate every node that has been receiving traffic from the intruder's node or run a new scan on every such node. Either intervention requires processing time, either from central control nodes or from the distributed nodes. The processing time required by each intervention on each cluster of nodes is shown in Table 4.2. The same intervention must be used throughout a cluster, but any mixture of central or distributed processing may be used on the thousands of nodes in a cluster. The opportunity cost of central processing time is $150/h and $70/h for distributed. No more than 16 hours of central

Table 4.2   Processing times for interventions against an intruder.

Cluster	Isolate processing time (h)		Scan processing time (h)	
	Central	Distributed	Central	Distributed
1	10	12	6	18
2	6	9	4	10
3	8	12	6	15

processing time and 33 hours of distributed time should be allocated to stopping the intruder. What is the least expensive way to stop the intruder?

Each cluster has an either/or constraint. Let x_i be the amount of central processing (hours) allocated to cluster $i = 1, 2, 3$ and z_i be the amount of distributed processing allocated. Also let $y_i = 1$ if cluster i is isolated and 0 if it is scanned. The mixed integer program to minimize processing cost is

$$\min 150(x_1 + x_2 + x_3) + 70(z_1 + z_2 + z_3)$$

$$\text{s.t.} \quad \frac{1}{10}x_1 + \frac{1}{12}z_1 \geq y_1$$

$$\frac{1}{6}x_1 + \frac{1}{18}z_1 \geq 1 - y_1$$

$$\frac{1}{6}x_2 + \frac{1}{9}z_2 \geq y_2$$

$$\frac{1}{4}x_2 + \frac{1}{10}z_2 \geq 1 - y_2$$

$$\frac{1}{8}x_3 + \frac{1}{12}z_3 \geq y_3$$

$$\frac{1}{6}x_3 + \frac{1}{15}z_3 \geq 1 - y_3$$

$$x_1 + x_2 + x_3 \leq 16$$

$$z_1 + z_2 + z_3 \leq 33$$

$$x_i, z_i \geq 0, \quad y_i \text{ binary.}$$

If $y_1 = 1$, the first constraint requires that sufficient central and distributed processing time be given to cluster 1 for the isolation intervention, while if $y_1 = 0$, the second constraint requires enough for it to be scanned. The next four constraints do the same for clusters 2 and 3. The last two constraints limit the total central and distributed processing time. The optimal solution is $\mathbf{x} = (0, 4, 0)$, $\mathbf{z} = (12, 0, 12)$, $\mathbf{y} = (1, 0, 1)$ for a cost of \$2280. This solution isolates clusters 1 and 3 and scans cluster 2.

4.3.1 Job Shop Scheduling

Another use of either/or constraints is in the *job shop scheduling problem*. We are given n jobs and m machines. Each job must be processed on a specified sequence of machines, each machine can only process one job at a time, and once a machine starts processing a job it must complete the job before it can start processing another job (interruption is not allowed). Typically, the objective is to minimize the *makespan*, which is the time until the last job is completed. For simplicity, consider the case where each job must be processed on all machines in order from machine 1 to machine m. We are given the processing times p_{ij} of job j

on machine i. To formulate as a mixed integer program, let x_{ij} be the start time of job j on machine i. The completion time of job j on the last machine is $x_{mj} + p_{mj}$, so the makespan is

$$\max_{j=1,\ldots,n} x_{mj} + p_{mj}.$$

Hence, the objective is

$$\min \left\{ \max_{j=1,\ldots,n} x_{mj} + p_{mj} \right\}. \tag{4.2}$$

This objective can be linearized as in Section 3.2. The precedence constraint

$$x_{ij} + p_{ij} \leq x_{i+1,j}, \quad i = 1, \ldots, m-1, \quad j = 1, \ldots, n \tag{4.3}$$

requires job j to finish processing on machine i before starting processing on the next machine. To write a constraint preventing a machine from processing two jobs at once, we must know which job it processes first. For example, if job 2 is scheduled before job 3 on machine 1, then

$$x_{12} + p_{12} \leq x_{13}, \tag{4.4}$$

but if job 3 is scheduled before job 2, the constraint is

$$x_{13} + p_{13} \leq x_{12}. \tag{4.5}$$

Either (4.4) or (4.5) must hold. Let

$$y_{i,j,k} = \begin{cases} 1, & \text{if job } j \text{ is processed before job } k \text{ on machine } i \\ 0, & \text{otherwise,} \end{cases}$$

where $j < k$. Then the either/or constraints are

$$x_{12} + p_{12} \leq x_{13} + M(1 - y_{123}),$$
$$x_{13} + p_{13} \leq x_{12} + My_{123},$$
$$y_{123} \text{ binary.}$$

Here M is a large number; any M larger than the possible makespan may be used, e.g. the total processing time of all the jobs. Then if $y_{123} = 1$, these constraints reduce to (4.4) and if $y_{123} = 0$, they reduce to (4.5). Thus, the mixed integer program has objective (4.2) and constraints (4.3) and

$$x_{ij} + p_{ij} \leq x_{ik} + M(1 - y_{i,j,k}),$$
$$x_{ik} + p_{ik} \leq x_{ij} + My_{i,j,k},$$
$$x_{ij} \geq 0, y_{i,j,k} \text{ binary,}$$

for $j < k$.

This formulation has $mn(n-1)/2$ binary variables, which makes it very challenging to solve. More sophisticated, nonlinear formulations are described in (Bertsimas and Weismantel, 2005, Section 2.5).

4.3.2 Linearization of Piecewise Linear Objectives

The minimax objective (4.2) and other piecewise linear functions could be linearized in Section 3.2 because they are convex (for minimization problems). When the function is not convex, we can use the following technique to express it in terms of linear functions and binary variables. The crucial role of convexity in solving optimization problems will be made clear in Theorem 6.2. The main point is that nonconvex problems can be much harder. Although we can linearize this nonconvex function, the "hardness" is reflected in the fact that we must introduce binary variables.

Consider the function $f(x)$ with linear segments between the points $(2, 4)$, $(4, 8)$, $(8, 5)$, $(10, 7)$ in Figure 4.1. A linear interpolation between the first two points is

$$x = 2\lambda_1 + 4\lambda_2, \ f(x) = 4\lambda_1 + 8\lambda_2, \ \lambda_1 + \lambda_2 = 1, \ \lambda_1, \lambda_2 \geq 0.$$

We can combine all three linear interpolations and express f as

$$x = 2\lambda_1 + 4\lambda_2 + 8\lambda_3 + 10\lambda_4, \ f(x) = 4\lambda_1 + 8\lambda_2 + 5\lambda_3 + 7\lambda_4,$$
$$\lambda_1 + \lambda_2 + \lambda_3 + \lambda_4 = 1, \ \lambda_i \geq 0,$$

however, we need the additional restriction that at most two consecutive λ_j can be nonzero. For example, if only λ_3 and λ_4 are nonzero, then $8 \leq x \leq 10$ and f is the linear interpolation between the last two points. Let $y_i = 1$ if x is in the ith interval and 0 otherwise, $i = 1, 2, 3$, where the intervals are $[2, 4)$, $[4, 8)$, and $[8, 10)$. Then we can express f, the function to be minimized, as

$$\min f(x) = 4\lambda_1 + 8\lambda_2 + 5\lambda_3 + 7\lambda_4$$
$$\text{s.t. } \lambda_1 \leq y_1$$
$$\lambda_2 \leq y_1 + y_2$$
$$\lambda_3 \leq y_2 + y_3$$
$$\lambda_4 \leq y_3$$
$$\lambda_1 + \lambda_2 + \lambda_3 + \lambda_4 = 1$$
$$y_1 + y_2 + y_3 = 1$$
$$\lambda_i \geq 0, \ y_i \text{ binary.}$$

The last constraint makes exactly one of the $y_i = 1$, indicating which interval x is in. The first four constraints allow only two consecutive $\lambda_i > 0$, corresponding to the interval x is in.

Figure 4.1 A piecewise linear function.

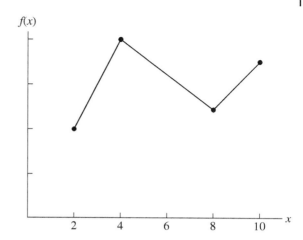

4.4 Additional Applications

Real applications often combine several of the linear and integer programming modeling approaches. This section gives three examples of larger models.

4.4.1 Supermarket Markdowns

Supermarkets routinely run promotions of selected products. In addition to increasing current sales, promotions can attract new customers and solidify relationships with suppliers. However, promotions could also motivate customers to wait for deals, hurt the store image, or hurt sales of similar items. In Cohen et al. (2017) and Cohen and Perakis (2020), a model is proposed where demand depends on the current price and price in the two previous weeks. Prices for the 13 weeks in a quarter are chosen to optimize profit given this demand. Other considerations are captured through the following constraints, taken from actual supermarket operations:

- No more than four promotions of an item in a quarter.
- Promotions may not be used in two consecutive weeks.

Consider a simpler model where demand $d(t, p(t))$ in week t depends only on the current price $p(t)$ and t. Assume all demand is met, so $d(t, p(t))$ is the quantity sold, and let c be the unit cost. Then the profit contribution of this item over the quarter is

$$\sum_{t=1}^{13}(p(t) - c)d(t, p(t)).$$

Note that profit is a nonlinear function of price. However, if price is selected from a "ladder" of a few prices, we can construct a linear integer program. Suppose we allow four prices, $p_0 > p_1 > p_2 > p_3$, where p_0 is full price and the others are markdowns, such as 80%, 70%, and 60% of p_0. Let $d_{kt} = d(t, p_k)$ be the demand in week t at price p_k. The decision variables are

$$x_{kt} = \begin{cases} 1, & \text{if the } k\text{th price is used in week } t, \\ 0, & \text{otherwise.} \end{cases}$$

In other words, if $x_{kt} = 1$ then the price in week t is p_k. The integer program to maximize profit is

$$\max \sum_{t=1}^{13} \sum_{k=0}^{3} (p_k - c) d_{kt} x_{kt}$$

$$\text{s.t.} \quad \sum_{k=0}^{3} x_{kt} = 1, \quad t = 1, \dots, 13$$

$$\sum_{t=1}^{13} \sum_{k=1}^{3} x_{kt} \leq 4$$

$$\sum_{k=1}^{3} x_{kt} + \sum_{k=1}^{3} x_{k,t+1} \leq 1, \quad t = 1, \dots, 12$$

$$x_{kt} \text{ binary.}$$

The first constraint requires one price to be chosen each week. The second constraint, which does not include $k = 0$, limits the total number of promotions to 4. The last constraint, again without $k = 0$, prevents promotions on consecutive weeks t and $t + 1$.

The demand function $d(t, p)$ is estimated using three years of historical data. A log–log regression model with quarterly seasonality and linear trend over time results in a demand function of the form

$$\ln d(t, p) = a + s_t + bt + \beta \ln p$$

or

$$d(t, p) = \exp(a + s_t + bt) p^{\beta}.$$

Here s_t is the seasonality parameter and only changes quarterly, i.e. $s_1 = s_2 = \cdots = s_{13}$ for the 13 weeks in the first quarter. Fitting a regression model to the historical data results in estimates of the parameters a, s, b, β. In Cohen et al. (2017), demand is also function of the two previous week's prices, capturing customer behavior such as waiting for promotions or continuing to buy a product that they bought when the price was low. This model requires the binary variables $y_{jklt} = 1$ if the jth price is used in week t, the kth price in week $t - 1$, and the lth price in week $t - 2$.

4.4.2 Warehouse Location and Transshipment

When designing a supply network, it can be desirable to simultaneously choose warehouse locations and transportation routes. As in the capacitated facility location problem, there are n possible locations for warehouses. As in the transshipment problem, there are m supply points and p demand points. Material can be shipped from a supply point to a demand point via any warehouse or can be shipped directly without using a warehouse; however, there is a fixed cost for each direct shipping route that is used. The inputs are

f_j = fixed cost of opening warehouse j (\$)
u_j = capacity of warehouse j (ton)
s_i = supply at supply point i (ton)
d_k = demand at demand point k (ton)
c_{ij}^I = inbound shipping cost from supply point i to warehouse j (\$/ton)
c_{jk}^O = outbound shipping cost from warehouse j to demand point k (\$/ton)
c_{ik}^D = direct shipping cost from supply point i to demand point k (\$/ton)
r_{ik} = fixed cost of direct shipping from supply i to demand k (\$).

We use the decision variables

x_{ij}^I = amount shipped from supply point i to warehouse j (ton)
x_{jk}^O = amount shipped from warehouse j to demand point k (ton)
x_{ik}^D = amount shipped direct from supply point i to demand point k (ton)
y_j = 1 if warehouse j is opened and 0 otherwise
z_{ik} = 1 if direct shipping is used from supply i to demand k and 0 Otherwise.

The mixed integer program to minimize cost is

$$\min \sum_{j=1}^{n} f_j y_j + \sum_{i=1}^{m}\sum_{k=1}^{p} r_{ik} z_{ik} + \sum_{i=1}^{m}\sum_{j=1}^{n} c_{ij}^I x_{ij}^I + \sum_{j=1}^{n}\sum_{k=1}^{p} c_{jk}^O x_{jk}^O + \sum_{i=1}^{m}\sum_{k=1}^{p} c_{ik}^D x_{ik}^D$$

$$\text{s.t. } \sum_{j=1}^{n} x_{ij}^I + \sum_{k=1}^{p} x_{ik}^D \le s_i, \quad i = 1, \dots, m$$

$$\sum_{i=1}^{m} x_{ij}^I \le u_j y_j, \quad j = 1, \dots, n$$

$$\sum_{i=1}^{m} x_{ij}^I = \sum_{k=1}^{p} x_{jk}^O, \quad j = 1, \dots, n$$

$$\sum_{j=1}^{n} x_{jk}^O + \sum_{i=1}^{m} x_{ik}^D = d_k, \quad k = 1, \dots, p$$

$$x_{ik}^D \le s_i z_{ik} \quad i = 1, \dots, m \quad k = 1, \dots, p$$

$$x_{ij}^I, x_{jk}^O, x_{ik}^D \ge 0, \ y_j, z_{ik} \text{ binary.}$$

The first two sums in the objective are the fixed costs of warehouses and direct shipping. The first constraint limits shipment out of a supply point, the second limits shipment into a warehouse to its capacity, or zero if it is not opened, the third enforces flow balance at a warehouse, the fourth requires demand to be met at a demand point, and the last requires a direct shipping route to be opened ($z_{ik} = 1$) before it can be used.

4.4.3 Locating Disaster Recovery Centers

After a disaster occurs in the United States, the Federal Emergency Management Agency (FEMA) opens disaster recovery centers (DRCs) to provide in-person assistance to those impacted. A study was done to assist them in deciding where to open DRCs and how to staff them more efficiently (Moline et al., 2019). DRCs are temporary and can only be opened in counties that have been declared disaster areas by the President. They are usually located in community centers, schools, or other public buildings. Individuals come to a DRC to receive guidance on how to obtain temporary housing, funds to repair their home or support small businesses, and other services. A model was developed to select towns in which to open a DRC to ensure that every expected visitor can reach a DRC within a maximum travel time of 60 minutes. Staffing levels are set so that the expected number of visitors at each DRC can be served.

The objective includes the fixed cost of opening a DRC, the labor costs for one week, and cost assigned to a visitor's travel time for one week. Demand usually peaks during the first two weeks and then decreases steadily. Because demand for future weeks is more uncertain, the model only considers the peak demand and a one-week horizon. Staffing would then be adjusted based on demand in future weeks. The model assumes that recipients will visit the closest DRC (in terms of travel time), but in fact recipients may choose to visit any DRC. A major challenge is forecasting demand. FEMA has a process for forecasting demand by county; these forecasts are then allocated to towns based on population. The inputs, and some of the values used in the study, are

n = number of towns
T_{ij} = travel time from town i to town j (hours)
$N(i)$ = set of towns within 60 minutes of town i
D_i = demand in town i (visitors/week)
S = service rate (visitors/hour) (1.75)
ρ = maximum utilization of staff (0.75)
$H_{\min}$ = minimum staff hours/week per DRC (90)
$H_{\max}$ = maximum additional staff hours/week per DRC

C_F = fixed cost of DRC (\$28 091)
C_L = labor cost (\$156/h)
C_T = travel cost (\$38.15/person per hour).

Travel times are symmetric and T_{ii} is the average travel time for recipients in town i to the DRC location in their town. We have added a maximum staff utilization of 75% at any DRC. Because of uncertain weekly demand and the "walk in" nature of visitors, staffing for 100% utilization would lead to intermittent overcrowding and long wait times. The decision variables are

x_i = additional staff time at DRC in town i (hours/week)
y_i = 1 if DRC is opened in town i and 0 otherwise
u_{ij} = recipients in town i who visit DRC in town j (visitors/week).

The mixed integer program to minimize cost is

$$\min \ (C_F + H_{\min}C_L) \sum_{i=1}^{n} y_i + C_L \sum_{i=1}^{n} x_i + C_T \sum_{i=1}^{n} \sum_{j \in N(i)} T_{ij} u_{ij}$$

$$\text{s.t. } x_i \leq H_{\max} y_i, \quad i = 1, \dots, n$$

$$\sum_{j \in N(i)} u_{ij} = D_i, \quad i = 1, \dots, n$$

$$\sum_{i \in N(j)} u_{ij} \leq \rho S(H_{\min} y_j + x_j), \quad j = 1, \dots, n$$

$$x_i, u_{ij} \geq 0, \ y_i \text{ binary}.$$

The first constraint only allows additional hours to be used at a DRC that is open. The second constraint requires all recipients to be allocated to a DRC, while the third limits the recipients allocated to a DRC to its weekly capacity. Note that the staff hours at DRC j are $H_{\min} y_j + x_j$. If $H_{\max}$ is set large enough, the potential staffing at a DRC is unlimited. It could be set to an actual limit on DRC staffing, in which case some recipients might be assigned to DRCs in other towns. As noted earlier, recipients choose which DRC to visit, so these limits may make the model unrealistic. Another option is to allow more than one DRC in a town by making y_i integer instead of binary.

Using data from tornados near Urbana, IL, in 2013, the model identified three DRCs to open out of 43 towns, compared with three that were actually opened, but spent only 22% as much due to less staffing. The demand forecast for week 1 used by FEMA, totaling about 350 visitors, was used in the model. The model set capacity significantly higher, 181% of that required by the actual visitors. In this relatively small scale disaster, the minimum required staff hours created excess capacity at some DRCs. Average travel time in the model was 41 minutes.

The model's DRC plan was markedly less expensive on each of three disaster data sets on which it was tested. For the two larger disasters, it also opened many fewer DRCs. One shortcoming of the model is that it does not include excess capacity to protect against uncertainty in demand. Likely of more concern to FEMA is that it ignores concerns other than efficiency that are important to government programs. Prior to this study, FEMA's decision process considered the location and size of the impacted population, their willingness to drive long distances or take public transport to a DRC, equitable access, the perception of fairness across different counties, and the need to establish a visible response. The issue of counties is particularly important. Local government officials in a county request federal assistance from FEMA. If FEMA does not open a DRC in their county, these officials and the recipients in their county will likely be dissatisfied. Since the county (and state) do not pay for the assistance, they have no incentive to reduce the number of DRCs. As noted in Moline et al. (2019), FEMA has partially adopted a different model that opens a DRC in each county to serve recipients in that county. This need to consider many different criteria and different stakeholders in a decision is typical of decision-making in the public sector. See Pollock and Maltz (1994) for a survey of public sector applications and Johnson (2012) for a discussion of approaches, focusing on the community level.

4.5 Traveling Salesperson and Cutting Stock Problems

A widely-studied discrete optimization problem is the *TSP*, where a salesperson must visit a set of cities and return to where they started. They seek to minimize the total cost of the route.

Example 4.6. (Business trip). A traveler from London must visit five towns on a business trip and wishes to minimize the length of her journey. The inter-city distances (in miles) are given in Table 4.3.

This is an example of a *symmetric TSP* because the cost (in this case, distance) of traveling between two cities is the same in either direction. There are many applications of the TSP, such as picking orders in a warehouse, routing a delivery truck, and scheduling jobs on a machine where there are sequence-dependent setup times for a job. In some applications, travel directly from one city to another might not be allowed. To accommodate this, it is convenient to use graph theory to represent the TSP.

As in Section 2.6, consider a directed graph with a set $\mathcal{V}$ of nodes and a set $\mathcal{A}$ of arcs. A *tour* is a sequence of arcs in $\mathcal{A}$, where the source of each arc is the previous destination, that visits each node exactly once. For example, if $\mathcal{V} = \{1, 2, 3, 4\}$, then

Table 4.3 Mileage between cities.

	London	Chester	Dover	Glasgow	Oxford	Plymouth
London	—	182	70	399	56	214
Chester	182	—	255	229	132	267
Dover	70	255	—	472	127	287
Glasgow	399	229	472	—	356	484
Oxford	56	132	127	356	—	179
Plymouth	214	267	287	484	179	—

$(1, 3), (3, 4), (4, 2), (2, 1)$ is a tour, as long as these arcs are in $\mathcal{A}$. Let c_{ij} be the cost of arc (i, j). In the example earlier, the cities are nodes, there is an arc (i, j) between every pair of cities, representing traveling from i to j, and c_{ij} is the distance from city i to city j.

The asymmetric TSP corresponds to a directed graph, where there may be an arc from i to j but not a reverse arc from j to i. For the symmetric TSP, we use an undirected graph, which has a set E of *edges* instead of arcs, namely,

$$E = \{(i, j) : i, j \in \mathcal{V}, \ i < j, \text{and there is an edge between } i \text{ and } j\}.$$

Note, for example, that $(2, 4)$ and $(4, 2)$ refer the same edge with the same cost $c_{24} = c_{42}$. We are using the convention that $(2, 4) \in E$ and $(4, 2)$ is not. The symmetric TSP is to find a set of edges with minimal total cost that form a tour. Here a tour is sequence of edges that visits each node exactly once.

To formulate the symmetric TSP as an integer program, let $x_{ij} = 1$ if edge $(i, j) \in E$ is in the tour and 0 otherwise. The total cost of a tour is

$$\sum_{(i,j) \in E} c_{ij} x_{ij}.$$

The sum is over edges in the graph. If the graph has an edge between every pair of vertices, we can write the sum more explicitly. For such a graph, there is an edge for every (i, j) such that $1 \leq i < j \leq n$, where $n = |\mathcal{V}|$ is the number of nodes. The sum over edges can be written

$$\sum_{i=1}^{n-1} \sum_{j=i+1}^{n} c_{ij} x_{ij}.$$

Here is one way to formulate the constraints. Each node must participate in two edges of the tour. For node i,

$$\sum_{j:(i,j) \in E} x_{ij} + \sum_{k:(k,i) \in E} x_{ki} = 2, \quad i \in \mathcal{V}.$$

The first sum includes edges with indices greater than i, the second with indices less than i. Not all sets of edges with exactly two edges containing each node form a tour. The other possibility is that they form two or more separate subtours. For example, $1 \to 2 \to 3 \to 1$ and $4 \to 6 \to 5 \to 7 \to 4$ are separate subtours. Thus, we also need constraints to prevent subtours. To prevent the first subtour, we can require that at most two of its edges be used:

$$x_{12} + x_{23} + x_{13} \le 2.$$

In general, to prevent a subtour on a set $S \subset V$ of vertices, we require that no more than $|S| - 1$ edges are chosen that have both endpoints in S. We need a *subtour elimination* constraint for every possible subset S of vertices:

$$\sum_{\substack{i,j \in S \\ (i,j) \in E}} x_{ij} \le |S| - 1, \text{ for all } S \subset V, \ 3 \le |S| \le n - 3.$$

It takes at least three vertices to form a subtour in S and in the other vertices, so we do not need constraints for $|S| = 2$ or $n - 2$. To summarize, one formulation of the symmetric TSP is

$$\min \sum_{(i,j) \in E} c_{ij} x_{ij}$$

$$\text{s.t.} \sum_{j:(i,j) \in E} x_{ij} + \sum_{k:(k,i) \in E} x_{ki} = 2, \ i \in V$$

$$\sum_{\substack{i,j \in S \\ (i,j) \in E}} x_{ij} \le |S| - 1, \text{ for all } S \subset V, \ 3 \le |S| \le n - 3 \qquad (4.6)$$

$$x_{ij} \text{ binary.}$$

This formulation has one binary variable for each edge and an exponential number of constraints, nearly 2^n. That may seem extravagant. Indeed, another formulation that has a polynomial number of constraints was introduced by Miller et al. (1960). However, formulation (4.6) is still relevant because it is a stronger formulation than one with a polynomial number of constraints. It is not practical to include all of the subtour elimination constraints. Instead, one approach is to solve the problem without these constraints, then check to see if there is a subtour. If there is, the constraint eliminating that subtour is added to the problem and it is resolved. This process is repeated until there are no subtours, at which point an optimal tour has been found. Other formulations of the TSP can be found in Applegate et al. (2006).

4.5.1 Cutting Stock Problem

For the TSP, we saw that a binary integer program with exponentially many constraints could be useful. One only adds a constraint to the problem when it is violated by the current solution. There are also problems where formulations with exponentially many variables are useful. The strategy for using these formulations is to start with only a small number of the variables (i.e. set the other variables to zero) and find a variable to add that might improve on the current solution. It must be possible to find such a variable quickly, since there are too many variables to check each one. One problem where this is possible is the *cutting stock problem*.

Example 4.7. A paper company makes large rolls of paper of width 70 in. They cut the paper into smaller widths to sell it. There is a demand for 40 rolls of width 17 in., 65 rolls of width 14 in., 80 rolls of width 11 in., and 75 rolls of width 8.5 in. How many large rolls are needed?

The key to formulating this problem is the choice of variables. Not all large rolls will be cut in the same way into final rolls. Instead, consider all possible patterns into which the large roll can be cut. For example, a 70 in. roll can be cut into two 17 in. rolls and three 11 in. rolls, with 3 in. of waste. We will denote this first pattern by $\mathbf{a}_1 = (2, 0, 3, 0)$, where the first component is for the largest cut roll, etc. Another pattern is five 14 in. rolls with no waste, denoted $\mathbf{a}_2 = (0, 5, 0, 0)$. The decision variables are the number of large rolls cut according to each pattern. For example, if 10 large rolls use pattern 1 and seven use pattern 2, then the number of rolls of each size cut is

$$10 \begin{bmatrix} 2 \\ 0 \\ 3 \\ 0 \end{bmatrix} + 7 \begin{bmatrix} 0 \\ 5 \\ 0 \\ 0 \end{bmatrix} = \begin{bmatrix} 20 \\ 35 \\ 30 \\ 0 \end{bmatrix},$$

i.e. 20 rolls of width 17 in., etc.

To generalize, if m different widths w_i, $i = 1, \ldots, m$ are sold, the jth pattern is the m-vector $\mathbf{a}_j$, where a_{ij} is the number of rolls of width w_i produced by pattern j, for $j = 1, \ldots, n$. Also let b_i be the demand for rolls of width w_i and W be the width of the large roll. Then $\mathbf{a} = (a_1, \ldots, a_m)$ is a feasible pattern if its components are nonnegative integers and

$$\sum_{i=1}^{m} a_i w_i \leq W.$$

Let x_j be the number of large rolls cut according to pattern j. The integer program to minimize the number of large rolls used is

$$\min \sum_{j=1}^{n} x_j$$

$$\text{s.t.} \sum_{j=1}^{n} a_{ij} x_j = b_i, \quad i = 1, \ldots, m$$

$$\mathbf{x} \geq \mathbf{0}, \text{integer.}$$

The constraints require that the number of rolls of each width produced equals demand for that width. Although "≥" constraints would be valid, they are not necessary; another pattern that produces less of a certain width (with this much additional waste) can be used to achieve equality with the same objective function value. The cutting stock problem can model many other situations, such as loading rolling freight of different sizes onto cargo aircraft.

There are m linear constraints and n variables. The number of feasible patterns n grows exponentially with W. As noted earlier, the solution strategy is to start with a small number of variables (one for each constraint), find a new feasible pattern that potentially improves on the current solution, add it to the problem, and find a new solution that uses this pattern. This technique is called *column generation* because adding a variable adds a column to the coefficient matrix $\mathbf{A}$. It can be found in any advanced book on linear programming.

Problems

1   For Exercise 2.22, suppose that Purewater must renew its leases to continue using each plant next year. Also, each distributor will charge a flat rate to contract for the next year. The costs and production capacity for next year, expressed per week, are shown in the following table. Purewater is deciding whether to continue using each plant and distributor next year. Assume that no other costs change if a plant is closed. Formulate and solve an integer program to minimize weekly cost.

	Plant 1	Plant 2	Plant 3	Distributor 1	Distributor 2
Cost ($)	4000	2200	1600	1200	2500
Capacity (cases)	2000	1500	1000	—	—

2   Four special operations teams are available to be deployed to a theater of operations where there is a need for six special operations missions. Exactly one team is to be assigned to each mission and each team can perform a maximum of two missions. Not every team can do every mission. The cost of deploying each team ($1000), the assessment of how likely they are to accomplish each mission, and the value of each mission ($1000) is shown in the following text. If a team is assigned to a mission, the benefit obtained is the value multiplied by the probability of success. Formulate and solve an integer program to maximize the net benefit.

| | Mission | | | | | | Deployment |
	1	2	3	4	5	6	cost
Team 1	0.9	0.7	0.7	0.8	—	0.6	50
Team 2	0.8	0.8	0.6	—	0.5	—	38
Team 3	—	0.9	0.5	0.7	0.7	0.4	42
Team 4	0.7	0.6	—	0.4	0.6	0.8	45
Value	80	100	120	90	70	150	—

3 For Exercise 2.11, suppose that if a process is used during the month there is a setup cost of $80 000 for process 1, $10 000 for process 2, and $15 000 for process 3. Formulate as an integer program and solve.

4 For Exercise 2.16, replace the unloading capacity constraint with the requirement that cranes must be rented to unload the containers. Each crane costs $1000 per month and can unload 200 containers per month. Up to four cranes can be rented each month. As suggested in Example 4.1, use integer variables for the number of cranes, but not the number of containers. Formulate as an integer program and solve.

5 For Exercise 2.13b, add the constraint that the number of tickets must be an integer. How much does the optimal number of tickets change when this constraint is added?

6 Six operations are being scheduled in three operating rooms (ORs) for the day. Because the ORs are configured differently, the total time (hours) that they must be reserved for an operation varies as shown in the following table.
   (a) Formulate an integer program to minimize the total time that must be reserved for all the operations. Do not include the setup time. Solve it by inspection.
   (b) If any operations are scheduled in an OR for the day, there is also a setup time, shown in the following table. Formulate and solve an integer program to minimize the longest that any of the ORs is reserved, including setup time.

	Operation						Setup
	1	**2**	**3**	**4**	**5**	**6**	**time**
OR 1	2	4	4.75	3	4.5	2	0.5
OR 2	3.25	4.5	5	3.25	4.25	3.5	1
OR 3	3	4.25	3.5	2.75	4.75	3.75	1.5

7 For Exercise 4.6b, how must your integer program be changed if
   (a) At most two operations can be scheduled in an OR?
   (b) Operations 1 and 2 must be in different ORs?
   (c) Operations 3 and 5 must be in the same OR?

8 WiFi is being installed in an office with eight rooms. To achieve full coverage, in addition to the main transmitter at the router, additional transmitters, called extenders, must be installed so that every point is within 40 feet of a transmitter. A candidate location has been identified in each room. From these locations, the distances to the other rooms are shown in following. Formulate and solve an integer program to provide full coverage with the minimum number of transmitters.

	Rooms							
	1	**2**	**3**	**4**	**5**	**6**	**7**	**8**
1	—	25	60	90	35	48	65	95
2		—	38	67	48	38	45	75
3			—	32	55	28	22	42
4				—	80	58	39	22
5					—	28	56	88
6						—	28	50
7							—	22
8								—

9 For the office in Exercise 4.8, TVs are also being installed. Although it is desirable to have more TVs, because they will be used to monitor different channels they wish to avoid noise interference. Thus, no two TVs should be in rooms with coverage distances, shown above, of less than 30 feet. Formulate and solve an integer program to maximize the number of TVs.

10 A holding company plans to sell at least $800 million of its subsidiaries in preparation for acquiring a new firm. The subsidiaries it could sell are listed below. To limit its exposure to companies it does not control, they wish to sell at least 80% of the shares in each subsidiary, or no more than 50%, so that they still have a controlling interest. The total of subsidiary 1 and 2 sales cannot be more than $700 million because they are in the same industry. The forecasted one year total return on each is also shown. Formulate and solve an integer program to determine how much of each subsidiary they should sell to maximize the return on their remaining holdings.

Subsidiary	Value ($ million)	Return (%)
1	450	6
2	310	5
3	380	4.25
4	180	4

11 For Example 2.4, the optimal solution showed that steel 2 was only 20% alloy 1 (12 tons were used), well below the limit of 40% alloy 1. What variables and constraints can be added to the linear program given there to require that if steel 2 uses any alloy 1, it must use at least 15 tons of alloy 1?

12 A company manufactures and distributes a lawn care product. It has plants in the cities listed below, which are also the major customer locations. However, it can decide which plants to operate each month. There is a fixed monthly cost of $120 000 of operating a plant, a capacity of 5000 lbs per month at each plant, and a shipping cost of $0.02 per pound per mile. Distances and monthly demand are listed below. Formulate and solve an integer program to minimize the cost of meeting demand.

	Chicago	Dallas	Denver	LA	Miami	Phoenix
Chicago	—	1205	1050	2112	1390	1729
Dallas	1205	—	801	1425	1332	1027
Denver	1050	801	—	1174	2065	1836
LA	2112	1425	1174	—	2757	398
Miami	1390	1332	2065	2757	—	2359
Phoenix	1729	1027	1836	398	2359	—
Demand (lb)	2800	1500	2300	1400	1700	2100

13 For the Disaster Recovery Center (DRC) example in Section 4.4:

(a) If travel cost is removed from the objective (set $C_T = 0$), does the model still consider travel time? If so, explain how.

(b) Write the constraints, showing the values of all constants, using the data in the table, $H_{max} = 400$, and the values given in Section 4.4.

(c) For this data, if only one DRC is opened, what town must it be in? If two DRCs are opened, what towns should they be in to minimize cost?

Town	Travel time (min)			Demand
	1	2	3	
1	—	30	80	310
2	30	—	50	100
3	80	50	—	240

14 (Based on Beasley et al. (2000)) When commercial aircraft land on the same runway at an airport, there is a minimum safe separation time between any two landings. The time is based on turbulence concerns and ranges from 5 to 15 minutes, depending on the aircraft types and which aircraft lands first. Each arriving aircraft has a target arrival time. The times might need to be adjusted to maintain the separation times; however, there are costs of deviating from the target time. The order in which they land may be changed. Let

a_i = target arrival time, aircraft $i = 1, \ldots, n$

s_{ij} = required separation time between aircraft i and $j, i \neq j$ (minutes)

c_i^E = cost of scheduling aircraft i earlier than its target time ($/minutes)

c_i^L = cost of scheduling aircraft i later than its target time ($/minutes).

Aircraft are numbered in the order of their target arrivals times, so $a_1 \leq a_2 \leq \cdots \leq a_n$.

(a) Define variables for the actual landing times and write constraints that maintain the required separation times. *Hint*: The required separation should be maintained between every pair of aircraft, not just consecutive aircraft.

(b) Define a variable for how early each plane is and one for how late each plane is. Write additional constraints and an objective function so that it equals the total early and late cost.

(c) There are also limits to how much the arrival time of an aircraft can be changed. Let

E_i = earliest arrival time, aircraft i (minutes)

L_i = latest arrival time, aircraft i (minutes).

Add the constraints that each aircraft must arrive between its earliest and latest times.

15 (Based on Udell and Toole (2019)) To design a residential rooftop solar system, first a set of possible panel locations is selected and the expected sunlight at different times – due to shading and sun angle – calculated. Then, a set of panels (i.e. locations) is chosen, and they are assigned to DC to AC inverters. A panel may have its own, less expensive inverter or a string of panels may share one. However, each panel on a string must use the same current, potentially reducing the energy produced: there is an optimal current depending on the amount of sunlight hitting each panel. To approximate the energy, each panel at each time is assumed to either produce at high energy (direct sunlight) or low energy (shaded). One of two currents is chosen; high current is better for high energy, low current for low energy. The goal is to meet the time-average energy requirement with minimum cost of the components used. Let

n = number of possible panel locations

T = number of times

H_{it} = energy of panel $i = 1, \ldots, n$ at time $t = 1, \ldots, T$ at high current

L_{it} = energy of panel i at time t at low current

E = total energy requirement (summed over all times)

c_i = cost of panel i

c^J = cost of inverter for a string

c^M = cost of a microinverter for a single panel

k = maximum number of panels per string.

(a) Suppose at most one string inverter is to be used, and that if one is used, no microinverters are used ($k \geq n$). Formulate an integer program to minimize the cost of meeting the total energy requirement.

(b) Now allow one string inverter and microinverters to be used together. Formulate an integer program to minimize the cost of meeting the total energy requirement.

16 (Based on Andrews et al. (2019)) Urban Outfitters uses omni-channel order fulfillment: e-commerce orders may be filled from distribution centers or using inventory in their stores nation-wide. All the inventory for a season can be thought of as available at the beginning of the season, distributed across these locations. When an e-order arrives, they must decide which location from which to fulfill each item in the order. The value of fulfilling an order is generally greater for distribution centers, but varies between stores because of their proximity to the order and the risk of a pick decline (not being able to fill the order because the item is lost or damaged, or the worker is not trained to pick it). Fulfilling an order from a store also has the disadvantage of not having that item to fill future store demand, which generally has greater value since there is no cost of picking and shipping. The value of fulfilling an order also changes over time due to price markdowns, etc. Consider just one item. Orders, including customer demand in stores, arrive one at a time, indexed $t = 1, \ldots, T$. Locations are indexed $i = 1, \ldots, n$, including stores and distribution centers. If an order is not filled, it is assigned to location 0. Let

R_{ti} = reward if order t is filled from location i

I_i = initial inventory at location i.

Note that R_{ti} depends on the location of the order. If it is a customer in store k, then $R_{ti} = 0, k \neq i$, so that an optimal solution will only fill the order from store k.

(a) The "offline" problem assumes that all orders and in-store demand are known, and decides which location to fulfill each order from. This fictitious problem gives an upper bound on the possible reward. Formulate a linear program to maximize total reward for the offline problem.

(b) The "online" problem assigns each order to a location, knowing only the rewards and current inventory. Let y_{ti} be the opportunity cost of filling order t from location $i > 0$ (using one unit of inventory). It is a variable that depends on the decisions for orders $1, \ldots, t - 1$, which affect current inventory at that location, expected future demand, and future R_{ti}. Suppose that y_{ti} has been computed; a method is described in Andrews et al. (2019) using a primal–dual algorithm. Then we should fill order t from the location i that maximizes the net reward $R_{ti} - y_{ti}$. Formulate an integer program to maximize the net reward just for order t (one order). In terms of the variables of this program, write an equation to update the current inventory. Let I_{ti} be inventory at location i after order t is filled.

(c) Now consider the online problem with two items. Let R_{ti}^j, y_{ti}^j, and I_j^{ti} denote reward, cost, and inventory for item $j = 1$ and 2. Suppose order t is for both items. There is a cost saving of S if both items are fulfilled from the same location. Formulate an integer program to maximize the net reward for order t.

17 For Exercise 2.25, Agua Para La Vida, add the requirement that if a certain diameter pipe is used on a link, at least 50 m of it must be used. Formulate an integer program to minimize the cost of the pipes.

18 For the TSP in Example 4.6:

(a) Write out the constraint for each node in (4.6).

(b) Solve (4.6) without the subtour elimination constraints.

(c) What are the subtours in the solution in (b)? Add their subtour elimination constraints to the integer program and solve. If the new solution has any subtours, list them.

5

Iterative Search Algorithms

Solving linear and integer programs, and the convex programs described in later chapters, requires iterative algorithms. This kind of "solution" is very different than in other areas of mathematics. In much of mathematics, the desired solution is a mathematical expression – a formula. The expressions are classified as closed form if they can be evaluated in a finite number of arithmetic operations, while expressions involving limits, such as infinite series and integrals, are not closed form. Beyond this, little attention is paid to how quickly the expression can be evaluated. For some of the optimization problems we consider, the exact optimal solution can be found in a finite number of operations – but this is nearly irrelevant. The focus is on finding efficient algorithms, so that the number of operations does not grow too quickly as the size of the problem increases. A problem class will be considered easy, or tractable, if there is a known efficient algorithm. Another distinction is that most solutions in mathematics are exact, while for some optimization problems there are efficient algorithms for finding approximate solutions, not exact solutions.

A familiar example of a closed form solution is the quadratic formula. Similarly, for a system $\mathbf{Ax} = \mathbf{b}$ of two linear equations in two unknowns, there is a formula for the solution. If there is a unique solution, it is

$$\mathbf{x} = \mathbf{A}^{-1}\mathbf{b} = \frac{1}{a_{11}a_{22} - a_{12}a_{21}} \begin{bmatrix} a_{22} & -a_{12} \\ -a_{21} & a_{11} \end{bmatrix} \begin{bmatrix} b_1 \\ b_2 \end{bmatrix}.$$

For larger linear systems, in principle there is a formula but it is more practical to describe it by an *algorithm*. An algorithm is a set of computational instructions that are specific enough to be followed for any inputs of a specified type. Basically, they are instructions that can be implemented as a computer program. For the linear system, there is an algorithm to row reduce the augmented matrix $[\mathbf{A}|\mathbf{b}]$. In this algorithm, we know how many steps are required once we know the size of the problem. The situation for solving an optimization problem is quite different. Practical methods for solving them are algorithms where we cannot say how many

Linear and Convex Optimization: A Mathematical Approach, First Edition. Michael H. Veatch.
© 2021 John Wiley & Sons, Inc. Published 2021 by John Wiley & Sons, Inc.
Companion website: www.wiley.com/go/veatch/convexandlinearoptimization

operations, or how many iterations, are required. They are iterative in the sense that we keep iterating until we find an optimal or near-optimal solution.

This chapter introduces some principles of algorithm design and assessment, focusing on iterative algorithms. Section 5.1 lays out terminology for iterative and other types of algorithms. The concepts of improving directions and local optimality conditions are covered in Section 5.2. Computational complexity of algorithms and what we mean by a correct algorithm are briefly introduced in Section 5.3.

5.1 Iterative Search and Constructive Algorithms

Although graphing the feasible region as we did in Chapter 1 is helpful for visualizing algorithms that move toward an optimal solution, it can be misleading when we think about algorithm design. An algorithm cannot, for example, list all the corner points – there are too many to do that efficiently. The algorithm cannot "see" the whole feasible region. In fact, the algorithms we present generally don't keep track of where they have been. The information stored is just the current solution and a small amount of ancillary information. In this sense, the algorithms use primarily *local* information about the problem. A good analogy for local information is to think of a bug crawling through the feasible region that can only see a short distance and must decide where to go next. For constrained problems, the algorithm may also use information on how far the current solution is from the constraints – which is not local information.

An *iterative search algorithm* for an optimization problem is one that starts with a feasible solution and applies some procedure to the current solution to find another one, which becomes the current solution. The procedure is repeated until an optimal solution is found, or it terminates for some other reason. Each time this procedure is performed is called an *iteration* of the algorithm. A popular example of an iterative search algorithm is Newton's method for finding a solution to an equation. To solve $f(x) = 0$, where f is a differentiable function of one variable, given an initial estimate x_0, an iteration consists of computing a new estimate

$$x_{t+1} = x_t - \frac{f(x_t)}{f'(x_t)}$$

from the current estimate x_t. Under certain conditions, the estimates converge rapidly to a solution. Here *convergence* means that $\lim_{t \to \infty} x_t = x^*$, where x^* is a solution; when the algorithm stops, the solution x_t found is not exact.

Returning to optimization problems, the iterative algorithms we consider will mostly be *improving search algorithms*: each iteration finds a feasible solution with a better, or in some cases equal, objective function value. A common-sense example is a hill climbing algorithm. Suppose we want to find the highest point on a large parcel of land that has only one hilltop and no level spots except the hilltop. Further, suppose that the boundaries can be ignored (think of the parcel as infinite, or assume the directions we walk, described below, never reach a boundary). Each location in the parcel is described by two coordinates, $\mathbf{x} = (x_1, x_2)$. An iteration is to face in the uphill direction (the opposite of the direction that an object would begin rolling down the hill) and walk in that direction until we are no longer going uphill. Because we only walk uphill, each iteration improves the objective function (our elevation). This algorithm will converge to the hilltop. Unlike Newton's method, an iteration of this algorithm uses more than local information at the current solution: determining the uphill direction uses local information, but determining how far to walk until we were no longer going uphill requires information about other points in the feasible region.

When we design an improving search algorithm, we should consider the following issues.

1. *Initial solution.* To start the algorithm, we must know an initial solution. The choice of an initial solution can have a major effect on the speed of the algorithm. Typically the initial solution is feasible. However, finding a feasible solution can be difficult. In this case, another algorithm might be used to find a feasible solution. Another approach is for the iterative search algorithm to start with an infeasible solution and simultaneously search for a solution that is both feasible and optimal. Some problems do not have feasible solutions, so the algorithm also needs to determine whether the problem is feasible.
2. *Stopping criterion.* Ideally, the algorithm will stop as soon as it reaches an optimal solution. Being able to quickly determine whether a solution is optimal is a key issue, addressed in the next section. Some algorithms converge to an optimal solution without reaching it. In this case a different stopping criterion is needed. Often these are based on a suboptimality bound, guaranteeing that the current objective function value differs from the optimal value by no more than the bound. In other cases, the algorithm stops when the change in the objective function becomes very small. As we have seen, not all optimization problems have an optimal solution. For such problems, the algorithm should be able to determine this and stop. We might also want an algorithm to stop after a certain time or number of iterations, and report the best solution found so far.
3. *Multiple optimal solutions.* We might also want the algorithm to determine if there are multiple optimal solutions and to describe all of them.

5.1.1 Constructive Algorithms

In contrast to iterative search algorithms, a constructive algorithm assigns some variables at each iteration and does not find a complete solution until the algorithm finishes. Typically these algorithms make a single pass through the variables; once a variable is set its value is not changed. For easy problems, a single-pass algorithm can find the optimal solution. A simple example is subset selection.

Example 5.1 (Subset selection). Given n different items with values $c_i > 0$, $i = 1, \ldots, n$, choose k of them to maximize the total value. Let $x_i = 1$ if object i is chosen and 0 otherwise. The binary integer program is

$$\max \quad \sum_{i=1}^{n} c_i x_i$$

$$\text{s.t.} \quad \sum_{i=1}^{n} x_i \leq k$$

$$x_i \text{ binary.}$$

The optimal solution is to choose the k most valuable items. If we renumber the items with the most valuable first, $c_1 \geq \cdots \geq c_n$, then $x_i = 1, i \leq k$ and $x_i = 0, i > k$. We can "construct" this solution by considering the items in order and choosing item i (setting $x_i = 1$) if there is still space in the subset of items.

If the items differ in size as well as value, the problem changes from very easy to hard.

Example 5.2 (Knapsack problem). Given n different books with values $c_i > 0$ and thicknesses a_i, for $i = 1, \ldots, n$, which books should be put in a knapsack of width b to maximize the total value? We assume $\sum_{i=1}^{n} a_i > b$; otherwise all the books would fit in the knapsack. A binary integer program formulation is

$$\max \quad \sum_{i=1}^{n} c_i x_i$$

$$\text{s.t.} \quad \sum_{i=1}^{n} a_i x_i \leq b \tag{5.1}$$

$$x_i \text{ binary.}$$

Any integer program in the form of (5.1) is called a *knapsack problem*. The key feature is that it has one constraint. For some values of c_i and a_i, the problem is hard in that no efficient algorithm is known. No single-pass algorithm will find the optimal solution. However, motivated by the first example, we might still try such an algorithm. If we do, in what order should the books be considered? Here are three possibilities:

1. *Greatest value.* Consider the books in order of decreasing value. Select a book if it fits.
2. *Smallest size.* Add books in order of increasing width until one does not fit.
3. *Greatest value per width.* Consider the books in order of decreasing $\frac{c_i}{a_i}$. Select a book if it fits.

None of these algorithms will always find an optimal solution or even a solution that is better than the other two. Option 1 is an example of a *greedy algorithm*: it makes the greatest possible improvement at each iteration. However, because it does not look ahead, this myopic approach may not find the optimal solution.

Another example of a constructive algorithm is Bellman's equation in dynamic programming, presented in Section 3.3. The algorithm is constructive in that decisions are assigned at each stage, working backwards from the last stage to the first.

5.1.2 Exact and Heuristic Algorithms

Because of the difficulty of optimization problems, we might not always be able to solve them exactly, or even converge to an optimal solution. *Exact algorithms*, which find optimal solutions, are not efficient enough to use on large instances of some types of optimization problems. For these problems, it is more realistic to use a *heuristic algorithm* that attempts to find an adequate solution quickly.

We have already seen three heuristic algorithms for the knapsack problem (Example 5.2). Although in general they do not find optimal solutions, they are very fast. After we sort the items, only a few computations need to be done for each item. In contrast, there is no efficient exact algorithm for this problem. As the number of items grows, the number of operations required grows exponentially and soon becomes prohibitive. Thus, for large numbers of items, an exact solution cannot be found.

Each of the three heuristics has some motivation. As already noted, the greatest value approach is greedy. The smallest size heuristic maximizes the number of books. Now consider the greatest value per width heuristic. If we remove the integer constraint from the knapsack problem, we obtain a linear program:

$$\max \ \sum_{i=1}^{n} c_i x_i$$

$$\text{s.t.} \ \sum_{i=1}^{n} a_i x_i \leq b \tag{5.2}$$

$$0 \leq x_i \leq 1.$$

Because we removed the integer constraint, the optimal solution to (5.1) is feasible for (5.2), implying that the optimal value of (5.2) is an upper bound on the optimal value of (5.1). This is an important principle: removing a constraint can only make the optimal value better, not worse. For this linear program, an algorithm isn't necessary; the greatest value per width heuristic suggests an optimal solution. Assume the items are in decreasing order of value per width: $c_1/a_1 \geq c_2/a_2 \geq \cdots \geq c_n/a_n$. As in the heuristic, add the items in order until an item doesn't fit, then add a fraction of that item, say, item r, to use the remaining space. After adding items $1, \ldots, r-1$, the remaining space in the knapsack is $b - \sum_{i=1}^{r-1} a_i$, where r satisfies

$$0 \leq b - \sum_{i=1}^{r-1} a_i < a_r.$$

An optimal solution to (5.2) is

$$x_k^* = \begin{cases} 1, & k < r, \\ \left(b - \sum_{i=1}^{r-1} a_i\right)/a_r, & k = r, \\ 0, & k > r. \end{cases} \tag{5.3}$$

Intuitively, (5.3) is optimal because if any increment of width is replaced by part of item $j > r$, that item has less value per width. We do not offer a formal proof, which is more involved.

Example 5.3 (Knapsack heuristic). Consider the knapsack problem

$$\begin{aligned} \max \quad & 38x_1 + 29x_2 + 16x_3 + 18x_4 + 6x_5 + 8x_6 \\ \text{s.t} \quad & 17x_1 + 14x_2 + 8x_3 + 10x_4 + 4x_5 + 6x_6 \leq 45 \\ & x_i \text{ binary.} \end{aligned}$$

The items are in decreasing order of value per width, so the greatest value per width heuristic adds items 1, 2, and 3, at which point the remaining space is $45 - 17 - 14 - 8 = 6$. Item 4 does not fit, but item 5 does, so it is added. The heuristic solution is $x = (1, 1, 1, 0, 1, 0)$ with value 89.

This solution is not optimal. There are two more units of space, so we can swap item 6 for item 5, giving the solution $x = (1, 1, 1, 0, 0, 1)$ with value 91, which is optimal. The linear programming relaxation has solution $x = (1, 1, 1, 0.6, 0, 0)$ with value 93.8, which is an upper bound on the optimal value. In fact, because the data are integers, the optimal value must be an integer, so the relaxation shows that the optimal value must be no more than 93.

When solving an integer program, having a good bound is very important. In the last example, if the relaxation produced a bound of, say, 91.5, and we considered the feasible solution $x = (1, 1, 1, 0, 0, 1)$, we could conclude that it is optimal because its value is 91.

5.1.3 Traveling Salesperson Heuristics

Another problem where heuristics are often needed is the traveling salesperson Problem (TSP) from Section 4.5. Given n cities, or nodes, with distance c_{ij} when traveling from city i to city j, the problem is to find a circuit, or tour, visiting all the cities with minimal total distance. We present several heuristics to illustrate some general concepts of heuristic algorithms. These heuristics do not make use of the integer program formulation of the problem, and so are rather different than our other topics. These heuristics are computationally efficient and can be used on very large instances. More involved heuristics have been successful in finding near-optimal tours on large examples, including some with over a million cities; see Applegate et al. (2006) and Lawler et al. (1985).

Nearest Neighbor
Our first heuristic is a constructive method: starting at a given city, first we choose the next city to visit, then the next, and so on until we complete the tour. The criterion for choosing the next city is greedy: choose the nearest city that has not been visited. If we are at city i and U is the set of cities that have not been visited, then we visit the city j that achieves the minimum in $\min_{j \in U} c_{ij}$. Note that no tour is found until the algorithm finishes. If we think of the decision variables as $x_k = k$th city visited, i.e. a permutation of the cities, then this is a single-pass algorithm where we set one variable at a time.

Cheapest Edge
A second constructive method applies to the symmetric TSP, where $c_{ij} = c_{ji}$. A tour that travels directly from i to j (in either direction) is said to contain edge (i, j). This heuristic makes a single pass through the edges, from shortest to longest, and chooses an edge whenever possible. Two criteria are used when choosing edges to insure that the edges chosen can be part of a tour:

- Do not choose the edge if, when combined with the edges already chosen, there are three edges incident to the same city. For example, if edges $(2, 5)$ and $(3, 5)$ have been chosen, edge $(5, 6)$ cannot be chosen because city 5 would appear three times. We say that node 5 has a *degree* of three in the graph consisting of these edges. A tour must have a degree of 2 at each node.
- Do not choose the edge if, when combined with the edges already chosen, it creates a subtour. For example, if edges $(2, 5)$, $(5, 3)$, and $(3, 4)$ have been chosen, edge $(4, 2)$ cannot be chosen because it creates the subtour $2 \to 5 \to 3 \to 4 \to 2$ that returns to city 2 without visiting city 1.

Again, no solution is found until the algorithm finishes.

Cheapest Insertion

Another constructive method for symmetric problems starts with a subtour and inserts a city, creating a larger subtour at each iteration until all cities are included. We insert a city for which the resulting subtour is as short as possible. If S is the set of cities and E is the set of edges in the current subtour, with $(i, j) \in E$ and $r \notin S$, we can insert r between i and j by removing edge (i, j) and adding edges (i, r) and (r, j). Thus, the selection criterion is to find $(i, j) \in E$ and $r \notin S$ for which $c_{ir} + c_{rj} - c_{ij}$ is minimal. For example, if the subtour is $2 \to 3 \to 4 \to 2$ and we insert city 5 between 2 and 3, the new subtour is $2 \to 5 \to 3 \to 4 \to 2$. We have added $c_{25} + c_{53} - c_{23}$ to the length of the subtour.

2-Opt

One iterative heuristic algorithm for symmetric problems starts with a tour (some other heuristic might be used to find the initial tour). At each iteration, two edges are removed from the tour and replaced by two other edges to create a shorter tour. This process is called a *2-interchange*. The procedure selects two edges in the tour that involve four different cities; such edges are nonadjacent in the tour. Removing them breaks the tour into two pieces, each one a path. There is only one way to reconnect them into a tour: the direction of travel is reversed on one of the pieces. For example, given the tour $1 \to 8 \to 7 \to 10 \to \cdots \to 5 \to 3 \to 4 \to 6 \to \cdots \to 1$, if we remove $(7, 8)$ and $(3, 4)$, the paths are $7 \to 10 \to \cdots \to 5 \to 3$ and $4 \to 6 \to \cdots \to 1 \to 8$ and we must add $(3, 8)$ and $(4, 7)$ to make a different tour.

The algorithm stops when there is no 2-interchange that reduces the length of the tour. Such a solution is called *2-optimal*. Unfortunately, a tour can be 2-optimal but not optimal. This situation helps explain why the TSP is so hard. Even if we guess an optimal tour, there is no fast way to check that it is optimal. We can check if it is 2-optimal fairly quickly, but that does not imply optimality. The number of possible 2-interchanges for a tour is fairly large, so the algorithm might check 2-interchanges until one is found that is shorter than the current tour and use that one, rather than checking all 2-interchanges and choosing the one that is shortest.

5.2 Improving Directions and Optimality

Iterative search algorithms try to find a better solution, given a current solution. This section describes general strategies for finding a better solution. For continuous problems, the strategy is to find a direction in $\mathbb{R}^n$ to move from the current solution that, at least for small distances, remains feasible and improves the objective function. For discrete problems, we will search over solutions that are "close" to the current solution to find a better solution. Of course, once an optimal solution is reached, it is impossible to find a better solution. We would like an optimality condition that allows us to easily check whether the current solution is optimal. We begin by defining an improving direction. Given a mathematical program

$$\max \quad f(\mathbf{x})$$
$$\text{s.t.} \quad \mathbf{A}\mathbf{x} \le \mathbf{b} \tag{5.4}$$
$$\mathbf{x} \ge \mathbf{0}$$

with current solution $\mathbf{x}$, an iterative algorithm could move in *direction* $\mathbf{d}$ to a new solution $\mathbf{x} + \lambda\mathbf{d}$, where the scalar $\lambda > 0$ is the *step size*.

> The vector $\mathbf{d}$ is an **improving direction** for maximizing f at $\mathbf{x}$ if $f(\mathbf{x} + \lambda\mathbf{d}) > f(\mathbf{x})$ for all $\lambda \in (0, \hat{\lambda})$ for some $\hat{\lambda} > 0$.

For differentiable functions, the gradient $\nabla f = (\frac{\partial f}{\partial x_1}, \dots, \frac{\partial f}{\partial x_n})$ can be used to determine improving directions.

Theorem 5.1: *If f is differentiable, then $\mathbf{d}$ is an improving direction at $\mathbf{x}$ for maximizing f if*

$$\nabla f(\mathbf{x})^T \mathbf{d} > 0$$

and for minimizing f if

$$\nabla f(\mathbf{x})^T \mathbf{d} < 0.$$

This standard theorem of calculus is closely related to linear approximation of a differentiable function. At $\mathbf{x}^0$, the linear approximation is

$$f(\mathbf{x}) \approx f(\mathbf{x}^0) + \nabla f(\mathbf{x})^T (\mathbf{x} - \mathbf{x}^0).$$

Notice that if we move in the direction of the gradient, $\mathbf{d} = \nabla f(\mathbf{x})$, then, assuming $\nabla f(\mathbf{x}) \ne \mathbf{0}$, $\nabla f(\mathbf{x})^T \nabla f(\mathbf{x}) > 0$ Thus, the gradient is an improving direction for a maximization problem. In fact, it is the direction of steepest increase for a function. Similarly, $-\nabla f(\mathbf{x})$ is an improving direction for a minimization problem.

Example 5.4 (Improving directions). For $\max f(x_1, x_2) = 2x_1x_2 + 2x_2 - x_1^2 - 2x_2^2$, we will find the improving directions at $(1, 0)$. At this point

$$\nabla f = \begin{bmatrix} -2x_1 + 2x_2 \\ 2x_1 - 4x_2 + 2 \end{bmatrix} = \begin{bmatrix} -2 \\ 4 \end{bmatrix}.$$

A direction **d** is improving if

$$\nabla f^T \mathbf{d} = [-2 \ 4] \begin{bmatrix} d_1 \\ d_2 \end{bmatrix} = -2d_1 + 4d_2 > 0,$$

$$d_2 > \frac{1}{2}d_1.$$

Geometrically, these are directions that make an acute angle with ∇f. One improving direction is shown in Figure 5.1. The improving directions lie in the open half plane to the left of the dashed line.

For this example, an iterative algorithm is not needed to find the optimal solution. One can simply set $\nabla f(\mathbf{x}) = \mathbf{0}$ and solve for $\mathbf{x}$, obtaining $\mathbf{x}^* = (1, 1)$. Using second derivatives, one can show that this point is in fact a maximum and therefore the optimal solution. We will be more interested in problems that we cannot easily solve algebraically.

5.2.1 Feasible Directions and Step Size

Once an improving direction is found, we would like to move some positive step size in that direction. However, that may not be possible because of the constraints. Requiring that the new solution is feasible sets bounds on λ. If it is possible to move a nonzero distance, i.e. if $\mathbf{x} + \lambda\mathbf{d}$ is feasible for some $\lambda > 0$, we call **d** a *feasible direction*. For linear constraints as in (5.4), we can substitute $\mathbf{x} + \lambda\mathbf{d}$ for $\mathbf{x}$ and solve each constraint for λ. The result will be one of the following cases.

1. $\lambda \leq 0$ or $\lambda = 0$. The constraint is active and shows that **d** is not a feasible direction.
2. $\lambda \leq \lambda_{max}$ for some $\lambda_{max} > 0$. The constraint is inactive and allows a positive step size.

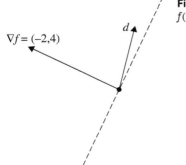

Figure 5.1 An improving direction for maximizing $f(\mathbf{x})$.

$\nabla f = (-2,4)$

d

3. $\lambda \geq \lambda_{\min}$ for some $\lambda_{\min} \leq 0$. The constraint does not bound the step size.
4. No bound on λ. The constraint is satisfied for all λ.

Because we assume $\mathbf{x}$ is feasible, all cases allow $\lambda = 0$. Since all constraints must be satisfied, we take the minimum of the upper bounds. For example, if the bounds are $\lambda \leq 5$, $\lambda \leq 8$, and $\lambda \geq -2$, then the largest feasible step size is $\lambda = 5$. If none of the constraints bound the step size then we set $\lambda = \infty$.

Example 5.5. For the constraints

$$
\begin{array}{rcrcl}
2x_1 & + & x_2 & \leq & 19, \\
3x_1 & + & 2x_2 & \leq & 52, \\
-x_1 & + & x_2 & \leq & 4, \\
& & x_j & \geq & 0,
\end{array}
$$

we will find the maximum step size in the direction $\mathbf{d} = (3, 2)$ from the point $(1, 5)$. Substituting

$$
\mathbf{x} + \lambda \mathbf{d} = \begin{bmatrix} 1 \\ 5 \end{bmatrix} + \lambda \begin{bmatrix} 3 \\ 2 \end{bmatrix}
$$

into the constraints,

$$
\begin{array}{rclcrcl}
2(1 + 3\lambda) & + & (5 + 2\lambda) & \leq & 19, & \lambda & \leq & 1.5, \\
3(1 + 3\lambda) & + & 2(5 + 2\lambda) & \leq & 52, & \lambda & \leq & 3, \\
-(1 + 3\lambda) & + & (5 + 2\lambda) & \leq & 4, & \lambda & \geq & 0, \\
& & 1 + 3\lambda & \geq & 0, & \lambda & \geq & -\frac{1}{3}, \\
& & 5 + 2\lambda & \geq & 0, & \lambda & \geq & -\frac{5}{2}.
\end{array}
$$

Thus, the largest feasible step size is $\lambda = 1.5$. At the new solution, $\mathbf{x} = (5.5, 8)$ and constraint 1 is active.

Only an active constraint can make a direction infeasible (Case 1). For linear constraints, we can describe which directions are feasible.

- For an active constraint $\mathbf{a}^T \mathbf{x} \leq b$, a feasible direction $\mathbf{d}$ satisfies $\mathbf{a}^T \mathbf{d} \leq 0$.
- For an active constraint $\mathbf{a}^T \mathbf{x} \geq b$, a feasible direction $\mathbf{d}$ satisfies $\mathbf{a}^T \mathbf{d} \geq 0$.
- For a constraint $\mathbf{a}^T \mathbf{x} = b$, a feasible direction $\mathbf{d}$ satisfies $\mathbf{a}^T \mathbf{d} = 0$.

Geometrically, a feasible direction must point into the half-space defined by a linear inequality constraint. For a linear equality constraint, it must lie in the hyperplane of the constraint.

It may be that there is no upper bound to the step size: $\mathbf{x} + \lambda \mathbf{d}$ is feasible for all $\lambda \geq 0$. Then the feasible region is unbounded. For a linear program, if $\mathbf{d}$ is also improving, then the objective function can improve without bound and the linear program is unbounded.

The linear programming algorithm in Chapter 9 uses the maximum step size, so that the constraint limiting the step size becomes active. However, as we shall see in Section 11.3 and Chapter 15, some algorithms intentionally stay in the interior of the feasible region. To stay in the interior when moving in a feasible direction, a smaller step size could be used. For nonlinear objective functions, finding a good step size must also consider the objective function. Although the objective improves for small steps in an improving direction, it may reverse direction as the step size increases. Thus, the step size in this direction that achieves the best value may be smaller than the maximum step size. Methods for finding a step size for convex programs are discussed in Chapter 15.

5.2.2 Optimality Conditions

The ability to quickly check whether a feasible solution is optimal is a key characteristic of an optimization problem. Easier optimization problems generally have such optimality conditions, while hard problems do not. Finding an optimality condition is a major concern when designing an algorithm and will be discussed for each type of problem we encounter. This section presents the general concept and some illustrative examples. A fundamental distinction is whether optimality of a solution can be checked using just local information.

For continuous problems, the concept of local information is based on the distance between two points. An ε-neighborhood of $\mathbf{x}$ in $\mathbb{R}^n$ is the set of all points whose distance from $\mathbf{x}$ is less than ε, i.e. $\{\mathbf{y} : \|\mathbf{x} - \mathbf{y}\| < \varepsilon\}$.

Local Optimal Solution   For the problem $\max_{\mathbf{x} \in S} f(\mathbf{x})$, $\mathbf{x}$ is a **local maximum** if it is feasible and has the largest objective function value of all feasible solutions in some ε-neighborhood of $\mathbf{x}$, i.e. $f(\mathbf{x}) \geq f(\mathbf{y})$ for all $\mathbf{y} \in S$ such that $\|\mathbf{x} - \mathbf{y}\| < \varepsilon$. For the problem $\min_{\mathbf{x} \in S} f(\mathbf{x})$, $\mathbf{x}$ is a **local minimum** if it is feasible and $f(\mathbf{x}) \leq f(\mathbf{y})$ for all $\mathbf{y} \in S$ such that $\|\mathbf{x} - \mathbf{y}\| < \varepsilon$.

Global Optimal Solution   For the problem $\max_{\mathbf{x} \in S} f(\mathbf{x})$, $\mathbf{x}$ is a **global maximum** if it is feasible and has the largest objective function value of all feasible solutions, i.e. $f(\mathbf{x}) \geq f(\mathbf{y})$ for all $\mathbf{y} \in S$. The point $\mathbf{x}$ is a **global minimum** if it is feasible and $f(\mathbf{x}) \leq f(\mathbf{y})$ for all $\mathbf{y} \in S$.

If there is an improving feasible direction at $\mathbf{x}$, then $\mathbf{x}$ cannot be a local optimum.

For unconstrained optimization of differentiable functions, there are optimality conditions from calculus. The first-order condition $\nabla f(\mathbf{x}) = \mathbf{0}$, used in Example 5.4, is a necessary condition for a local optimal solution. Note that there is one equation, $\frac{\partial f}{\partial x_i} = 0$, for each variable; if there is only one variable, this condition reduces to $f'(x) = 0$. Second-order conditions, involving second derivatives, can be used to strengthen the conditions. For a function of one variable, if $f'(x) = 0$ and $f''(x) < 0$ then x is a local maximum.

We would like to have similar optimality conditions for problems with constraints. As a simple example, consider the one-variable problem $\max f(x)$, $x \in [a, b]$. For x to be a local maximum, a necessary condition is

$$f'(x) = 0 \text{ if } x \in (a, b),$$
$$f'(x) \leq 0 \text{ if } x = a,$$
$$f'(x) \geq 0 \text{ if } x = b.$$

The proof is based on linear approximation. For example, if $f'(a) > 0$ then there are points just to the right of a with larger function values and a cannot be a local maximum. General approaches are given in Chapter 7 for linear programs and Chapter 14 for nonlinear programs.

All of these conditions are for local optima. For general functions, there is no guarantee that a local optimum will be a global optimum. Without this guarantee, an optimization problem is truly difficult. If we use an improving search algorithm, the algorithm will find a local optimum, but that may not be a (global) optimal solution. Even worse, there will be no optimality condition that we can check to discover that it is not optimal. Completely different algorithms, from the field of global optimization, are needed to find global optima for these problems; we do not consider them.

We can extend the concepts of local and global optimality to problems with discrete variables. However, we need to define a measure of distance between two solutions and specify what counts as "local." These may vary from problem to problem. First, consider the subset selection problem in Example 5.1, where $x_i = 1$ if object i is chosen and 0 otherwise. Define the distance between two solutions as the number of objects for which they disagree. This distance can be computed using the $\mathcal{L}_1$ *norm*, defined as

$$\|\mathbf{x}\|_1 = \sum_{i=1}^{n} |x_i|.$$

For example, if $\mathbf{x} = (1, 1, 1, 0, 0, 0)$ and $\mathbf{y} = (1, 1, 0, 0, 1, 1)$, then $\|\mathbf{x} - \mathbf{y}\|_1 = 3$, indicating that they differ in three components. Given a solution $\mathbf{x}$, its 2-neighborhood consists of all solutions that differ from $\mathbf{x}$ in at most two components. As in the TSP heuristics, we call a solution 2-optimal if it is a local optimum over its 2-neighborhood.

Theorem 5.2: *For the subset selection problem, 2-optimality implies optimality.*

Proof: For this problem we know the optimal solution is to choose the k most valuable items. However, to illustrate the concept of 2-optimality, we will not use this fact in the proof. Suppose $\mathbf{y}$ is optimal and consider a feasible but not optimal solution $\mathbf{x}$. We will show that $\mathbf{x}$ is not 2-optimal. Let $X = \{i : x_i = 1\}$ be the objects in the solution $\mathbf{x}$ and Y be the objects in $\mathbf{y}$. If $\|\mathbf{x}\|_1 < k$, then add any object to X (change some $x_i = 0$ to $x_i = 1$); this solution is better than $\mathbf{x}$, feasible, and in the 2-neighborhood of $\mathbf{x}$, so $\mathbf{x}$ is not 2-optimal. If, instead, $\|\mathbf{x}\|_1 = \|\mathbf{y}\|_1 = k$, let m be the most valuable object in Y that is not in X,

$$c_m = \max_{i \in Y \setminus X} c_i,$$

and l be the least valuable object in X that is not in Y,

$$c_l = \min_{i \in X \setminus Y} c_i.$$

Now, because the sum of the values in X is less than the sum of values in Y and each set has the same size, we must have

$$c_l \leq \bar{c}_x < \bar{c}_y \leq c_m,$$

where $\bar{c}_x$ is the average value of the objects in X and similarly for Y. The last inequality, for example, is true because the maximum of a set of numbers is at least as large as its mean. Thus, adding m and removing l from X gives a feasible solution $\mathbf{x} + \mathbf{e}_m - \mathbf{e}_l$ that is better than $\mathbf{x}$ and in the 2-neighborhood of $\mathbf{x}$, i.e. $\mathbf{x}$ is not 2-optimal. □

For the TSP problem in Section 5.1, the situation is the opposite: 2-optimality does not imply optimality, nor does k-optimality for any k. In this problem, local optimality does not guarantee global optimality, which is consistent with it being a very hard problem. Many other discrete optimization problems can be categorized in the same way: those for which local optimality implies global optimality are problems for which efficient algorithms are known. Thorough treatments are given in Papadimitriou and Steiglitz (1982) and Wolsey and Nemhauser (2014).

5.3 Computational Complexity and Correctness

Analysis of algorithms focuses on two main questions. Does the algorithm produce the correct answer? How efficient is it? This section specifies what we mean by correctness and introduces a theory of computational complexity to understand efficiency.

We would like an iterative algorithm to reach an optimal solution after a finite number of iterations, or at least converge to optimal as the number of iterations grows. If the problem does not have an optimal solution, the algorithm should stop (in a finite number of iterations), having verified that the problem does not have an optimal solution. An algorithm that does this for all problems of a certain type is called *correct* or *exact*. Algorithms that are not guaranteed to find an optimal solution might still be valuable heuristics.

Why might an iterative algorithm not terminate in a finite number of iterations? Assuming the algorithm is deterministic and that the next solution depends only on the current solution, if it revisits a solution that it visited on an earlier iteration, then it will cycle repeatedly through a sequence of solutions. The cycle might consist of just one solution, in which case it is stuck at that solution. In some algorithms, the next solution $\mathbf{x}^{k+1}$ depends on the current solution $\mathbf{x}^k$ and additional information, say, $\mathbf{w}^k$. It can be described as a mapping of the state $(\mathbf{x}^k, \mathbf{w}^k)$ into a new state $(\mathbf{x}^{k+1}, \mathbf{w}^{k+1}) = \Gamma(\mathbf{x}^k, \mathbf{w}^k)$. In this setting, the algorithm only cycles if it revisits a state. We will see that the simplex algorithm can visit the same solution $(x_1, \ldots, x_n)$ repeatedly without getting stuck because it uses a set of current variables, called the basis, as additional state information. If the same solution and basis were revisited, the simplex algorithm would cycle. If the feasible region is unbounded, an algorithm might not terminate because it visits larger and larger solutions; this behavior is called diverging. Even on a bounded feasible region, other undesirable behavior is possible. One famous example is that Newton's method for solving an equation can exhibit chaos, meaning that the sequence of estimates x^k does not converge or even approach a cycle.

The ultimate measure of the efficiency of an algorithm is how long it takes to run for a specific problem on specific computer hardware. One approach that is popular in the optimization community is to run a collection of test problems on different algorithms using the same computers. This captures all features of the software that implements the algorithm. Programming details, such as the data structure, amount of memory used and when it is accessed, avoiding duplicate computations, and the use of multiple processors, can have a substantial impact on run time. However, we are interested in comparing algorithms at a more abstract level without considering how they are implemented and the implications of hardware.

The simplest measure used for the speed of an algorithm is the number of arithmetic operations performed. Arithmetic operations include addition, multiplication, division, and comparisons. This measure ignores other instructions, such as conditional or read/write statements. It also assumes that each operation takes the same time, when in fact operations on large integers or high-precision floating point numbers take longer.

Another crucial issue ignored by the operation count is the use of multiple processors. Some algorithms, such as the traditional method of multiplying matrices, can be easily divided up among multiple processors and run in parallel. For such algorithms, using n processors achieves a speedup factor of nearly n. Other algorithms, such as Gauss–Jordan elimination for solving a system of linear equations, are not easily parallelized and their speedup factor may not grow linearly in n. Unfortunately, the optimization algorithms presented in this book are not easily parallelized. However, there has been extensive work on parallel algorithms for optimization, particularly for network optimization problems and dynamic programming. See Bertsekas and Tsitsiklis (2015) for a comprehensive treatment.

The amount of memory required by different algorithms for the same problem may also vary enough to be of concern. However, for the algorithms we consider the memory requirement is comparable with the size of the input data, so we will not use memory requirements to compare the algorithms.

Once we decide to use the operation count to describe the time required by an algorithm, the remaining issue is how to summarize the efficiency of an algorithm over the variety of problems of different sizes that it can solve. It is not practical to determine the number of iterations required for each specific problem. Instead, one can often determine the maximum possible number of iterations as a function of the problem size, or a useful bound on this maximum. If the algorithm converges to an optimal solution, one tries to bound the rate of convergence. Using the maximum, or worst-case, number of iterations gives a good prediction of typical performance for some algorithms; for others it is quite pessimistic. However, there is no good theory for calculating average behavior, so worst-case performance is used. Next, we describe the kinds of computational complexity results that can be obtained.

5.3.1 Computational Complexity

For simple algorithms, the exact number of operations can be found. To compute the dot product of two vectors in $\mathbb{R}^n$, n multiplications and $n - 1$ additions are required, so the total number of operations is $2n - 1$. For more complex algorithms, an exact count may be difficult but an approximate count is useful. When solving an $n \times n$ system of linear equations using Gauss–Jordan elimination, we perform up to n^2 row operations on the augmented matrix. At the beginning, they require up to n multiplications and n additions, but later in the algorithm some entries are zero and fewer operations are required. We could find an exact count, but a useful bound is $2n(n^2) = 2n^3$. In fact, in computational complexity, we are most concerned about the order, or rate of growth. The function $2n^3$ increases cubically with n.

Order of Growth Notation   Consider nonnegative functions $f(n)$ and $g(n)$, defined for $n > 0$.
We say $f(n) = O(g(n))$ if $f(n) \leq kg(n)$ for all $n \geq N$ for some $k, N > 0$.
We say $f(n) = \Omega(g(n))$ if $f(n) \geq kg(n)$ for all $n \geq N$ for some $k, N > 0$.
We say $f(n) = \Theta(g(n))$ if both $f(n) = O(g(n))$ and $f(n) = \Omega(g(n))$.

The symbol $O(\cdot)$ is called "big O" notation. If $f(n) = O(n^2)$, we say f is "big O of n^2" or order n^2. If $f(n) = \Theta(g(n))$, we say f and g are the same order. For example, $2n^3 = \Theta(n^3)$.

As another example, consider $f(n) = n + 10 \ln n$. The fastest growing term is n, so f is order n, i.e. $n + 10 \ln n = \Theta(n)$. To demonstrate this from the definition, we can verify that

$$n \leq n + 10 \ln n \leq 2n$$

for $n \geq 100$. Because "big O" is an upper bound, weaker bounds such as $n + 4n \ln n = O(n^2)$ are correct, but the tighter bound $n + 4n \ln n = O(n \ln n)$ is more informative.

For optimization problems, to describe how run time depends on the size of the problem, suppose we use one number, n, to represent the problem size, such as the number of objects in the knapsack problem. Linear programs have two obvious measures of size, the number of variables and the number of functional constraints. However, to simplify the discussion, we assume these are combined into a single measure. It is customary to refer to all optimization problems of a certain form as a *problem* and specific numerical examples as *instances* of the problem. Thus, Example 5.3 is an instance of the knapsack problem, which consists of all instances of the form (5.1). The set of all linear programs also comprises a problem.

We will use the worst-case operation count to measure the running time of an algorithm. Each instance has a running time. Let $T(n)$ be the worst-case running time for all instances of size n. For many discrete optimization problems, such as the shortest path problem, there are algorithms where $T(n)$ is low order, such as $O(n)$, $O(n^2)$, or $O(n^3)$. We could also say they are $\Theta(n)$, etc. These are often grouped together as polynomial time.

An algorithm is **polynomial time** if its running time $T(n) = O(n^k)$ for some positive integer k.

However, for other important problems, the best known algorithms have running times that are $\Theta(2^{cn})$ for some constant $c > 0$. This means that their running time increases exponentially in the size of the problem. Such algorithms are called *exponential time.* If we don't know the exact value of c, we might have a lower bound and say the running time is $\Omega(2^{cn})$ for some c.

While the distinction between, say, an $O(n^2)$ and $O(n^3)$ algorithm has some impact on the size of problem that can be solved, the difference between an $O(n^3)$ and an exponential time algorithm is much more dramatic. As $n \to \infty$, exponential functions grow faster than any polynomial function. Thus, in this ordering, polynomial time algorithms are all viewed as relatively fast and efficient, while exponential time algorithms are viewed as slow. Much of the work in computational complexity seeks to determine whether or not there is a polynomial time algorithm for a problem, and to show that a given algorithm is polynomial time. When there are polynomial time algorithms for a problem and large instances – or short running times – are important, then these algorithms are compared and faster ones sought.

Example 5.6 (Running time and problem size). Suppose we have a computing budget of 10^5 seconds (a little over a day) on a machine that runs at 4GHz but averages 4 cycles per operation (we are still ignoring instructions other than arithmetic operations). Then the budget allows for $10^5 \times 10^9 = 10^{14}$ operations. For an algorithm with $T(n) = 2n^3$, this budget can solve any problem with size n satisfying

$$2n^3 \leq 10^{14},$$

or $n \leq 36{,}840$. In contrast, an exponential time algorithm with $T(n) = 2(2^n)$ can solve any problem with size $n \leq 45$.

Generally, having lots of computing power makes a big difference for polynomial time algorithms but not for exponential time algorithms. As noted earlier, solving an $n \times n$ system of linear equations using Gauss–Jordan elimination is $O(n^3)$. Finding the inverse of an $n \times n$ matrix also requires an $O(n^3)$ algorithm. The standard method of multiplying two $n \times n$ matrices is $O(n^3)$, although there are somewhat faster methods.

Problems

1 Solve by hand the subset selection problem

$$\begin{aligned}
\max \quad & 9x_1 + 11x_2 + 8x_3 + 21x_4 + 7x_5 + 20x_6 + 5x_7 \\
\text{s.t.} \quad & x_1 + x_2 + x_3 + x_4 + x_5 + x_6 + x_7 \leq 4 \\
& x_j \text{ binary.}
\end{aligned}$$

2 For the knapsack problem

$$\text{max} \quad 34x_1 + 42x_2 + 25x_3 + 15x_4 + 9x_5 + 8x_6$$
$$\text{s.t.} \quad 10x_1 + 14x_2 + 8x_3 + 6x_4 + 4x_5 + 3x_6 \le 28$$
$$x_j \text{ binary.}$$

(a) Use the greatest value heuristic to find a solution.
(b) Use the smallest size heuristic to find a solution.
(c) Use the greatest value per width heuristic to find a solution.
(d) Solve the linear programming relaxation by hand. Does this show that any of the heuristic solutions are optimal?
(e) Find the optimal solution.

3 For the knapsack problem

$$\text{max} \quad 6x_1 + 8x_2 + 5x_3 + 22x_4 + 5x_5$$
$$\text{s.t.} \quad 4x_1 + 5x_2 + 3x_3 + 12x_4 + 2x_5 \le 17$$
$$x_j \text{ binary.}$$

(a) Use the greatest value per width heuristic to find a solution.
(b) Solve the linear programming relaxation by hand. Does this show that the solution to (a) is optimal?

4 A machine is usually configured to run type 1 jobs, but each day it must also run type 2, 3, 4, and 5 jobs. The setup time in seconds between each type of job is shown in the table below. Choosing the order in which to set up the job types as quickly as possible is a traveling salesperson problem.

Job type	1	2	3	4	5
1	—	500	200	185	205
2	500	—	305	360	340
3	200	305	—	320	165
4	185	360	320	—	302
5	205	340	165	302	—

(a) Use the nearest neighbor heuristic, starting at 1 and 2, to find solutions.
(b) Use the cheapest edge heuristic to find a solution.
(c) Use the cheapest insertion heuristic, starting at 1, to find a solution.
(d) For the solution in (b), use a 2-interchange to find a better solution.

5 For the TSP in Example 4.6:

(a) Use the nearest neighbor algorithm, starting at London, to find a solution.

(b) Use the cheapest edge heuristic to find a solution.

(c) Use the cheapest insertion heuristic, starting at London, to find a solution.

6 For the objective max $f(x_1, x_2) = x_1^3 - 4x_1x_2^2 + x_2^3$, find the improving directions at

(a) $(1, 1)$

(b) $(2, 0)$.

7 For the objective min $f(x_1, x_2) = 4x_1^2 + 4x_1x_2 + 3x_2^2 - 6x_2$ at $(2, -1)$, determine whether each direction is improving or not

(a) $\mathbf{d} = (1, -4)$

(b) $\mathbf{d} = (0, 2)$

(c) $\mathbf{d} = (-1, 3)$.

8 For the objective max $5x_1 - 2x_2 + 4x_3$ at $(3, 3, 0)$, determine whether each direction is improving or not

(a) $\mathbf{d} = (2, 4, 1)$

(b) $\mathbf{d} = (-2, 2, 3)$.

9 For the constraints

$$2x_1 + 4x_2 \leq 18,$$
$$5x_1 - 3x_2 \leq 15,$$
$$x_j \geq 0.$$

Find the maximum feasible step size in the direction $\mathbf{d}$ from the point $\mathbf{x}$, where

(a) $\mathbf{d} = (1, 2)$ and $\mathbf{x} = (3, 1)$

(b) $\mathbf{d} = (3, -1)$ and $\mathbf{x} = (0, 2)$

(c) $\mathbf{d} = (3, 0)$ and $\mathbf{x} = (3, 3)$.

10 For the constraints

$$x_1 - 3x_2 + 5x_3 \leq 30,$$
$$3x_1 + 4x_2 \qquad \geq 12,$$
$$x_j \geq 0.$$

find the maximum feasible step size in the direction $\mathbf{d}$ from the point $\mathbf{x}$, where

(a) $\mathbf{d} = (1, 2, 1)$ and $\mathbf{x} = (4, 1, 2)$

(b) $\mathbf{d} = (2, -4, 0)$ and $\mathbf{x} = (10, 5, 1)$

(c) $\mathbf{d} = (2, -2, -1)$ and $\mathbf{x} = (1, 3, 4)$.

11 Suppose **x** is a local maximum for a knapsack problem (5.1). In particular, suppose it is 2-optimal.
  (a) Is **x** guaranteed to be an optimal solution? If not, give an example where it is not.
  (b) Now assume $a_i = 1$, which is the subset selection problem. Is **x** guaranteed to be an optimal solution? If not, give an example where it is not.

12 For a problem with three binary variables, find the 2-neighborhood of the solution $(1, 1, 0)$. Is $(0, 1, 1)$ in this 2-neighborhood?

13 For the problem $\min f(x)$, $x \in [a, \infty)$ where f is differentiable, find a necessary condition for a local minimum.

14 Finding the row-reduced form of a $m \times n$ matrix requires at most $\frac{m(m-1)}{2}$ row operations, each of which requires at most $2n$ arithmetic operations. What is the complexity of this algorithm in $O(\cdot)$ notation?

15 Given a computation budget of 10^{12} operations, find the largest n for which any problem of size n can be solved if the worst-case running time of the algorithm is
  (a) $T(n) = 8n^3$
  (b) $T(n) = 3^n$.

6

Convexity

For continuous optimization, the "easy" problems are those where local optimality implies global optimality. Improving search algorithms for such problems will not get stuck at the wrong local optimum. We would like conditions on the objective function and the constraints that guarantee this. The condition on the objective function used in this chapter is an extension of the condition used when minimizing a function of one variable: if the function is convex, a local minimum is also a global minimum. An analogous condition, also called convexity, is introduced for the feasible region. The following example shows why it is needed.

Example 6.1. Consider the problem

$$\max \quad z = 30x + 2y$$
$$\text{s.t.} \quad y \leq x^3 - 15x^2 + 48x + 80$$
$$x \leq 10$$
$$x, y \geq 0.$$

If we sweep the objective function line in the increasing direction, Figure 6.1 shows that the line with $z = 322$ has a point of tangency to the boundary of the feasible region at $\mathbf{x}^* = (3, 116)$. It is a local maximum because all points in the feasible region near $\mathbf{x}^*$ lie below this line. In other words, if we sweep the line any further, it leaves this part of the feasible region. However, the optimal solution is $(10, 60)$ with an optimal value $z = 420$.

In this example, the local maximum arises because the first constraint, viewed as a function of x, changes from concave to convex. Can we describe when a local maximum can occur in a more general way? Notice that the line left the feasible region at $x = 3$ but reentered it at $x = 9$. If we only consider regions where lines cannot do this, the problem of local extrema will be eliminated – at least for linear objective functions. That is exactly how convexity of a feasible region will be defined.

Linear and Convex Optimization: A Mathematical Approach, First Edition. Michael H. Veatch.
© 2021 John Wiley & Sons, Inc. Published 2021 by John Wiley & Sons, Inc.
Companion website: www.wiley.com/go/veatch/convexandlinearoptimization

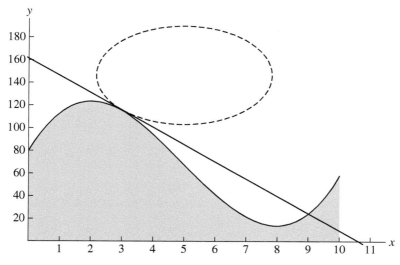

Figure 6.1 Feasible region and isocontours for Example 6.1.

The same condition on the feasible region works for nonlinear objectives if they have the correct concavity. To understand why, change the objective in Example 6.1 to min $f(x,y) = (x-5)^2 + \frac{1}{225}(y-146)^2$. The level curves of this function are ellipses; the curve $f(x,y) = 8$ is tangent to the boundary of the feasible region at the point (3,116) as shown in Figure 6.1. Notice that, because of the direction of curvature of the level curve, it lies above the tangent line, making (3,116) a local minimum of this problem. In fact, the entire region inside the ellipse, where $f(x,y) \geq 8$, lies above tangent line. As long as the feasible region does not cross this line, it is impossible for it to intersect the ellipse and we would know (without examining other feasible solutions) that (3,116) was optimal.

There are other conditions that are sometimes used, but the convexity conditions play a central role in optimization because they are often easy to check. We begin in Section 6.1 by describing convexity for feasible regions, that is, for sets. Although the conditions can be stated in terms of the constraints, it is useful – and more general – to define convexity for sets. Section 6.2 turns to convexity of functions and presents the primary result of this chapter, on when local optimality implies global optimality.

6.1 Convex Sets

A convex set is one that contains the line segment between any two points in the set. The line segment between $\mathbf{x}, \mathbf{y} \in \mathbb{R}^n$ can be written parametrically

Figure 6.2 The first set is convex. The second is not.

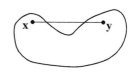

as $y + \lambda(x - y)$, or $\lambda x + (1 - \lambda)y$, where the scalar $\lambda \in [0, 1]$. For example, if $x = (7, 0, -2)$ and $y = (1, 3, 4)$, the line segment between them can be written

$$\lambda \begin{bmatrix} 7 \\ 0 \\ -2 \end{bmatrix} + (1 - \lambda) \begin{bmatrix} 1 \\ 3 \\ 4 \end{bmatrix} = \begin{bmatrix} 1 \\ 3 \\ 4 \end{bmatrix} + \lambda \begin{bmatrix} 6 \\ -3 \\ 6 \end{bmatrix}.$$

Choosing $\lambda = \frac{1}{3}$, this expression becomes $(1 + 2, 3 - 1, 4 - 2) = (3, 2, 2)$, which is one-third of the distance from x to y.

> The set $S \in \mathbb{R}^n$ is **convex** if for all $x, y \in S$ and $\lambda \in [0, 1]$, $\lambda x + (1 - \lambda)y \in S$.

Figure 6.2 shows an example of a set that is convex and one that is not. To show that a set is not convex, we pick two points in the set and a value of λ for which the intermediate point is not in the set.

Example 6.2. Show that $S = \{(x, y) : xy \le 1, x \ge 0, y \ge 0\}$ is not convex.

Pick the points $(0, 4)$ and $(4, 0)$. Then $0.5(0, 4) + 0.5(4, 0) = (2, 2)$ is not in S.

Example 6.3. Show that the hypercube $S = \{x \in \mathbb{R}^n : 0 \le x_i \le 1, i = 1, \ldots, n\}$ is convex.

For any $x, y \in S$ and $\lambda \in [0, 1], 0 \le \lambda x_i + (1 - \lambda)y_i \le \lambda + (1 - \lambda) = 1, i = 1, \ldots, n$. Thus, $\lambda x + (1 - \lambda)y \in S$.

Example 6.4. Show that $S = \{(x_1, x_2) : x_1^2 + x_2 \le 4, x_1 \ge 0, x_2 \ge 0\}$ is convex.

For any $x, y \in S$ and $\lambda \in [0, 1], x_1 \ge 0$ and $y_1 \ge 0$, so $\lambda x_1 + (1 - \lambda)y_1 \ge 0$ and similarly for the second nonnegativity constraint. It remains to show that

$$[\lambda x_1 + (1 - \lambda)y_1]^2 + \lambda x_2 + (1 - \lambda)y_2 \le 4.$$

Expanding the left side,

$$\lambda^2 x_1^2 + 2\lambda(1 - \lambda)x_1 y_1 + (1 - \lambda)^2 y_1^2 + \lambda x_2 + (1 - \lambda)y_2$$
$$= \lambda(x_1^2 + x_2) + (1 - \lambda)(y_1^2 + y_2)$$

$$- \lambda(1 - \lambda)x_1^2 + 2\lambda(1 - \lambda)x_1y_1 - \lambda(1 - \lambda)y_1^2$$
$$\leq 4 - \lambda(1 - \lambda)(x_1 + y_1)^2$$
$$\leq 4. \tag{6.1}$$

As the last example illustrates, checking convexity from the definition can be tedious. Another method, based on convex functions, is described in Section 6.2. Notice that each constraint was checked for convexity separately. This approach is valid because the feasible region is the intersection of the solution sets for the individual constraints and the intersection of convex sets is another convex set (see Exercise 6.7).

We will often be interested in solution sets and feasible regions that are unbounded.

A set S is **bounded** if there exists a constant M such that $\|\mathbf{x}\| \leq M$ for all $\mathbf{x} \in S$. Otherwise, S is **unbounded**.

A realistic optimization problem should not have an unbounded solution. However, although actual decision variables are bounded, it is often simpler from a modeling perspective to allow an unbounded region than to include all the constraints that would make it bounded. The following terminology applies to these sets.

Let $\mathbf{a}$ be a nonzero vector in $\mathbb{R}^n$.
  A **hyperplane** is a set of the form $\{\mathbf{x} \in \mathbb{R}^n : \mathbf{a}^T\mathbf{x} = k\}$.
  A **half-space** is a set of the form $\{\mathbf{x} \in \mathbb{R}^n : \mathbf{a}^T\mathbf{x} \leq k\}$.

Because they are defined by one constraint, hyperplanes and half-spaces have dimension $n - 1$. For example, a hyperplane in $\mathbb{R}^3$ is a plane; it has dimension 2. In general, the dimension of the solution set of a system of linear equations $\{\mathbf{x} \in \mathbb{R}^n : \mathbf{Ax} = \mathbf{b}\}$ is defined as the dimension of the corresponding subspace $\{\mathbf{x} \in \mathbb{R}^n : \mathbf{Ax} = \mathbf{0}\}$, i.e. the number of vectors in a basis. These solution spaces are known as *affine sets*. The definition of an affine set is that it contains the line through any two points in the set. Since convex sets are only required to contain the line segment between the two points, all affine sets are convex. Also, all affine sets are unbounded, except for single points (which have dimension 0).

Now consider the general case of a system of linear inequalities.

A **polyhedron** is a set of the form $\{\mathbf{x} \in \mathbb{R}^n : \mathbf{Ax} \leq \mathbf{b}\}$.
A **polytope** is a bounded polyhedron.

This terminology is not universal. The intersection of any number of half-spaces and hyperplanes is a polyhedron. They are called the *defining hyperplanes* of the polyhedron. Exercises 6.8 and 6.9 show that hyperplanes and half-spaces are convex. These facts and the intersection property (Exercise 6.7) lead to the following convexity result.

Theorem 6.1: *Polyhedral sets are convex.*

Thus, the feasible region of any linear program is convex. Polytopes have another nice property: they are determined by their corner points. Although we won't make the notion of a corner point precise until the next chapter, the following definitions are closely related. The point $\lambda\mathbf{x} + (1 - \lambda)\mathbf{y}$ is a weighted average of the points $\mathbf{x}$ and $\mathbf{y}$. The line segment between the points can be thought of as all weighted averages of them. The following definitions extend the concept to any number of points.

For the set of points $\mathbf{v}_1, \ldots, \mathbf{v}_k \in \mathbb{R}^n$ and scalars α_i satisfying $\sum_{i=1}^k \alpha_i = 1$ and $\alpha_i \geq 0$, $i = 1, \ldots, k$, $\sum_{i=1}^k \alpha_i \mathbf{v}_i$ is a **convex combination** of $\mathbf{v}_1, \ldots, \mathbf{v}_k$.
 The **convex hull** $\text{conv}\{\mathbf{v}_1, \ldots, \mathbf{v}_k\}$ is the set of all convex combinations of $\mathbf{v}_1, \ldots, \mathbf{v}_k$.

In Figure 6.3, $\mathbf{x} = 0.5\mathbf{v}_1 + 0.3\mathbf{v}_2 + 0.2\mathbf{v}_3$ is a convex combination of the three corner points of the triangle. The triangle is the convex hull of $\{\mathbf{v}_1, \mathbf{v}_2, \mathbf{v}_3\}$. One can show that $\text{conv}\{\mathbf{v}_1, \ldots, \mathbf{v}_k\}$ is convex and that it is the smallest convex set containing $\mathbf{v}_1, \ldots, \mathbf{v}_k$ (see Exercise 6.10).

This construction works for any polytope: it is the convex hull of its corner points. If a polyhedral set is unbounded, it is clearly not the convex hull of its corner points, which is a bounded set. However, as we will see in the next chapter, it can be constructed from its corner points and certain directions that describe where it is unbounded. The following definition lays the groundwork.

The vector $\mathbf{d} \neq 0$ is an **unbounded direction** of S if for all $\mathbf{x} \in S, \mathbf{x} + \lambda\mathbf{d} \in S$ for all $\lambda > 0$.

Figure 6.3 The point $\mathbf{x}$ is a convex combination of $\mathbf{v}_1, \mathbf{v}_2, \mathbf{v}_3$.

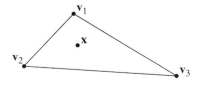

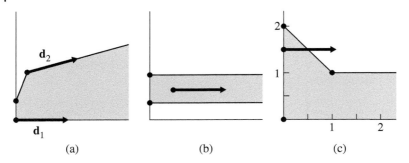

Figure 6.4 Unbounded sets. Only the first two have unbounded directions.

In Figure 6.4, the first set can be described by three corner points and two unbounded directions, one lying in each of the rays that are boundaries of the set. However, every nonnegative linear combination of these two directions is also an unbounded direction. The second set has only one unbounded direction $\mathbf{d} = (1, 0)$ (plus nonnegative multiples of it). The third set has no unbounded directions. Note that $(1, 0)$ is not an unbounded direction; for example, $(0, 1.5) + \lambda(1, 0)$ is not in the set for large λ. The last example is not a convex set. If an unbounded set is convex, then it has at least one unbounded direction.

The set of unbounded directions of a convex set is a particularly simple form of unbounded set, known as a cone. This concept will be useful when we describe optimality conditions in the next two chapters.

A set $C \in \mathbb{R}^n$ is a **cone** if, for every $\mathbf{x} \in C$, $\lambda\mathbf{x} \in C$ for all $\lambda \geq 0$. A polyhedron of the form $C = \{\mathbf{x} \in \mathbb{R}^n : \mathbf{Ax} \geq \mathbf{0}\}$ is a **polyhedral cone**.

We will only be interested in polyhedral cones. Figure 6.5 shows an example; the positive orthant in $\mathbb{R}^3$ is another. Polyhedral cones start at the origin and have at most one corner point. As polyhedra, they are convex. To summarize, given a set

Figure 6.5 A polyhedral cone.

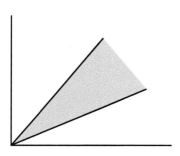

of vectors $\mathbf{v}_1, \ldots, \mathbf{v}_k \in \mathbb{R}^n$ we can define the following sets of linear combinations $\sum_{i=1}^{k} \alpha_i \mathbf{v}_i$:

- The set of all linear combinations is a subspace of $\mathbb{R}^n$.
- The set of all affine combinations, with $\sum_{i=1}^{k} \alpha_i = 1$, is an affine set.
- The set of all conic combinations, with $\alpha_i \geq 0$, $i = 1, \ldots, k$, is a polyhedral cone.
- The set of all convex combinations, with $\sum_{i=1}^{k} \alpha_i = 1$ and $\alpha_i \geq 0$, $i = 1, \ldots, k$, is the convex hull of $\mathbf{v}_1, \ldots, \mathbf{v}_k$.

6.2 Convex and Concave Functions

There is a close connection between convex sets and convex and concave functions, to which we now turn. Informally, a function is convex if all secant lines lie above (or on) the function and concave if all secant lines lie below (or on) the function; see Figure 6.6. Many functions are convex on some interval and concave on another; such functions are neither convex nor concave The same concept applies to functions of more than one variable.

A function f on a convex set S is **convex** if for all $\mathbf{x}, \mathbf{y} \in S$ and $\lambda \in [0, 1]$, $f(\lambda \mathbf{x} + (1 - \lambda)\mathbf{y}) \leq \lambda f(\mathbf{x}) + (1 - \lambda)f(\mathbf{y})$.

A function f on a convex set S is **concave** if for all $\mathbf{x}, \mathbf{y} \in S$ and $\lambda \in [0, 1]$, $f(\lambda \mathbf{x} + (1 - \lambda)\mathbf{y}) \geq \lambda f(\mathbf{x}) + (1 - \lambda)f(\mathbf{y})$.

If these inequalities are strict for all $x \neq y$ and $\lambda \in (0, 1)$, then f is **strictly convex** or **strictly concave**.

Example 6.5. Show that the function $g(x_1, x_2) = x_1^2 + x_2$ used in Example 6.4 is convex.

For any $\mathbf{x}, \mathbf{y} \in \mathbb{R}^2$ and $\lambda \in [0, 1]$, we must show that

$$[\lambda x_1 + (1 - \lambda)y_1]^2 + \lambda x_2 + (1 - \lambda)y_2 \leq \lambda(x_1^2 + x_2) + (1 - \lambda)(y_1^2 + y_2).$$

Figure 6.6 The first function is convex. The second is concave.

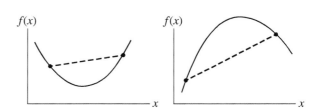

Substituting (6.1) for the left side,

$$\lambda(x_1^2 + x_2) + (1 - \lambda)(y_1^2 + y_2) - \lambda(1 - \lambda)x_1^2 + 2\lambda(1 - \lambda)x_1 y_1 - \lambda(1 - \lambda)y_1^2$$
$$\leq \lambda x_1^2 + \lambda x_2 + (1 - \lambda)y_1^2 + (1 - \lambda)y_2,$$

which simplifies to

$$-x_1^2 + 2x_1 y_1 - y_1^2 = -(x_1 - y_1)^2 \leq 0.$$

We usually do not need to apply the definition to check that a function is convex or concave because there are tests based on the second derivative, discussed in Chapter 14. Now consider an inequality constraint.

Lemma 6.1: *If f is a convex function on a convex set S, then $L = \{\mathbf{x} : f(\mathbf{x}) \leq k\}$ is a convex set. If f is a concave function on S, then $U = \{\mathbf{x} : f(\mathbf{x}) \geq k\}$ is a convex set.*

Proof: We prove the case where f is convex. For $\mathbf{x}, \mathbf{y} \in L$, $f(\mathbf{x}) \leq k$ and $f(\mathbf{y}) \leq k$. Then for $\lambda \in [0, 1]$,

$$f(\lambda \mathbf{x} + (1 - \lambda)\mathbf{y}) \leq \lambda f(\mathbf{x}) + (1 - \lambda)f(\mathbf{y}) \leq k,$$

which means that $\lambda \mathbf{x} + (1 - \lambda)\mathbf{y} \in L$. □

The proof is taken from (Luenberger and Ye, 2016, Proposition 7.3). This lemma allows the convexity of a feasible region to be checked by checking the convexity/concavity of its inequality constraints. Now consider the objective function. For unconstrained optimization problems, convexity of the objective function is sufficient to guarantee that a local minimum is also a global minimum. We are ready to state the analogous result for constrained optimization problems.

Theorem 6.2: *Let S be a convex set.*
If f is convex and $\mathbf{x}^*$ is a local minimum of $\min_{\mathbf{x} \in S} f(\mathbf{x})$ then $\mathbf{x}^*$ is optimal.
If f is concave and $\mathbf{x}^*$ is a local maximum of $\max_{\mathbf{x} \in S} f(\mathbf{x})$ then $\mathbf{x}^*$ is optimal.

Proof: We prove the convex case using proof by contradiction. Let $\mathbf{x}^*$ be a local minimum and suppose there is a $\mathbf{y} \in S$ with $f(\mathbf{y}) < f(\mathbf{x}^*)$. Since S is convex, $\lambda \mathbf{x}^* + (1 - \lambda)\mathbf{y} \in S$ for all $\lambda \in [0, 1]$. Then

$$f(\lambda \mathbf{x}^* + (1 - \lambda)\mathbf{y}) \leq \lambda f(\mathbf{x}^*) + (1 - \lambda)f(\mathbf{y})$$
$$< \lambda f(\mathbf{x}^*) + (1 - \lambda)f(\mathbf{x}^*)$$
$$= f(\mathbf{x}^*).$$

Now, for any neighborhood of $\mathbf{x}^*$ we can choose a λ close to 1 so that $\lambda \mathbf{x}^* + (1 - \lambda)\mathbf{y}$ is in the neighborhood. But this point has a better function value than $\mathbf{x}^*$, contradicting $\mathbf{x}^*$ being a local minimum. □

To apply Theorem 6.2 to linear programs, we need the following lemma. The proof is Exercise 6.12.

Lemma 6.2: *All linear functions are both convex and concave.*

Combining this lemma with Theorem 6.1 establishes the following corollary of Theorem 6.2.

Corollary 6.1: *If $\mathbf{x}^*$ is a local optimum of a linear program, then $\mathbf{x}^*$ is optimal.*

When using an improving search algorithm, Theorem 6.2 is important because it prevents stopping at a local optimum that is not a global optimum. The proof also indicates what the improving directions are: for any feasible $\mathbf{x}$ and $\mathbf{y}$ with $f(\mathbf{x}) > f(\mathbf{y})$ (when minimizing), $\mathbf{d} = \mathbf{y} - \mathbf{x}$ is an improving direction. Another way to think of Theorem 6.2 is that optimality conditions are possible for these problems. It reduces optimality to a local condition that can be checked at $\mathbf{x}^*$ (i.e. in any neighborhood of $\mathbf{x}^*$). When solving a linear program graphically by sweeping the line where the objective function is constant, an optimal solution is found when the line touches the feasible region S at $\mathbf{x}^*$ but does not enter the interior of S in neighborhoods of $\mathbf{x}^*$. Because S is convex, it lies in one of the half spaces defined by the line (Figure 6.7). In higher dimensions, a hyperplane where the objective function is constant plays the same role and is called a *supporting hyperplane* of S. The next chapter states optimality conditions for linear programs based on this idea.

6.2.1 Combining Convex Functions

Certain combinations of convex functions are also convex. For example, sums and nonnegative multiples of convex functions on the same domain are also convex, while a negative multiple of a convex function is concave. These properties can be seen in the second derivative tests for convexity as well.

While second derivative tests are very useful, a different method is needed to establish concavity for the piecewise linear functions encountered in Chapter 3 because these functions are not differentiable.

Figure 6.7 The line $\mathbf{c}^T\mathbf{x} = k$ only intersects S at the point shown.

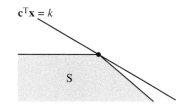

$\mathbf{c}^T\mathbf{x} = k$

S

Lemma 6.3: *If $f_1, \ldots, f_k$ are convex functions on a convex set S and*

$$g(\mathbf{x}) = \max\{f_1(\mathbf{x}), \ldots, f_k(\mathbf{x})\}$$

then g is convex. If $f_1, \ldots, f_k$ are concave functions on a convex set S and

$$h(\mathbf{x}) = \min\{f_1(\mathbf{x}), \ldots, f_k(\mathbf{x})\}$$

then h is concave.

Proof: For any $\mathbf{x}, \mathbf{y} \in S$ and $\lambda \in [0, 1]$ and for some i,

$$\begin{aligned}
g(\lambda\mathbf{x} + (1 - \lambda)\mathbf{y}) &= f_i(\lambda\mathbf{x} + (1 - \lambda)\mathbf{y}) \\
&\le \lambda f_i(\mathbf{x}) + (1 - \lambda)f_i(\mathbf{y}) \\
&\le \lambda g(\mathbf{x}) + (1 - \lambda)g(\mathbf{y}),
\end{aligned}$$

which is the definition of convexity of g. The same proof applies to h with the inequalities reversed. □

For maximization problems, piecewise linear objective functions that can be written as the maximum of linear functions were converted to linear programs. This is an instance of the dictum there is no free lunch. A convex objective function was converted to a linear program by adding variables, but Theorem 6.2 applies to both. Non-convex problems generally cannot be converted to convex problems. In particular, Theorem 6.2 does not apply to the problem

$$\max_{\mathbf{x} \in S} \; g(\mathbf{x})$$

$$\text{s.t. } g(\mathbf{x}) = \max\{f_1(\mathbf{x}), \ldots, f_k(\mathbf{x})\}.$$

Even if $f_1, \ldots, f_k$ are convex functions, g is not necessarily convex.

In Examples 6.4 and 6.5, the solution set of the constraint $g(x_1, x_2) = x_1^2 + x_2 \le 4$ was shown to be convex by showing that g is a convex function. Another approach is to solve for the second variable, $x_2 \le h(x_1) = 4 - x_1^2$, and show that h is concave. For a constraint in the opposite direction, $x_2 \ge f(x_1)$, we would need to show that f is convex. This reduces the problem from a function of two variables to a function of one variable. To generalize, consider the graph of $y = f(\mathbf{x})$ for $\mathbf{x} \in \mathbb{R}^n$. The graph is in $n + 1$ dimensions. Another connection between convex sets and convex functions is the following.

Corollary 6.2: *If f is a convex function, then $\{(x_1, \ldots, x_n, y) : y \ge f(\mathbf{x})\}$ is a convex set.*

This set is called the *epigraph* of f; see Figure 6.8. Conversely, if f is concave then the points below the graph of f are a convex set, sometimes called the *hypograph*.

Figure 6.8 Epigraph of $f(x)$.

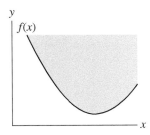

Problems

1 Find a formula for the line segment between $(2, -4, 0)$ and $(-1, 4, 6)$. Find the point on this line segment that is $\frac{3}{4}$ of the distance from the first point to the second.

2 Find a formula for the convex hull of $(1, 1)$, $(1, 5)$, $(4, 4)$, and $(2, 3)$. Use only the points that are necessary.

3 Determine graphically if the set $\{(x, y) : y \leq e^{-x}, x \geq 0, y \geq 0\}$ is convex. If not, find two points that violate the definition.

4 Determine graphically if the set $\{(x, y) : y \geq e^x, y \geq e^{-x}, y \leq e\}$ is convex. If not, find two points that violate the definition.

5 Is the following region is convex? If not, find two points that violate the definition.

$$x_1 + 2x_2 + x_3 \leq 4,$$
$$2x_1 + 2x_2 \quad\quad \leq 4,$$
$$x_1 + 2x_2 + x_3 \leq 3,$$
$$x_j \geq 0, \text{ integer.}$$

6 Show that the region

$$x_1 - x_2 \leq 0,$$
$$-2x_1 + x_2 \leq 0$$

is convex.

7 Let S and T be convex sets in $\mathbb{R}^n$. Show that $U = S \cap T$ is a convex set.

8 Show that the hyperplane $\{\mathbf{x} \in \mathbb{R}^n : \mathbf{a}^T\mathbf{x} = b\}$ is a convex set.

9   Show that the half-space $\{x \in \mathbb{R}^n : a^T x \leq b\}$ is a convex set.

10   (a) Show that $\text{conv}(v_1, \dots, v_k)$ is a convex set.
      (b) Suppose that $v_1, v_2, v_3 \in S \subset \mathbb{R}^n$ and S is convex. Show that, for $\alpha_i \geq 0$ and $\alpha_1 + \alpha_2 + \alpha_3 = 1$, the convex combination $x = \alpha_1 v_1 + \alpha_2 v_2 + \alpha_3 v_3 \in S$.
      (c) Assuming that (b) holds for convex combinations of any number of points, show that $\text{conv}(v_1, \dots, v_k)$ is the smallest convex set containing $v_1, \dots, v_k$.

11   Find an unbounded direction of the region

$$
\begin{aligned}
2x_1 - 6x_2 &\geq -18, \\
-x_1 + 3x_2 &\geq -6, \\
x_1 + x_2 &\geq 6.
\end{aligned}
$$

12   Show that the linear function $f(x) = a^T x$ is both convex and concave.

13   Show that if f_1 and f_2 are linear functions in $\mathbb{R}^n$, then $g(x) = \max\{f_1(x), f_2(x)\}$ is convex.

14   Show that if f_1 and f_2 are linear functions in $\mathbb{R}^n$, then $h(x) = \min\{f_1(x), f_2(x)\}$ is concave.

15   Prove Corollary 6.2.

7

Geometry and Algebra of LPs

In this chapter we explore how the general principles of improving search algorithms from Chapter 5 can be applied to linear programs. In particular, we will (i) construct improving feasible directions, (ii) find a convenient optimality condition, and (iii) show that we only need to consider "corner points" when looking for optimal solutions. These results take advantage of the special structure of linear programs, namely, that the feasible region is a polyhedral set (from Chapter 6) and the objective function is linear.

We saw in Chapter 1 that, for examples of linear programs with two variables, the optimal solution occurred at a corner point. One idea for an improving search algorithm would be to move from a corner point along an edge to another corner point. We can visualize the same approach in three dimensions, where the feasible region is a polyhedron. In this chapter, these intuitive concepts are shown to be correct in higher dimensions. We derive a formula for computing the edge directions, checking whether they are improving, and using them to check optimality. Studying polyhedra and their corner points has both a geometric and algebraic aspect. The geometric concepts and graphs are important to develop understanding, but algebra is needed to do computations. More specifically, the feasible region as a set is a geometric concept; it is "representation free." The algebraic approach is to give a representation of the region using linear equations and inequalities, e.g. $\mathbf{Ax} \leq \mathbf{b}$, $\mathbf{x} \geq \mathbf{0}$. Properties that are true for any representation of a polyhedron are truly geometric properties; properties that may vary depending on the representation are algebraic.

The main theoretical result in this chapter is that, for polyhedra that do not contain a line, the search for optimal solutions to a linear program only needs to consider corner points. It is proven for general linear programs in Section 7.2 after describing corner points geometrically and algebraically in Section 7.1. After that, attention focuses on one form of linear program, the canonical form. Although algorithms could be constructed for the various forms of linear programs, it is more convenient to convert them to this form, so that a simpler and more streamlined algorithm can be used. Section 7.3 introduces this form and a method for constructing corner points. It also addresses the issues of degeneracy and adjacency of corner points in canonical form. Section 7.4 develops formulas for edge directions and an optimality condition based on them, again in canonical form. All of these will be needed for the algorithm presented in Chapter 9. The optional Section 7.5 returns to a geometric view, presenting an optimality condition and representation theorem for general polyhedra.

7.1 Extreme Points and Basic Feasible Solutions

Recall that a constraint is active at $\mathbf{x}$ if it is satisfied with equality. For problems with two variables, each corner point is the intersection of two lines and has two linearly independent active constraints. There may be additional active constraints, but only two can be linearly independent. Typically, a corner point is a vertex of a polygon. To find the point, we could solve the two constraint equations. This section extends these geometric and algebraic notions to any number of variables. In $\mathbb{R}^n$, n linearly independent active constraints have a unique solution; some of these are corner points.

Example 7.1 (Finding corner points). For the feasible region

$$1.5x + y \le 16,$$
$$x + y \le 12,$$
$$y \le 10,$$
$$x, y \ge \ 0$$

the first two constraints are linearly independent because their coefficient matrix has full rank. Requiring them to be active gives the equations

$$\begin{bmatrix} 1.5 & 1 \\ 1 & 1 \end{bmatrix} \begin{bmatrix} x \\ y \end{bmatrix} = \begin{bmatrix} 16 \\ 12 \end{bmatrix}.$$

The solution $(8, 4)$ is a corner point, which is feasible. However, if we pick the first and third constraints to be active, they intersect at $(4, 10)$, which is infeasible.

To look for corner points in $\mathbb{R}^n$, we first impose the equality constraints, then choose enough additional constraints to be active so that there are n linearly independent active constraints. These constraints have a unique solution; however, it may not be feasible because some of the inactive constraints could be violated.

> **Basic Solution**   The point $x \in \mathbb{R}^n$ is a **basic solution** of a polyhedral set if (i) All equality constraints are active and (ii) There are n linearly independent active constraints. A **basic feasible solution** (bfs) is a basic solution that satisfies all constraints.

For example, suppose a polyhedral set in $\mathbb{R}^5$ has three equality constraints, but only two are linearly independent. Then the solution space of these constraints has dimension $5 - 2 = 3$ and three linearly independent inequality constraints must be chosen as active to obtain a basic solution.

Example 7.2 (Feasible and infeasible basic solutions).   In Example 7.1 there are five inequality constraints, so there are at most $_5C_2 = 10$ basic solutions. Figure 7.1 shows that there are actually 9, because the two constraints $y \geq 0$ and $y \leq 10$ are linearly dependent and do not intersect. Five of them are bfs's; these are the corner points.

In fact, corner points and bfs's are equivalent. To show this, we first replace the informal concept of corner point with the following definition, which also applies to other convex sets.

Figure 7.1   Basic solutions for Example 7.1.

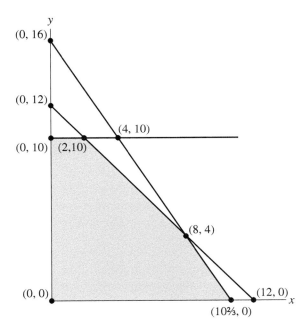

> **Extreme Point**   For a convex set S, $x \in S$ is an **extreme point** if it does not lie on a line segment between two distinct points in S.

For example, let $S = \{x \in \mathbb{R}^3 : 0 \le x_j \le 1\}$, the unit cube. Then $(\frac{1}{4}, 1, 1)$ is not an extreme point because it can be written as a convex combination of $\mathbf{y} = (0, 1, 1)$ and $\mathbf{z} = (1, 1, 1)$, namely, $\frac{3}{4}\mathbf{y} + \frac{1}{4}\mathbf{z}$, showing that it is on the line segment between $\mathbf{y}$ and $\mathbf{z}$. It is somewhat more tedious to show that a point *is* an extreme point, e.g. that $\mathbf{z}$ is an extreme point of the cube. We generally won't need to use the definition of an extreme point, as we will use the equivalent definition of a bfs.

Theorem 7.1:   *For a polyhedral set, extreme points are the same as basic feasible solutions.*

A proof is given in (Bertsimas and Tsitsiklis 1997, Theorem 2.3). See also Exercise 7.7.

7.1.1   Degeneracy

Theorem 7.1 establishes that an extreme point in $\mathbb{R}^2$ must have at least two active constraints, three in $\mathbb{R}^3$, etc. If it has more, they cannot be linearly independent and we call the point degenerate.

> **Degenerate Basic Solution**   For a set of linear constraints in $\mathbb{R}^n$, a basic solution is **degenerate** if it has more than n active constraints. Otherwise it is **nondegenerate**.

Figure 7.2 illustrates degenerate basic solutions in $\mathbb{R}^2$. The point C is a degenerate bfs. The points A and B are nondegenerate bfs's and D is a nondegenerate basic solution.

Although a bfs can also be defined geometrically, basic and degenerate solutions cannot. Their definition hinges on the algebraic representation. For example, we can always add more constraints without changing the feasible region, giving a different representation with more basic solutions. If a bfs is nondegenerate, we can add a constraint through that point and make it degenerate. Given a representation of a polyhedron, it is straightforward to check if a point is degenerate. We check whether each constraint is active; if more than n constraints are active (including all equality constraints) it is degenerate.

A final geometric concept will be useful. In two or three dimensions, an improving search algorithm can move along an edge of the feasible region

Figure 7.2 The point *C* is a degenerate basic feasible solution.

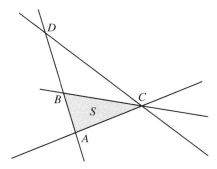

from one extreme point to another. Such extreme points are *adjacent* because they are connected by an edge. However, to generalize to any dimension, it is more convenient to use an equivalent algebraic concept. As long as they are nondegenerate, adjacent extreme points share all but one active constraint. For example, in Figure 7.1, the points $(2, 10)$ and $(8, 4)$ share one active constraint.

Adjacent Extreme Point   Distinct extreme points in $\mathbb{R}^n$ that share $n - 1$ linearly independent active constraints are **adjacent**.

The notion of an edge also extends to any dimension; see Section 7.5. Informally, we can think of moving to an adjacent extreme point as staying on the boundary of the feasible region, rather than moving through the interior, as long as this region has a dimension greater than one.

7.2 Optimality of Extreme Points

In this section we show that we only need to consider extreme points when searching for optimal solutions to linear programs. The proof is quite simple for bounded feasible regions, given the result from Section 6.1 that such a region is the convex hull of its extreme points. In other words, any point in a bounded polyhedron can be represented as a convex combination of its extreme points. For unbounded regions, we also need to consider unbounded directions, defined in Section 6.1.

Theorem 7.2   (*Representation theorem*):   *Let S be a polyhedral set with one or more extreme points* $\mathbf{v}_1, \dots, \mathbf{v}_k$. *Then* $\mathbf{x} \in S$ *if and only if*

$$\mathbf{x} = \sum_{i=1}^{k} \alpha_i \mathbf{v}_i + \mathbf{d} \tag{7.1}$$

for some $\{\alpha_i\}$ satisfying $\sum_{i=1}^{k} \alpha_i = 1$, $\alpha_i \geq 0$ and $\mathbf{d}$ is either an unbounded direction of S or $\mathbf{d} = \mathbf{0}$.

This theorem says that any point in such a polyhedron can be represented as a convex combination of its extreme points plus an unbounded direction. A stronger version of this theorem is stated in Section 7.4 that specifies how find $\mathbf{d}$; it is proven in (Bertsimas and Tsitsiklis, 1997, Theorem 4.15). Although we do not have a way to check the assumption that S has an extreme point, this will not be a problem. Any linear program that is feasible can be transformed to have non-negative variables, and in this form it has an extreme point. Now we are ready to prove the main result.

Theorem 7.3 *(**Fundamental theorem of linear programming**): If a linear program is feasible and has at least one extreme point, then either an extreme point is optimal or the linear program is unbounded.*

Proof: Consider the linear program $\max_{\mathbf{x} \in S} \mathbf{c}^T\mathbf{x}$, where S is a polyhedron with at least one extreme point. If S has an improving unbounded direction $\mathbf{d}$, then $\mathbf{c}^T\mathbf{d} > 0$ and the linear program is unbounded. Now suppose $\mathbf{c}^T\mathbf{d} \leq 0$ for any unbounded direction. Using the notation in (7.1), order the extreme points so that $\mathbf{v}_k$ has the largest objective function value, $\mathbf{c}^T\mathbf{v}_1 \leq \mathbf{c}^T\mathbf{v}_2 \leq \cdots \leq \mathbf{c}^T\mathbf{v}_k$. By Theorem 7.2, for any $\mathbf{x} \in S$

$$\mathbf{c}^T\mathbf{x} = \mathbf{c}^T \left(\sum_{i=1}^{k} \alpha_i \mathbf{v}_i + \mathbf{d} \right)$$
$$= \sum_{i=1}^{k} \alpha_i \mathbf{c}^T \mathbf{v}_i + \mathbf{c}^T \mathbf{d}$$
$$\leq \sum_{i=1}^{k} \alpha_i \mathbf{c}^T \mathbf{v}_k$$
$$= \mathbf{c}^T \mathbf{v}_k,$$

i.e. $\mathbf{v}_k$ is optimal. $\qquad\qquad\square$

This result has shaped the algorithms created to solve linear programs. While linear programs have continuous variables, they also have a discrete aspect, in that there are a finite number of extreme points. As noted earlier, the Fundamental Theorem applies to all linear programs that we will solve. For completeness, we now describe when polyhedra have extreme points. The next example shows that a linear program without extreme points can still have optimal solutions.

Figure 7.3 Feasible region for Example 7.3.

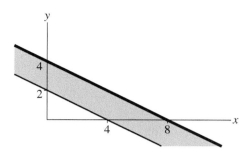

Example 7.3. Consider the linear program

$$\max x + 2y$$
$$\text{s.t. } x + 2y \le 8$$
$$x + 2y \ge 4.$$

As shown in Figure 7.3, it has an unbounded feasible region with no extreme points. However, it has an optimal solution: any point on the upper boundary has an objective function value of 8.

Notice that the feasible region in Figure 7.3 contains a line. In fact, this is equivalent to not having an extreme point.

> A polyhedron S **contains a line** if there exists a point $\mathbf{x} \in S$ and a nonzero vector $\mathbf{d}$ such that $\mathbf{x} + \lambda \mathbf{d} \in S$ for all λ.

In this definition the vector $\mathbf{d}$ is an unbounded direction of S; since $\lambda < 0$ is allowed, so is $-\mathbf{d}$. Containing a line requires S to be unbounded in two opposite directions.

Theorem 7.4: *Let S be a nonempty polyhedron. Then S has an extreme point if and only if it does not contain a line.*

Proof: First we prove that if $S = \{\mathbf{x} \in \mathbb{R}^n : \mathbf{a}_i^T \mathbf{x} \le b_i, \ i = 1, \ldots, m\}$ does not contain a line then it has an extreme point. Suppose S does not have an extreme point. Choose $\mathbf{x} \in S$ to have the maximum number of linearly independent active constraints, which must be less than n. Let $I = \{i : \mathbf{a}_i^T \mathbf{x} = b_i\}$ be the set of active constraint indices at $\mathbf{x}$. Because the vectors $\mathbf{a}_i$, $i \in I$ do not span $\mathbb{R}^n$, there exists a vector $\mathbf{d} \ne \mathbf{0}$ that is orthogonal to each of them, $\mathbf{a}_i^T \mathbf{d} = 0$, for $i \in I$. Consider the line $\mathbf{y} = \mathbf{x} + \lambda \mathbf{d}$ parameterized by λ. All constraints active at $\mathbf{x}$ are active on this line. One implication is that both $\mathbf{d}$ and $-\mathbf{d}$ are feasible directions at $\mathbf{x}$. However, because S does not contain a line, some constraint $j \notin I$ will eventually become violated as λ varies. At the point where it is about to be violated it becomes active, so $\mathbf{a}_j^T(\mathbf{x} + \lambda^* \mathbf{d}) = b_j$ for some λ^*.

Next we show that $\mathbf{a}_j$ for this newly active constraint is not a linear combination of the $\mathbf{a}_i$, $i \in I$. Note that $\mathbf{a}_j^T\mathbf{x} \neq b_j$ (because $j \notin I$) and $\mathbf{a}_j^T(\mathbf{x} + \lambda^* \mathbf{d}) = b_j$, so that $\mathbf{a}_j^T\mathbf{d} \neq 0$. But since $\mathbf{a}_i^T\mathbf{d} = 0$, for $i \in I$, every linear combination of the $\mathbf{a}_i$, $i \in I$ is also orthogonal to $\mathbf{d}$. Thus, $\mathbf{a}_j$ is not a linear combination of these vectors, and the number of linearly independent active constraints at $\mathbf{x} + \lambda^*\mathbf{d}$ is greater than at $\mathbf{x}$, which is a contradiction.

To prove the other direction, suppose S contains the line $\mathbf{x} + \lambda\mathbf{d}$ for some $\mathbf{x} \in S$. We will show that S also contains the line $\mathbf{x}' + \lambda\mathbf{d}$ for all $\mathbf{x}' \in S$ (equivalently, $\mathbf{d}$ and $-\mathbf{d}$ are unbounded directions of S). But if $\mathbf{x}'$ lies on a line, it is not an extreme point. For a given $\lambda > 0$, consider the convex combination $\mathbf{z}_\lambda = \frac{\lambda}{\lambda+1}\mathbf{x}' + \frac{1}{\lambda+1}(\mathbf{x} + \lambda\mathbf{d})$, which is in S by convexity. Because S is closed (defined in Appendix A.2), $\lim_{\lambda \to \infty} \mathbf{z}_\lambda = \mathbf{x}' + \mathbf{d}$ is also in S. Repeating this argument for multiples of $\mathbf{d}$ shows that $\mathbf{d}$ and $-\mathbf{d}$ are unbounded directions of S. $\qquad\square$

Clearly, the feasible region of a linear program with nonnegative variables does not contain a line. We conclude that the Representation Theorem and Fundamental Theorem apply to linear programs with nonnegative variables.

7.3 Linear Programs in Canonical Form

The main calculation in the algorithm for linear programs in Chapter 9 is moving from a bfs to an adjacent bfs. This step is made simpler by only considering inequality constraints of the form $\mathbf{x} \geq \mathbf{0}$. Other inequality constraints can be converted to equality by adding slack variables as in Section 3.1. For simplicity, we focus on maximization problems. First we define a form with all inequality constraints.

Standard Form of Linear Programs

$$\max \quad \mathbf{c}^T\mathbf{x}$$
$$\text{s.t.} \quad \mathbf{A}\mathbf{x} \leq \mathbf{b}$$
$$\mathbf{x} \geq \mathbf{0}.$$

Here the variables are $\mathbf{x} = (x_1, \ldots, x_n)$, $\mathbf{c} = (c_1, \ldots, c_n)$, $\mathbf{b} = (b_1, \ldots, b_m)$, and $\mathbf{A} = [a_{ij}]$ is an $m \times n$ matrix. There are m functional constraints (rows), n variables (columns), and n nonnegativity constraints. Denote the jth column of $\mathbf{A}$ by $\mathbf{a}_j$. Now we define the equality form.

Canonical Form of Linear Programs

$$\max \ \mathbf{c}^T\mathbf{x}$$

$$\text{s.t.} \quad \mathbf{Ax} = \mathbf{b}$$

$$\mathbf{x} \geq \mathbf{0}.$$

Example 7.4. Consider the standard form linear program

$$\begin{aligned}
\max \quad & 8x + 10y \\
\text{s.t.} \quad & 3x - 5y \leq 15 \\
& x + 4y \leq 12 \\
& 2x + 10y \leq 28 \\
& x, y \geq 0.
\end{aligned}$$

Adding slack variables converts it to the canonical form

$$\begin{aligned}
\max \quad & 8x + 10y \\
\text{s.t.} \quad & 3x - 5y + s_1 && = 15 \\
& x + 4y && + s_2 && = 12 \\
& 2x + 10y && && + s_3 = 28 \\
& x, y, s_1, s_2, s_3 \geq 0,
\end{aligned}$$

or in matrix form

$$\max \quad \begin{bmatrix} 8 & 10 & 0 & 0 & 0 \end{bmatrix} \begin{bmatrix} x \\ y \\ s_1 \\ s_2 \\ s_3 \end{bmatrix}$$

s.t.

$$\begin{bmatrix} 3 & -5 & 1 & 0 & 0 \\ 1 & 4 & 0 & 1 & 0 \\ 2 & 10 & 0 & 0 & 1 \end{bmatrix} \begin{bmatrix} x \\ y \\ s_1 \\ s_2 \\ s_3 \end{bmatrix} = \begin{bmatrix} 15 \\ 12 \\ 28 \end{bmatrix}$$

$$x, y, s_1, s_2, s_3 \geq 0.$$

Notice that converting from standard to canonical form introduces an identity submatrix in $\mathbf{A}$.

7.3.1 Basic Solutions in Canonical Form

The definition of a basic solution in Section 7.1 applies to any polyhedron. When the linear program is in canonical form, choosing a set of linearly independent active constraints is simpler. To further simplify, we assume that the rows of $\mathbf{A}$ are linearly independent. If not, row reducing $[\mathbf{A}|\mathbf{b}]$ will produce a set of linearly independent rows. This is called the *full rank* assumption and results in rank$(\mathbf{A}) = m \leq n$. There are at least as many variables as functional constraints. Basic solutions must have all m equality constraints and $n - m$ nonnegativity constraints active, i.e. $n - m$ variables set to zero. That leaves m variables. A given set of m variables corresponds to a basic solution if the constraints with the other variables deleted are linearly independent, or equivalently, if the columns of $\mathbf{A}$ for these m variables are linearly independent. These m columns are a basis for $\mathbb{R}^m$. Variables for these columns are called *basic* variables and the others *nonbasic* variables. The set of basic variables is also called a *basis*. There are $n - m$ nonbasic variables, set to zero, and m basic variables, found by solving constraints. A bfs (in canonical form) is a basic solution that is nonnegative.

It will be convenient to write the linear program in matrix form partitioned into basic and nonbasic variables. For a given set of basic variables, let

$\mathcal{B} = \{B(1), \ldots, B(m)\}$ be the indices of the basic variables

$\mathcal{N} = \{N(1), \ldots, N(n - m)\}$ be the indices of nonbasic variables

$\mathbf{x}_B = (x_{B(1)}, \ldots, x_{B(m)})$ be the basic variables

$\mathbf{x}_N = (x_{N(1)}, \ldots, x_{N(n-m)})$ be the nonbasic variables

$\mathbf{B} = [\mathbf{a}_{B(1)}| \ldots |\mathbf{a}_{B(m)}]$ be the basic columns of $\mathbf{A}$

$\mathbf{N} = [\mathbf{a}_{N(1)}| \ldots |\mathbf{a}_{N(n-m)}]$ be the nonbasic columns of $\mathbf{A}$, $m \times (n - m)$

$\mathbf{c}_B =$ basic components of $\mathbf{c}$

$\mathbf{c}_N =$ nonbasic components of $\mathbf{c}$.

We call the $m \times m$ invertible matrix $\mathbf{B}$ a *basis matrix*. The canonical form linear program is then

$$\max \quad z = \mathbf{c}_B^T \mathbf{x}_B + \mathbf{c}_N^T \mathbf{x}_N$$
$$\text{s.t.} \quad \mathbf{B} \mathbf{x}_B + \mathbf{N} \mathbf{x}_N = \mathbf{b}$$
$$\mathbf{x}_B \geq \mathbf{0}, \ \mathbf{x}_N \geq \mathbf{0}.$$

Setting $\mathbf{x}_N = \mathbf{0}$, the basic solution is

$$z = \mathbf{c}_B^T \mathbf{B}^{-1} \mathbf{b}$$
$$\mathbf{x}_B = \mathbf{B}^{-1} \mathbf{b}.$$

If $\mathbf{x}_B \geq \mathbf{0}$, this basic solution is a bfs. We can also solve in terms of $\mathbf{x}_N$:

$$\mathbf{x}_B = \mathbf{B}^{-1}\mathbf{b} - \mathbf{B}^{-1}\mathbf{N}\mathbf{x}_N,$$

$$z = \mathbf{c}_B^T(\mathbf{B}^{-1}\mathbf{b} - \mathbf{B}^{-1}\mathbf{N}\mathbf{x}_N) + \mathbf{c}_N^T\mathbf{x}_N$$

$$= \mathbf{c}_B^T\mathbf{B}^{-1}\mathbf{b} + (\mathbf{c}_N^T - \mathbf{c}_B^T\mathbf{B}^{-1}\mathbf{N})\mathbf{x}_N.$$

These relationships will be useful later, although when $\mathbf{x}_N \neq \mathbf{0}$ it is generally not a basic solution.

The order of the components in $\mathbf{x}_B$ is determined by the basic indices $B(1), \ldots, B(m)$. For example, the basic variables could be x_1, x_4, x_5 but $B(1) = 5$, $B(2) = 1$, $B(3) = 4$, meaning that $\mathbf{x}_B = (x_5, x_1, x_4)$. This notation controlling the order of the basis will come in handy in Section 9.4. Often we can avoid this issue by simply referring to the first component of $\mathbf{x}_B$ as x_5 rather than $(\mathbf{x}_B)_1$. Although we could write the constraints

$$[\mathbf{B} \mid \mathbf{N}] \begin{bmatrix} \mathbf{x}_B \\ \mathbf{x}_N \end{bmatrix} = \mathbf{b},$$

notice that $[\mathbf{B} \mid \mathbf{N}]$ is not necessarily the same as $\mathbf{A}$ because the columns have been reordered to put basic columns first.

Example 7.5. For Example 7.1 with objective function $4x + 3y$ (see Figure 7.1), the basis (x, y, s_3) has

$$\mathbf{B} = \begin{bmatrix} 1.5 & 1 & 0 \\ 1 & 1 & 0 \\ 0 & 1 & 1 \end{bmatrix}, \quad \mathbf{N} = \begin{bmatrix} 1 & 0 \\ 0 & 1 \\ 0 & 0 \end{bmatrix}, \quad \mathbf{c}_B = \begin{bmatrix} 4 \\ 3 \\ 0 \end{bmatrix}, \quad \mathbf{c}_N = \begin{bmatrix} 0 \\ 0 \end{bmatrix}$$

$$\mathbf{x}_B = \begin{bmatrix} x \\ y \\ s_3 \end{bmatrix} = \begin{bmatrix} 1.5 & 1 & 0 \\ 1 & 1 & 0 \\ 0 & 1 & 1 \end{bmatrix}^{-1} \begin{bmatrix} 16 \\ 12 \\ 10 \end{bmatrix} = \begin{bmatrix} 8 \\ 4 \\ 6 \end{bmatrix}, \quad z = \begin{bmatrix} 4 & 3 & 0 \end{bmatrix} \begin{bmatrix} 8 \\ 4 \\ 6 \end{bmatrix} = 44.$$

Note that we must know the order of the variables in the basis in order to interpret $\mathbf{x}_B$ and expand it to $\mathbf{x} = (8, 4, 0, 0, 6)$. The variables (x, y, s_2) also form a basis with

$$\mathbf{B} = \begin{bmatrix} 1.5 & 1 & 0 \\ 1 & 1 & 1 \\ 0 & 1 & 0 \end{bmatrix}.$$

However, for this basis $\mathbf{x}_B = (4, 10, -2)$. Since $s_2 = -2 < 0$, this basic solution is not feasible.

Example 7.6. If we remove the third constraint from Example 7.5, we can visualize the feasible region by removing the horizontal constraint from Figure 7.1,

Table 7.1 Basic solutions for Example 7.6.

Basic variables	Solution type	Solution
(s_1, s_2)	bfs	$(0, 0, 16, 12)$
(x, s_1)	Basic solution	$(12, 0, -2, 0)$
(x, s_2)	bfs	$(\frac{32}{3}, 0, 0, \frac{4}{3})$
(y, s_1)	bfs	$(0, 12, 4, 0)$
(y, s_2)	Basic solution	$(0, 16, 0, -4)$
(x, y)	bfs	$(8, 4, 0, 0)$

leaving four bfs and two other basic solutions. In canonical form, the problem is

$$\begin{aligned}
\max \quad & 4x_1 + 3x_2 \\
\text{s.t.} \quad & 1.5x_1 + x_2 + s_1 && = 16 \\
& x_1 + x_2 && + s_2 && = 12 \\
& x_2 && + s_3 = 10 \\
& x_j, s_i \geq 0
\end{aligned}$$

The basic solutions are listed in Table 7.1.

Recall from Section 7.1 that a bfs is degenerate if it has more than n active constraints. In canonical form, these extra constraints are the nonnegativity constraints on basic variables, so these basic variables are zero.

Example 7.7. The problem

$$\begin{aligned}
\max \quad & x_1 + x_2 + x_3 + x_4 + x_5 \\
\text{s.t.} \quad & 2x_1 + 2x_2 + 3x_3 && + 2x_5 = 8 \\
& x_1 + 3x_2 && + 4x_4 + x_5 = 4 \\
& 2x_2 + 4x_3 + 4x_4 + x_5 = 4 \\
& x_j \geq 0
\end{aligned}$$

with basis (x_1, x_4, x_5) has $\mathbf{x}_B = (0, 0, 4)$, or $\mathbf{x} = (0, 0, 0, 0, 4)$. Because two of the basic variables equal zero, this basis is degenerate. Other bases containing x_5 will have the same bfs. For example, (x_2, x_3, x_5) is a basis with this bfs.

When a basic solution has less than m positive components, some of the basic variables equal zero and one cannot determine the basis from the basic solution. Any set of m variables that includes all of the positive ones will be a basis corresponding to this solution, provided these columns of $\mathbf{A}$ are linearly independent. Generally, then, multiple bases correspond to one degenerate basic solution. Some authors use a different convention and call these separate basic solutions, so that there is a one-to-one correspondence between bases and basic solutions.

Example 7.7 is in canonical form but does not have an identity submatrix in the columns of the slack variables in $\mathbf{A}$, which it would have had if we started in standard form. This distinction will be important when we solve linear programs in Chapter 9.

7.4 Optimality Conditions

Now we are ready to derive an optimality condition for linear programs in canonical form. We saw in Chapter 6 that, because of convexity, any local optimum of a linear program is also a global optimal solution. Combined with the local optimality condition in Chapter 5, we know that any feasible solution with no improving feasible direction is optimal. For a problem in canonical form, a direction $\mathbf{d}$ is improving if $\mathbf{c}^T \mathbf{d} > 0$ (because it is a maximization problem) and feasible if $\mathbf{Ad} = \mathbf{0}$ and $d_j \geq 0$ for all j with $x_j = 0$. Putting these together, a feasible solution is optimal if and only if the following system has no solution $\mathbf{d}$:

$$\mathbf{Ad} = \mathbf{0}, \ d_j \geq 0 \text{ for all } j \text{ with } x_j = 0, \tag{7.2}$$
$$\mathbf{c}^T \mathbf{d} > 0.$$

This condition is hard to check. The Fundamental Theorem and its proof suggest that if we are at a bfs, checking directions leading to other bfs's and certain unbounded directions is sufficient. In fact, we will show that only $n - m$ directions, either pointing to an adjacent bfs or an unbounded direction, need be checked, one for each nonbasic variable. First we find formulas for these directions.

Applying (7.2) to a bfs $\mathbf{x}$ with basis matrix $\mathbf{B}$ we obtain

$$\mathbf{Ad} = \mathbf{Bd}_B + \mathbf{Nd}_N = \mathbf{0}, \tag{7.3}$$
$$\mathbf{d}_B = -\mathbf{B}^{-1} \mathbf{Nd}_N.$$

Since $\mathbf{x}_N = \mathbf{0}$, (7.2) also requires $\mathbf{d}_N \geq \mathbf{0}$. We will use the standard basis for $\mathbb{R}^{n-m}$ to generate all such $\mathbf{d}_N$. For each nonbasic index $k \in \mathcal{N}$, define the *simplex direction* $\mathbf{d}^{(k)}$ by $d_k^{(k)} = 1, d_j^{(k)} = 0, j \in \mathcal{N} \backslash \{k\}$ and

$$\mathbf{d}_B^{(k)} = -\mathbf{B}^{-1} \mathbf{Nd}_N^{(k)} = -\mathbf{B}^{-1} \mathbf{a}_k. \tag{7.4}$$

The name refers to the simplex algorithm of Chapter 9. Moving in this direction increases x_k, keeping the other nonbasic variables at 0 and maintaining feasibility by adjusting the basic variables. It satisfies (7.3), with

$$\mathbf{Ad}^{(k)} = \mathbf{Bd}_B^{(k)} + \mathbf{Nd}_N^{(k)} = -\mathbf{BB}^{-1} \mathbf{a}_k + \mathbf{a}_k = \mathbf{0}.$$

If $\mathbf{x}_B > 0$ (which is true if $\mathbf{x}$ is a nondegenerate bfs), then $\mathbf{d}^{(k)}$ is a feasible direction. Choosing a feasible direction amounts to choosing $\mathbf{d}_N \geq \mathbf{0}$, since $\mathbf{d}_B$ is then determined. Thus, any feasible direction $\mathbf{d}$ can be written as a nonnegative linear combination of simplex directions, namely,

$$\mathbf{d} = \sum_{k \in \mathcal{N}} \lambda_k \mathbf{d}^{(k)},$$

where $\lambda_k \geq 0$. For example, for basic variables $\mathcal{B} = \{1, 2\}$ and nonbasics $\mathcal{N} = \{3, 4\}$ suppose the simplex directions are $\mathbf{d}^{(3)} = (-8, 3, 1, 0)$ and $\mathbf{d}^{(4)} = (2, 5, 0, 1)$. Then the feasible direction with $\mathbf{d}_N = (1, 3)$ must be $\mathbf{d} = \mathbf{d}^{(3)} + 3\mathbf{d}^{(4)} = (-2, 18, 1, 3)$. Because $\mathbf{d}^{(3)}$ and $\mathbf{d}^{(4)}$ satisfy (7.3), $\mathbf{d}$ does also.

Now, the simplex direction $\mathbf{d}^{(k)}$ is nonimproving if

$$\mathbf{c}^T \mathbf{d}^{(k)} = \mathbf{c}_B^T \mathbf{d}_B^{(k)} + \mathbf{c}_N^T \mathbf{d}_N^{(k)}$$
$$= \mathbf{c}_k - \mathbf{c}_B^T \mathbf{B}^{-1} \mathbf{a}_k \leq 0. \tag{7.5}$$

If (7.5) holds for all nonbasic indices k, then it also holds for nonnegative linear combinations of these directions, so all feasible directions are nonimproving and the bfs $\mathbf{x}$ (with basis matrix $\mathbf{B}$) is optimal. We have proven the following theorem.

Theorem 7.5 *(Optimality in canonical form): Consider a bfs with basis matrix $\mathbf{B}$ for a linear program in canonical form. If $\mathbf{c}_N^T - \mathbf{c}_B^T \mathbf{B}^{-1} \mathbf{N} \leq \mathbf{0}$ then there is no improving feasible simplex direction and the bfs is optimal.*

The converse is not always true. If the bfs is degenerate, a basic variable x_j could be zero. Then we have ignored the requirement $d_j \geq 0$ in (7.2), so some of the $\mathbf{d}^{(k)}$ might not be feasible directions. As a result, a bfs might be optimal even if (7.5) does not hold for this k.

In summary, for a basic solution of a linear program in canonical form, the two requirements for its basis matrix $\mathbf{B}$ to be optimal are

1. *Feasibility.* $\mathbf{x}_B = \mathbf{B}^{-1}\mathbf{b} \geq \mathbf{0}$.
2. *Optimality.* $\mathbf{c}_N^T - \mathbf{c}_B^T \mathbf{B}^{-1} \mathbf{N} \leq \mathbf{0}$.

Although these conditions allow us to check whether a basis is optimal, we still need an algorithm to efficiently search for bases to check.

7.5 Optimality for General Polyhedra

The simplex directions $\mathbf{d}^{(k)}$ used in the optimality condition have an intuitive geometric interpretation: starting at an extreme point, they move along an edge of the feasible region. Now we make this idea precise for general polyhedra. In Figure 7.4 the bfs $\mathbf{x}$ has two edge directions, shown as arrows. One lies on the edge bounded by the adjacent bfs $\mathbf{y}$; the other lies on an unbounded edge, which

Figure 7.4   Edge directions for the bfs **x**.

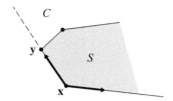

is an unbounded direction. In general, a bfs will have these two types of edge directions. This example does not have any equality constraints; the same idea applies in canonical form, but only very small examples can be visualized.

An **edge** of a polyhedron $S \in \mathbb{R}^n$ is a non-singleton set of points in S that lie on the intersection of $n - 1$ linearly independent defining hyperplanes.

Defining hyperplanes are just the constraints (at equality). The intersection of $n - 1$ linearly independent constraints is a line in $\mathbb{R}^n$; an edge is the intersection of a line with the convex set S, so it is either a line, a ray, or a line segment. If the polyhedron has a bfs **x**, then it does not contain a line, leaving just rays and line segments as edges. However, not every set of $n - 1$ linearly independent constraints at a bfs has an edge: if the bfs is degenerate, the intersection with S could be a point. In this case there is a simplex direction but no edge. We will call **d** an *edge direction* of a bfs **x** for the polyhedron S if $\mathbf{x} + \lambda \mathbf{d}$ lies on an edge for all $\lambda \in [0, \varepsilon]$ for some $\varepsilon > 0$. With these definitions, we are ready to extend Theorem 7.5 to general polyhedra.

Theorem 7.6   *(Optimality for general polyhedra):*   *A bfs for a linear program is optimal if and only if it has no improving edge direction.*

Proof: Edge directions are feasible, so if there is an improving edge direction the bfs cannot be optimal.

   For the other direction, consider the problem of maximizing $\mathbf{c}^T\mathbf{x}$ subject to $\mathbf{x} \in S$, where S is a polyhedral set in $\mathbb{R}^n$, and suppose the bfs **x** is not optimal. Omit all constraints that are not active at **x**. Call the expanded feasible region C (see Figure 7.4). The set of edge directions of C containing **x** is the same as for S because they are not affected by inactive constraints. Then there is a $\mathbf{y} \in S$ with $z = \mathbf{c}^T\mathbf{y} > \mathbf{c}^T\mathbf{x}$ and the polyhedron

$$P = C \cap \{\mathbf{x} \in \mathbb{R}^n : \mathbf{c}^T\mathbf{x} = z\}$$

is not empty (it contains **y**). Since C has a bfs, it does not contain a line and neither does P, so P has at least one bfs $\mathbf{x}'$. At $\mathbf{x}'$ there are n linearly independent active constraints, $n - 1$ of which must be defining hyperplanes of C. Thus, $\mathbf{x}'$ lies on an edge of C. But $\mathbf{x}$ also lies on this (and every) edge of C because all constraints defining C are active at $\mathbf{x}$. Then $\mathbf{d} = \mathbf{x}' - \mathbf{x}$ is an improving edge direction of C (and S). □

7.5.1 Representation of Polyhedra

The ability to represent every point in a polyhedron in terms of a finite number of points is an important result of linear programming theory; Theorem 7.6 and many other results follow from it. We close this section by introducing this theory. It is a stronger version of the Representation Theorem (Theorem 7.2). First we need some definitions. Recall from Section 6.1 that a polyhedral cone is a polyhedron of the form $C = \{\mathbf{x} \in \mathbb{R}^n : \mathbf{Ax} \geq \mathbf{0}\}$. When a cone is shifted away from the origin it is called an *affine cone*. The proof of Theorem 7.6 constructed an affine cone C by omitting all the inactive constraints at an extreme point of the polyhedron S (see Figure 7.4).

The set of unbounded directions of a polyhedron form the nonzero elements of a polyhedral cone; they are also called the *recession cone* of the polyhedron. We use cones to describe the set of unbounded directions of a polyhedron. A positive multiple of an unbounded direction is also an unbounded direction; all positive multiples form a ray.

Extreme Ray A nonzero element $\mathbf{x}$ of a polyhedral cone in $\mathbb{R}^n$ is on an **extreme ray** if there are $n - 1$ linearly independent active constraints at $\mathbf{x}$. An extreme ray of the set of unbounded directions of a polyhedron (its recession cone) is called an **extreme ray of the polyhedron**.

Figure 7.5 shows examples. The extreme rays of a polyhedral cone are the same as its edges, except that the extreme rays do not include the origin. The following theorem states that every point in a polyhedral set with at least one extreme point can be represented as a convex combination of extreme points plus a nonnegative linear combination of extreme rays.

Theorem 7.7 **(Resolution theorem):** *Let S be a polyhedral set with at least one extreme point. Let $\mathbf{v}_1, \ldots, \mathbf{v}_k$ be the extreme points and $\mathbf{d}_1, \ldots, \mathbf{d}_r$ contain one unbounded direction from each extreme ray. Then $\mathbf{x} \in S$ if and only if*

$$\mathbf{x} = \sum_{i=1}^{k} \alpha_i \mathbf{v}_i + \sum_{j=1}^{r} \lambda_j \mathbf{d}_j$$

for some $\{\alpha_i\}$ satisfying $\sum_{i=1}^{k} \alpha_i = 1$, $\alpha_i \geq 0$ and some $\lambda_j \geq 0$.

A proof is given in (Bertsimas and Tsitsiklis, 1997, Theorem 4.15). For bounded polyhedra, there are no unbounded directions $\mathbf{d}_j$ so the second sum disappears.

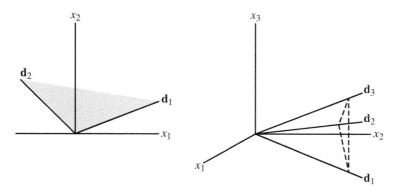

Figure 7.5 Cones and their extreme rays $\mathbf{d}_j$.

In this case, the polyhedron is the convex hull of its extreme points. While Theorem 7.2 refers to an arbitrary unbounded direction, Theorem 7.7 refers only to the extreme rays, of which there are finitely many. Polyhedra and cones are said to be finitely generated sets for this reason.

One corollary of the Resolution Theorem is that any feasible direction can be written as a nonnegative linear combination of edge directions, leading immediately to the fact that optimality can be checked using edge directions (Theorem 7.6).

Problems

For Exercises 7.1–7.4, find all extreme points of the set.

1
$$
\begin{aligned}
x_1 + 2x_2 &\leq 12, \\
-x_1 + x_2 &\leq 3, \\
x_1 &\leq 7, \\
x_1, x_2 &\geq 0.
\end{aligned}
$$

2
$$
\begin{aligned}
3x_1 + 5x_2 &= 35, \\
x_1, x_2 &\geq 0.
\end{aligned}
$$

3
$$
\begin{aligned}
x_1 - x_2 &\leq 0, \\
2x_1 - x_2 &\leq 4, \\
-2x_1 + x_2 &\leq 2, \\
x_1, x_2 &\geq 0.
\end{aligned}
$$

4
$$2x_1 + 3x_2 + 4x_3 = 12,$$
$$x_1 \geq 2,$$
$$x_2, x_3 \geq 0.$$

5 For the constraints in Exercise 7.1:
(a) Find all basic solutions. List the linearly independent active constraints for each basic solution.
(b) Which of the basic solutions are feasible?

6 For the constraints:
$$x_2 + x_3 + 2x_4 \leq 10,$$
$$x_1 + x_2 + x_3 + x_4 \leq 12,$$
$$0 \leq x_i \leq 4, \quad i = 1, 2, 3, 4.$$

(a) Show that $(4, 4, 4, 0)$ is a degenerate basic feasible solution.
(b) Show that none of the constraints are redundant.

7 Prove this part of Theorem 7.1: If $\mathbf{x}$ is an extreme point then it is a bfs.

8 Consider the linear program
$$\begin{aligned} \max \quad & 2x + 5y \\ \text{s.t.} \quad & -x + 2y \leq 5 \\ & x + y = 7 \\ & x, y \geq 0. \end{aligned}$$

(a) Convert to canonical form.
(b) Write the answer to (a) in matrix form.

9 Convert the following linear program to canonical form:
$$\begin{aligned} \max \quad & 3x + 2y \\ \text{s.t.} \quad & x - y \geq -2 \\ & 5x + y \leq 8 \\ & x \leq 1.5 \\ & x, y \geq 0. \end{aligned}$$

10 Consider the linear program
$$\begin{aligned} \max \quad & 4x + 6y + z \\ \text{s.t.} \quad & 3x + y - z \geq 8 \\ & x + z \leq 4 \\ & y + z = 6 \\ & x, y, z \geq 0. \end{aligned}$$

(a) Convert to canonical form.

(b) Write the answer to (a) in matrix form.

11 Convert the following linear program to canonical form:

$$
\begin{aligned}
\min \quad & x + y - z \\
\text{s.t.} \quad & 5x - 4y + z \geq 12 \\
& 2x + y = 5 \\
& -x + 2y + z \geq 5 \\
& x, y, z \geq 0.
\end{aligned}
$$

12 For the constraints in Exercise 7.1

(a) Add slack variables to convert the constraints to canonical form.

(b) For each extreme point found in Exercise 7.1, find the corresponding extreme point in canonical form. Identify the basic variables for each extreme point.

8

Duality Theory

When dealing with a maximizing linear program, any feasible solution gives a lower, or suboptimal, bound on the optimal value. These may be much easier to find than the optimal solution and value. What about upper, or superoptimal, bounds? In this chapter we associate with each linear program another linear program, called the dual. There is a rich theory of relationships between the two. We begin in Section 8.1 by showing how bounds can be expressed as a linear program and defining the dual. The two main theoretical results are in Section 8.2, which relates their objective function values, and Section 8.3, which relates their optimal solutions. The final two sections establish connections between these duality theorems and more general concepts that predate linear programming. Section 8.4 shows that the dual is an extension of the method of Lagrange multipliers for solving optimization problems with equality constraints; it also provides an interpretation of the dual in terms of prices or penalties for violating a constraint. Section 8.5 presents Farkas' lemma and its connection with the duality theorem. One use of this lemma is to give another optimality condition for a linear program.

We only prove the key duality theorem (Theorem 8.3) under the assumption that an optimal basic feasible solution (bfs) is nondegenerate. Degeneracy raises additional issues. Duality theory has many applications in linear programming. Its relationship to sensitivity analysis is discussed in Section 10.2 and its use in algorithms is covered in Chapter 11.

8.1 Dual of a Linear Program

Starting with any linear program, called the primal, we will associate with it another linear program, called the dual. To motivate the definition of the dual, suppose we wish to find upper bounds to the following problem.

Linear and Convex Optimization: A Mathematical Approach, First Edition. Michael H. Veatch.
© 2021 John Wiley & Sons, Inc. Published 2021 by John Wiley & Sons, Inc.
Companion website: www.wiley.com/go/veatch/convexandlinearoptimization

Example 8.1. Consider the problem

$$z^* = \max\ 3x_1 + 8x_2 + 5x_3 +\ x_4$$
$$\text{s.t.}\quad 6x_1 + 4x_2 + 2x_3 + 4x_4 \le 12$$
$$x_1 + 2x_2 + 2x_3 -\ x_4 \le\ 8 \qquad\qquad (8.1)$$
$$x_j \ge\ 0.$$

Notice that if we multiply the first constraint by $\frac{5}{2}$, all of its coefficients are larger than the objective function coefficients. Thus, for any feasible solution,

$$z = 3x_1 + 8x_2 + 5x_3 + x_4 \le 15x_1 + 10x_2 + 5x_3 + 10x_4 \le \left(\frac{5}{2}\right)12 = 30$$

and $z^* \le 30$. The first inequality is valid because $x_j \ge 0$. We cannot create a bound from the second constraint because of the negative coefficient. However, we can form more bounds using nonnegative linear combinations of the two constraints. For example,

$$(6x_1 + 4x_2 + 2x_3 + 4x_4) + 2(x_1 + 2x_2 + 2x_3 - x_4)$$
$$= 8x_1 + 8x_2 + 6x_3 + 2x_4 \le 12 + 8(2) = 28,$$

giving the bound $z^* \le 28$.

In fact, any nonnegative linear combination of the constraints with coefficients as large as the objective function gives a bound. We can write all nonnegative linear combinations as

$$y_1(6x_1 + 4x_2 + 2x_3 + 4x_4) + y_2(x_1 + 2x_2 + 2x_3 - x_4)$$
$$= (6y_1 + y_2)x_1 + (4y_1 + 2y_2)x_2 + (2y_1 + 2y_2)x_3 + (4y_1 - y_2)x_4 \qquad (8.2)$$
$$\le 12y_1 + 8y_2,$$

where $y_1, y_2 \ge 0$ and $x_j \ge 0$. The coefficients are as large as those in the objective function if

$$6y_1 +\ y_2 \ge 3,$$
$$4y_1 + 2y_2 \ge 8,$$
$$2y_1 + 2y_2 \ge 5,$$
$$4y_1 -\ y_2 \ge 1.$$

If these constraints hold, then $z^* \le 12y_1 + 8y_2$. The smallest such bound is the optimal value of a linear program! We have shown that

$$z^* \le \min\ 12y_1 + 8y_2$$
$$\text{s.t.}\quad 6y_1 +\ y_2 \ge 3$$
$$4y_1 + 2y_2 \ge 8$$
$$2y_1 + 2y_2 \ge 5 \qquad\qquad (8.3)$$
$$4y_1 -\ y_2 \ge 1$$
$$y_i \ge 0.$$

Notice that the process of forming linear combinations of the rows of (8.1) creates a constraint in (8.3) for each column of (8.1). The variables or rows of (8.1) correspond to constraints or columns of (8.3) and vice versa.

Any linear program has a corresponding bounding linear program, known as its dual. Suppose we start with a linear program in standard form,

$$\max \ \mathbf{c}^T\mathbf{x}$$
$$\text{s.t. } \mathbf{A}\mathbf{x} \le \mathbf{b} \tag{8.4}$$
$$\mathbf{x} \ge \mathbf{0}.$$

Let $\mathbf{A}_i$ be the ith row of $\mathbf{A}$ and $\mathbf{a}_j$ the jth column. The ith constraint can be written $\mathbf{A}_i\mathbf{x} \le b_i$. The dual is constructed exactly as previously. Multiplying the constraints by y_i and adding,

$$y_1\mathbf{A}_1\mathbf{x} + \cdots + y_m\mathbf{A}_m\mathbf{x} \le y_1 b_1 + \ldots + y_m b_m = \mathbf{b}^T\mathbf{y}.$$

Requiring the coefficient of x_j to be as large as the objective function coefficient gives the constraint

$$\mathbf{y}^T\mathbf{a}_j \ge \mathbf{c}_j, \quad j = 1, \ldots, n$$

or

$$\mathbf{y}^T\mathbf{A} \ge \mathbf{c}^T.$$

Any $\mathbf{y} \ge \mathbf{0}$ satisfying these constraints gives the bound $\mathbf{b}^T\mathbf{y}$ and the smallest bound is the optimal value of

$$\min \ \mathbf{b}^T\mathbf{y}$$
$$\text{s.t. } \mathbf{A}^T\mathbf{y} \ge \mathbf{c} \tag{8.5}$$
$$\mathbf{y} \ge \mathbf{0}.$$

We have used $(\mathbf{y}^T\mathbf{A})^T = \mathbf{A}^T\mathbf{y}$ to rewrite the constraints. As in Example 8.1, the optimal value of (8.5) is an upper bound on the optimal value of (8.4).

Dual of a Linear Program   Given a linear program (8.4) in standard form, the **dual** is (8.5). The original linear program is known as the **primal**.

Any linear program can be converted to an equivalent linear program in standard form and then the definition of the dual applied. Here "equivalent" means not only that it has the same optimal value (possibly multiplied by -1 to convert a minimization to a maximization) but also that the optimal solution, if it exists, can be recovered from the standard form problem. In particular, the canonical form

$$\max \ \mathbf{c}^T\mathbf{x}$$
$$\text{s.t. } \mathbf{A}\mathbf{x} = \mathbf{b}$$
$$\mathbf{x} \ge \mathbf{0}$$

is equivalent to

$$\max \quad \mathbf{c}^T \mathbf{x}$$

$$\text{s.t.} \quad \begin{bmatrix} \mathbf{A} \\ -\mathbf{A} \end{bmatrix} \mathbf{x} \leq \begin{bmatrix} \mathbf{b} \\ -\mathbf{b} \end{bmatrix}$$

$$\mathbf{x} \geq \mathbf{0},$$

which has dual

$$\min \quad \begin{bmatrix} \mathbf{b}^T & -\mathbf{b}^T \end{bmatrix} \begin{bmatrix} \mathbf{y}^+ \\ \mathbf{y}^- \end{bmatrix}$$

$$\text{s.t.} \quad \begin{bmatrix} \mathbf{A}^T & -\mathbf{A}^T \end{bmatrix} \begin{bmatrix} \mathbf{y}^+ \\ \mathbf{y}^- \end{bmatrix} \geq \mathbf{c}$$

$$\mathbf{y}^+, \mathbf{y}^- \geq \mathbf{0}.$$

Substituting $\mathbf{y} = \mathbf{y}^+ - \mathbf{y}^-$ and letting $\mathbf{y}$ be unrestricted in sign, the dual can be simplified to

$$\min \quad \mathbf{b}^T \mathbf{y}$$

$$\text{s.t.} \quad \mathbf{A}^T \mathbf{y} \geq \mathbf{c}$$

$$\mathbf{y} \text{ u.r.s.}$$

Thus, equality constraints in the primal result in unrestricted variables in the dual. Starting with other forms, a complete set of rules for forming duals can be obtained. To state these rules more concisely, we first note that duality is a symmetric relationship.

Theorem 8.1: *For any linear program with dual (D), the dual of (D) is equivalent to the original linear program.*

Proof: Convert the linear program into standard form (8.4). Its dual (8.5), with objective function z_D, is equivalent to

$$\text{(D)} \quad \max \quad -z_D = -\mathbf{b}^T \mathbf{y}$$

$$\text{s.t.} \quad -\mathbf{A}^T \mathbf{y} \leq -\mathbf{c}$$

$$\mathbf{y} \geq \mathbf{0}.$$

Taking the dual of (D),

$$\text{(DD)} \quad \min \quad -\mathbf{c}^T \mathbf{x}$$

$$\text{s.t.} \quad -\mathbf{A}\mathbf{x} \geq -\mathbf{b}$$

$$\mathbf{x} \geq \mathbf{0}.$$

Multiplying the objective by -1 again to get back to z_D, (DD) is equivalent to (8.4). □

Table 8.1 Dual relationships.

Maximize	Minimize
≤ constraint	≥ 0 variable
≥ constraint	≤ 0 variable
= constraint	Unrestricted variable
≥ 0 variable	≥ constraint
≤ 0 variable	≤ constraint
Unrestricted variable	= constraint

Since the dual of the dual is the primal, we are free to label either problem as the primal in Table 8.1.

Rules for Forming a Dual

1. Assign a dual variable to each constraint in the primal.
2. Write a constraint in the dual for each variable in the primal. The coefficients are the transpose of the coefficient matrix of the primal, i.e. columns of the matrix in the primal become rows in the dual.
3. Objective function coefficients in the primal become constraint right-hand sides in the dual.
4. Constraint right-hand sides in the primal become objective function coefficients in the dual.
5. Using Table 8.1, the direction of each constraint in the primal determines whether its dual variable is positive, negative, or unrestricted.
6. Using Table 8.1, the sign restriction of each primal variable determines the direction of its constraint in the dual.

Example 8.2. To find the dual of the problem

$$\max\ -3x_1 + 2x_2 + 5x_3$$
$$\text{s.t.}\ -x_1 + 3x_2 + 4x_3 = 12$$
$$2x_1 - x_2 + 3x_3 \geq 6$$
$$x_3 \leq 4$$
$$x_1 \geq 0,\ x_2 \leq 0,\ x_3\ \text{u.r.s.}$$

using Table 8.1, the "=" in the first constraint makes y_1 unrestricted in sign, the "≥" in the second constraint makes $y_2 \leq 0$, and the "≤" in the third constraint makes

$y_3 \geq 0$. Also, the sign restriction $x_1 \geq 0$ makes the first dual constraint "$\geq$", $x_2 \leq 0$ makes the second dual constraint "$\leq$", and x_3 being unrestricted in sign makes the third dual constraint "$=$". Hence, the dual is

$$
\begin{aligned}
\min \quad & 12y_1 + 6y_2 + 4y_3 \\
\text{s.t.} \quad -y_1 + 2y_2 \quad & \geq -3 \\
3y_1 - y_2 \quad & \leq 2 \\
4y_1 + 3y_2 + y_3 & = 5 \\
y_1 \text{ u.r.s., } y_2 \leq 0, \ y_3 & \geq 0
\end{aligned}
$$

The argument that the optimal value of the dual is a bound on the optimal value of the primal can be extended to other forms of linear programs. For example, if the second constraint in (8.1) changed from "$\leq$" to "$\geq$", then we must restrict $y_2 \leq 0$ for inequality (8.2) to hold and obtain an upper bound on z^*. Rather than checking all the cases, a general derivation will be given in Section 8.2, where the notion of a bound is called weak duality.

8.2 Duality Theorems

We saw in Section 8.1 that, for a standard form problem, the optimal value of the dual is an upper bound on the optimal value of the primal problem. In this section we extend this relationship to all linear programs and strengthen it. We will call the maximization problem the primal and call its data **c**, **A**, and **b** as in (8.4). Thus, the primal looks like (8.4) and the dual like (8.5) except that the constraints may have any direction and the variables any sign restrictions. The quantity of interest in this section is the difference in their objective function values. We saw that the value of the dual is an upper bound, so

$$
z_D - z_P = \mathbf{b}^T \mathbf{y} - \mathbf{c}^T \mathbf{x} \geq 0 \tag{8.6}
$$

when **x** and **y** are *optimal* for the primal and dual. By the definition of optimality, (8.6) also holds for any feasible solutions, since suboptimal primal values will be smaller and suboptimal dual values will be larger. That fact is called *weak duality*. A deeper question is how good the bound is. If the quantity in (8.6) is positive at optimal solutions, there is an *duality gap* and the bound is imperfect; if it is zero, there is no duality gap and the bound is perfect: solving the dual tells us the optimal value of the primal. This question is settled by the Strong Duality Theorem.

Theorem 8.2 *(Weak duality):* *For any primal that is a maximization problem, if* **x** *is a feasible solution for the primal and* **y** *is a feasible solution for the dual, then* $\mathbf{c}^T \mathbf{x} \leq \mathbf{b}^T \mathbf{y}$.

Proof: Write the difference in objective function values

$$\mathbf{y}^T\mathbf{b} - \mathbf{c}^T\mathbf{x} = (\mathbf{y}^T\mathbf{b} - \mathbf{y}^T\mathbf{A}\mathbf{x}) + (\mathbf{y}^T\mathbf{A}\mathbf{x} - \mathbf{c}^T\mathbf{x}),$$
$$= \mathbf{y}^T(\mathbf{b} - \mathbf{A}\mathbf{x}) + (\mathbf{y}^T\mathbf{A} - \mathbf{c}^T)\mathbf{x}, \tag{8.7}$$
$$= \sum_{i=1}^{m} y_i(b_i - \mathbf{A}_i\mathbf{x}) + \sum_{j=1}^{n}(\mathbf{y}^T\mathbf{a}_j - c_j)x_j.$$

Each term in this expression is a variable in one problem multiplied by the slack or surplus in the corresponding constraint of the other problem. The variable x_j is multiplied by $\mathbf{y}^T\mathbf{a}_j - c_j$, which is the surplus in constraint j in the dual. From Table 8.1 we see that, because $\mathbf{x}$ and $\mathbf{y}$ are feasible, these two quantities must have the same sign or equal zero. For example, if $x_j \geq 0$, it is a "$\geq$" constraint and the surplus is nonnegative. Hence, their product is nonnegative.

Similarly, y_i is multiplied by $b_i - \mathbf{A}_i\mathbf{x}$, which is the slack in constraint i in the primal. Again, because $\mathbf{x}$ and $\mathbf{y}$ are feasible, these two quantities must have the same sign. Thus, (8.7) is nonnegative. $\square$

Weak duality is not a deep result. Indeed, the proof for some forms of problems is very simple. If the primal is in canonical form, then

$$\mathbf{y}^T\mathbf{b} = \mathbf{y}^T\mathbf{A}\mathbf{x} \geq \mathbf{c}^T\mathbf{x},$$

with the last inequality following from the dual constraint $\mathbf{y}^T\mathbf{A} \geq \mathbf{c}^T$ and the primal constraint $\mathbf{x} \geq \mathbf{0}$. Weak duality has the following useful consequences.

Corollary 8.1: *If $\mathbf{x}^*$ is a feasible solution for the primal problem, $\mathbf{y}^*$ is a feasible solution for its dual, and $\mathbf{c}^T\mathbf{x}^* = \mathbf{b}^T\mathbf{y}^*$, then $\mathbf{x}^*$ and $\mathbf{y}^*$ are optimal solutions for the primal and dual, respectively.*

Proof: Assume the primal is a maximization problem. For any feasible solution $\mathbf{x}$ to the primal, by weak duality $\mathbf{c}^T\mathbf{x} \leq \mathbf{b}^T\mathbf{y}^* = \mathbf{c}^T\mathbf{x}^*$, which shows that $\mathbf{x}^*$ is optimal for the primal. By primal–dual symmetry, $\mathbf{y}^*$ is also optimal. $\square$

Corollary 8.2: *For any linear program*

(a) If a primal problem is unbounded then its dual is infeasible.
(b) If the dual of a problem is unbounded then the primal problem is infeasible.

Proof: (a) Assume the primal is a maximization problem. Suppose there is a feasible solution $\mathbf{y}$ to the dual. Because the primal problem is unbounded, there is a feasible solution $\mathbf{x}$ for the primal with objective function value $\mathbf{c}^T\mathbf{x} > \mathbf{b}^T\mathbf{y}$, contradicting weak duality. Hence, the dual is infeasible. Part (b) follows from (a) by primal–dual symmetry. The proof for minimization problems is similar. $\square$

Next we present the central result of duality theory. For brevity, the proof assumes that at least one optimal bfs is nondegenerate.

Theorem 8.3 *(Strong duality):* *If a linear program has an optimal solution, then so does its dual and their optimal values are equal.*

Proof: First we prove the result for a problem in canonical form

$$\max \quad \mathbf{c}^T \mathbf{x}$$
$$\text{s.t.} \quad \mathbf{A}\mathbf{x} = \mathbf{b}$$
$$\mathbf{x} \geq \mathbf{0}$$

with dual

$$\min \quad \mathbf{b}^T \mathbf{y}$$
$$\text{s.t.} \quad \mathbf{A}^T \mathbf{y} \geq \mathbf{c}.$$

By Theorem 7.3, the primal has an optimal bfs $\mathbf{x}^*$. As in Section 7.3, we partition $\mathbf{A}$ into basic columns $\mathbf{B}$ and nonbasic columns $\mathbf{N}$, and $\mathbf{c}$ into basic and nonbasic components $\mathbf{c}_B$ and $\mathbf{c}_N$, with $\mathbf{x}_B^* = \mathbf{B}^{-1}\mathbf{b}$ and $\mathbf{x}_N^* = \mathbf{0}$. Assume that $\mathbf{x}^*$ is nondegenerate. Then the optimality condition of Theorem 7.5 is also a necessary condition, and

$$\mathbf{c}_N^T - \mathbf{c}_B^T \mathbf{B}^{-1} \mathbf{N} \leq \mathbf{0}.$$

The same inequality holds with equality for basic variables:

$$\mathbf{c}_B^T - \mathbf{c}_B^T \mathbf{B}^{-1} \mathbf{B} = \mathbf{c}_B^T - \mathbf{c}_B^T = \mathbf{0}.$$

Combining these,

$$\mathbf{c}^T - \mathbf{c}_B^T \mathbf{B}^{-1} \mathbf{A} \leq \mathbf{0}. \tag{8.8}$$

The key observation is that this expanded primal optimality condition is also a dual feasibility condition. Let $\mathbf{y}^T = \mathbf{c}_B^T \mathbf{B}^{-1}$. Then

$$\mathbf{c}^T - \mathbf{y}^T \mathbf{A} \leq \mathbf{0} \text{ or}$$
$$\mathbf{A}^T \mathbf{y} \geq \mathbf{c}$$

and $\mathbf{y}$ is feasible for the dual. Further,

$$\mathbf{y}^T \mathbf{b} = \mathbf{c}_B^T \mathbf{B}^{-1} \mathbf{b} = \mathbf{c}_B^T \mathbf{x}_B^* = \mathbf{c}^T \mathbf{x}^*.$$

By Corollary 8.1, $\mathbf{y}$ is optimal for the dual, so the primal and dual have the same optimal values.

If the problem is not in canonical form, it can be converted to an equivalent problem in canonical form with the same optimal value. Call these problems (P) and (P_{can}). We claim that their duals, (D) and (D_{can}), also have the same optimal value. The proof earlier shows that (P_{can}) and (D_{can}) have the same optimal value and we conclude that (P) and (D) have the same optimal value. To establish this claim, first note that any linear program can be converted to canonical form by a sequence of transformations that

(a) Replace an unrestricted in sign variable with the difference of two nonnegative variables.
(b) Replace a nonpositive variable with a nonnegative variable that is its negative.
(c) Replace an inequality constraint with an equality constraint, adding a nonnegative slack variable.

It is easily verified that none of these transformations change the optimal value. □

The Strong Duality Theorem settles the question posed above about a duality gap: there is none. If a linear program has an optimal solution, then its dual doesn't just bound its optimal value, it has the same optimal value. Further, the proof is constructive. Given an optimal basis for the primal problem, $\mathbf{y}^T = \mathbf{c}_B^T \mathbf{B}^{-1}$ is optimal for the dual. Since the primal–dual relationship is symmetric, one way to solve a linear program is to solve its dual (assuming that we can find an optimal solution that is also basic), then convert the dual solution to an optimal bfs for the primal problem.

The proof also establishes, for a primal problem in canonical form, that primal optimality of a bfs is the same as feasibility of a corresponding dual solution. This equivalence applies to other forms as well. For example, starting in standard form and adding slack variables, the coefficient matrix is $[\mathbf{A}|\mathbf{I}]$ and the objective function costs are $(\mathbf{c}, \mathbf{0})$. Substituting these into the optimality condition (8.8), we have

$$\mathbf{c}^T - \mathbf{c}_B^T \mathbf{B}^{-1} \mathbf{A} \leq \mathbf{0},$$
$$\mathbf{c}_B^T \mathbf{B}^{-1} \geq \mathbf{0}.$$

Substituting $\mathbf{y}^T = \mathbf{c}_B^T \mathbf{B}^{-1}$ gives the dual constraints

$$\mathbf{A}^T \mathbf{y} \geq \mathbf{c},$$
$$\mathbf{y} \geq \mathbf{0}.$$

The equivalence between primal optimality and dual feasibility, as well as between primal feasibility and dual optimality, will be expanded in Section 8.3. More implications of strong duality are explored in Chapter 10.

If we classify linear programs as having an optimal solution, unbounded objective function values, or infeasible, Corollary 8.2 and Theorem 8.3 tell us that certain combinations for the primal and its dual are impossible. These are shown in Table 8.2. Because the primal–dual relationship is symmetric, the table must

Table 8.2 Possibilities when solving the primal and the dual.

	Optimal solution	Unbounded	Infeasible
Optimal solution	✓		
Unbounded			✓
Infeasible		✓	✓

be symmetric, so we have not labeled which is the primal and which is the dual. From the table, we can see that if both problems have feasible solutions, then both problems have optimal solutions.

The case where both problems are infeasible can in fact occur. Feasibility of a linear program includes as a special case existence of solutions to systems of linear equations. Consider the form

$$\max \ \mathbf{c}^T \mathbf{x}$$
$$\text{s.t.} \ \mathbf{Ax} = \mathbf{b}$$

which has the dual

$$\min \ \mathbf{b}^T \mathbf{y}$$
$$\text{s.t.} \ \mathbf{A}^T \mathbf{y} = \mathbf{c}.$$

If $\mathbf{A}$ does not have full rank, we can choose $\mathbf{b}$ to not be in the row space of $\mathbf{A}$ and $\mathbf{c}$ to not be in the column space of $\mathbf{A}$. Then both problems are infeasible.

For a primal problem in standard form, the direction of the constraint inequalities is reversed in the dual. We would not be surprised, then, if an example where the primal feasible region is bounded had an unbounded dual feasible region. For example, if all entries in $\mathbf{A}$ are positive, then the dual constraints $\mathbf{A}^T \mathbf{y} \geq \mathbf{c}, \mathbf{y} \geq \mathbf{0}$ define an unbounded region. Interestingly, this is true for all linear programs. Clark's theorem states that unless both problems are infeasible, at least one of them must have an unbounded feasible region (Clark, 1961). Since an unbounded problem must have an unbounded feasible region, Table 8.2 already verifies Clark's theorem for problems with no optimal solutions.

8.3 Complementary Slackness

For a problem in canonical form, a dual solution $\mathbf{y}^T = \mathbf{c}_B^T \mathbf{B}^{-1}$ from the proof of Theorem 8.3 can be associated with any basic solution for the primal. These complementary basic solutions have several uses. For now, we examine a relationship between when a variable is zero and when the corresponding constraint is active. In the dual constraints $\mathbf{y}^T\mathbf{A} \geq \mathbf{c}^T$, the constraints corresponding to primal basic variables are $\mathbf{y}^T\mathbf{B} \geq \mathbf{c}_B^T$. Using the complimentary dual solution,

$$\mathbf{y}^T\mathbf{B} = \mathbf{c}_B^T\mathbf{B}^{-1}\mathbf{B} = \mathbf{c}_B^T.$$

Thus, these dual constraints are active and have no slack. This relationship is called *complementary slackness*, because either a variable is zero (nonbasic) or its corresponding dual constraint is active. It holds for optimal solutions of any primal–dual pair.

Theorem 8.4 *(Complementary slackness):* *Let* $\mathbf{x}^*$ *and* $\mathbf{y}^*$ *be optimal solutions to the primal and the dual problem, respectively, and let* s_i^* *and* w_j^* *be the slack in the ith primal and jth dual constraints, respectively. Then*

$$x_j^* w_j^* = 0, \quad j = 1, \ldots, n,$$
$$y_i^* s_i^* = 0, \quad i = 1, \ldots, m.$$

Proof: Use the notation (8.4) for the primal and (8.5) for the dual, but allow any direction of constraints and restrictions on the sign of the variables. From (8.7),

$$\mathbf{y}^T\mathbf{b} - \mathbf{c}^T\mathbf{x} = \mathbf{y}^T(\mathbf{b} - \mathbf{A}\mathbf{x}) + (\mathbf{y}^T\mathbf{A} - \mathbf{c}^T)\mathbf{x},$$
$$= \sum_{i=1}^{m} y_i s_i + \sum_{j=1}^{n} w_j x_j.$$

As in the proof of Theorem 8.2, if $\mathbf{x}$ and $\mathbf{y}$ are feasible for the primal and dual, respectively, then x_j and w_j have the same sign and y_i and s_i have the same sign. Hence, each term $w_j x_j$ and $y_i s_i$ in the sum is nonnegative. For optimal solutions $\mathbf{x}^*$ and $\mathbf{y}^*$, by strong duality $\mathbf{y}^{*T}\mathbf{b} - \mathbf{c}^T\mathbf{x}^* = 0$, which implies that each term is zero. □

Notice that in this notation there is a slack variable for every constraint, so each problem has $m + n$ variables. For equality constraints, any feasible solution has a slack of zero, so their complementary slackness condition is automatically satisfied. In particular, if the primal problem is in canonical form, the first condition is automatically satisfied.

We can view complementary slackness as an optimality condition. In fact, if we add primal and dual feasibility, it becomes a necessary and sufficient optimality condition. Given a bfs, we can use the complementary slackness conditions to try to check if the solution is optimal by constructing a dual solution.

Example 8.3. We will show that $x = (4, 6)$ is optimal for the standard form problem

$$\begin{array}{ll} \max & 9x_1 + 30x_2 \\ \text{s.t.} & x_1 \qquad\quad \le 8 \\ & \qquad 4x_2 \le 24 \\ & 3x_1 + 4x_2 \le 36 \\ & \qquad x_j \ge 0. \end{array}$$

The dual is

$$\begin{array}{ll} \min & 8y_1 + 24y_2 + 36y_3 \\ \text{s.t.} & y_1 \qquad\quad + 3y_3 \ge 9 \\ & \qquad 4y_2 + 4y_3 \ge 30 \\ & \qquad\quad y_i \ge 0. \end{array}$$

The slacks in the primal constraints are $s_1 = 8 - 4 = 4$, $s_2 = 24 - 4(6) = 0$, and $s_3 = 36 - 3(4) - 4(6) = 0$, so we set $y_1 = 0$. Since $x_1, x_2 > 0$, we know that both dual constraints are active, or $w_1 = w_2 = 0$. We can solve these constraints for the remaining dual variables:

$$\begin{array}{rl} 3y_3 = & 9, \\ 4y_2 + 4y_3 = & 30 \end{array}$$

with solution $y_2 = \frac{9}{2}, y_3 = 3$. Since these solutions are feasible and satisfy complementary slackness, they are optimal. We can also check that the primal and dual objective function values are equal: $9(4) + 30(6) = 8(0) + 24(\frac{9}{2}) + 36(3) = 216$.

We were able to solve for the dual solution from the primal solution x in Example 8.3 because there were exactly two active primal constraints. Including the nonnegativity constraints, there are $5 - 2 = 3$ inactive constraints, each corresponding to a nonzero components of (x, s). That forced three components of (y, w) to be zero; the other two could be found from the two dual constraints (converted to equalities). We can do this for any problem with nonnegative variables as long as x is a nondegenerate bfs: (x, s) will have m nonzeros and n zeros; the complimentary slackness conditions lead to m zeros in (y, w), and the n dual constraints can be solved for the remaining n variables. The equations use the basic columns of A, which by definition are linearly independent.

8.4 Lagrangian Duality

There is another derivation of the dual of a linear program which is related to the Lagrange multiplier method for solving optimization problems with equality constraints. This approach gives another meaning to dual variables as "prices" and introduces the idea of relaxing the constraints of a linear program, replacing them with penalties in the objective function. The result will be an interpretation of dual variables that describes their meaning in the primal problem, not just the dual.

As an example of the Lagrange multiplier method, suppose we have 40 feet of fence and want to find the rectangle with the largest area out of all rectangles with a total length of three of its sides equal to 40 feet. We write it as a constrained optimization problem in two variables:

$$\text{max} \quad xy$$
$$\text{s.t.} \quad 2x + y = 40.$$

The method introduces a Lagrange multiplier λ_i for each equality constraint. Since there is only one multiplier in this problem, call it λ. The Lagrangian is formed by adding a complementary slackness term to the objective function:

$$L(x, y, \lambda) = xy + \lambda(40 - 2x - y).$$

For a fixed λ, we can minimize the Lagrangian over x and y (with no constraints) by setting $\partial L/\partial x = y - 2\lambda$ and $\partial L/\partial y = x - \lambda$ to zero. The solution $x = \lambda$ and $y = 2\lambda$ is in fact a maximum (over x and y) of the Lagrangian even though it is not convex. Using the constraint of the original problem, $\lambda = 10$ and $(x, y) = (10, 20)$ are the dimensions in feet of the rectangle with largest area.

The key idea in this example is that instead of a constraint on the amount of fencing, we allow the constraint to be violated but introduce a penalty for using more fencing. The Lagrange multiplier λ is the penalty per foot of fencing; in economic terms it plays the role of a price. When the price is properly chosen, the unconstrained maximum of the Lagrangian is also optimal for the original problem. In a problem with more than one constraint, some or all of them can be relaxed, moving them to the objective function.

Now we extend the Lagrangian approach to a linear program, which has inequality constraints. Consider the standard form primal problem

$$\text{max} \quad \mathbf{c}^T\mathbf{x}$$
$$\text{s.t.} \quad \mathbf{Ax} \leq \mathbf{b}$$
$$\mathbf{x} \geq \mathbf{0}$$

and let $\mathbf{x}^*$ be an optimal solution. We introduce a Lagrangian relaxation that "dualizes" the functional constraints by moving them to the objective function:

$$g(\lambda) = \max_{\mathbf{x} \geq 0} \mathbf{c}^T\mathbf{x} + \lambda^T(\mathbf{b} - \mathbf{Ax}).$$

Here λ_i is the Lagrange multiplier for the constraint $\mathbf{A}_i\mathbf{x} \leq b_i$. We would like to create an upper bound for the primal using the relaxation. To do this, we must restrict $\lambda \geq \mathbf{0}$. For any $\lambda \geq \mathbf{0}$ and any $\mathbf{x}$ feasible for the primal,

$$g(\lambda) \geq \mathbf{c}^T\mathbf{x}^* + \lambda^T(\mathbf{b} - \mathbf{Ax}^*) \geq \mathbf{c}^T\mathbf{x}^* \geq \mathbf{c}^T\mathbf{x}. \tag{8.9}$$

The first inequality holds because $\mathbf{x}^*$ is feasible for the Lagrangian relaxation, the second because it is feasible for the primal. Notice that (8.9) is just weak duality for the Lagrangian relaxation. The tightest bound is

$$\min_{\lambda \geq 0} g(\lambda) = \min_{\lambda \geq 0} \max_{\mathbf{x} \geq 0} \mathbf{c}^T\mathbf{x} + \lambda^T(\mathbf{b} - \mathbf{A}\mathbf{x}). \tag{8.10}$$

Problem (8.10) is a Lagrangian dual of the primal. We will show that it is equivalent to the dual, as defined previously. Rewriting the Lagrangian relaxation as

$$g(\lambda) = \max_{\mathbf{x} \geq 0}(\mathbf{c}^T - \lambda^T\mathbf{A})\mathbf{x} + \lambda^T\mathbf{b}, \tag{8.11}$$

we see that it will be unbounded, with an optimal value of ∞, unless $\mathbf{c}^T - \lambda^T\mathbf{A} \leq \mathbf{0}$. Since (8.10) minimizes over λ, we can add this constraint without changing the optimal solution. Furthermore, if $\mathbf{c}^T - \lambda^T\mathbf{A} \leq \mathbf{0}$, then $\mathbf{x} = \mathbf{0}$ optimizes (8.11) and $g(\lambda) = \lambda^T\mathbf{b}$. Substituting into (8.10) gives

$$\min \quad \lambda^T\mathbf{b}$$
$$\text{s.t.} \quad \mathbf{c}^T - \lambda^T\mathbf{A} \leq \mathbf{0}$$
$$\lambda \geq \mathbf{0},$$

which is the dual.

Example 8.4. For the primal problem in Example 8.3, dualizing the three constraints gives the Lagrangian relaxation

$$g(\lambda) = \max_{\mathbf{x} \geq 0} 9x_1 + 30x_2 + \lambda_1(8 - x_1) + \lambda_2(24 - 4x_2) + \lambda_3(36 - 3x_1 - 4x_2).$$

The Lagrangian dual is

$$\min_{\lambda \geq 0} \max_{\mathbf{x} \geq 0} 9x_1 + 30x_2 + \lambda_1(8 - x_1) + \lambda_2(24 - 4x_2) + \lambda_3(36 - 3x_1 - 4x_2)$$
$$= \min_{\lambda \geq 0} \max_{\mathbf{x} \geq 0} (9 - \lambda_1 - 3\lambda_3)x_1 + (30 - 4\lambda_2 - 4\lambda_3)x_2 + 8\lambda_1 + 24\lambda_2 + 36\lambda_3.$$

Requiring the coefficients of x_1 and x_2 to be nonpositive, the maximum occurs at $\mathbf{x} = \mathbf{0}$ and we obtain the linear program

$$\min \quad 8\lambda_1 + 24\lambda_2 + 36\lambda_3$$
$$\text{s.t.} \quad 9 - \lambda_1 \qquad\quad - 3\lambda_3 \leq 0$$
$$30 \qquad - 4\lambda_2 - 4\lambda_3 \leq 0$$
$$\lambda_i \geq 0$$

which is the same as the dual in Example 8.3.

Properties of the Lagrangian Dual

1. The Lagrangian dual of a linear program is equivalent to the dual of the linear program. Although the Lagrangian relaxation $g(\lambda)$ is not linear (because it is infinite for some λ), once the constraints are added it is linear and equal to the objective of the dual.
2. The Lagrange multipliers are the dual variables.
3. The Lagrange multiplier interpretation shows that dual variables can be interpreted as prices associated with the primal constraints.

We will further develop the price interpretation of dual variables in Chapter 10. Lagrangian duality is more general than linear programming duality in two ways. First, we can choose just some of the constraints to dualize. Nonnegativity constraints can also be dualized. Second, as in the area maximization example, the objective function does not have to be linear – nor do the constraints; see Chapter 14.

8.5 Farkas' Lemma and Optimality

When we stated the optimality condition for any feasible solution of a linear program in Section 7.4, we noted that it is difficult to check; the "easy" optimality condition only applies to basic feasible solutions. Now we provide another optimality condition for any feasible solution. It interprets Farkas' lemma, which dates from 1894, for linear programming. Although not directly useful as an optimality condition, Farkas' lemma is interesting for other reasons. First, it gives another geometric interpretation of the condition that there is no improving feasible direction. Second, it can be used to prove strong duality.

To better visualize feasible directions, consider the inequality-constrained problem

$$\max \quad \mathbf{c}^T \mathbf{x}$$
$$\text{s.t.} \quad \mathbf{A}_{\text{full}} \mathbf{x} \leq \mathbf{b}.$$

Standard form problems can be written this way by including the nonnegativity constraints in $\mathbf{A}_{\text{full}}$. For a feasible solution $\mathbf{x}$, throw out the rows of $\mathbf{A}_{\text{full}}$ for inactive constraints, leaving $\mathbf{A}$. Adapting (7.2), an improving feasible direction for this problem is a vector **d** that satisfies

$$\mathbf{Ad} \le 0 \quad \text{and} \quad \mathbf{c}^T \mathbf{d} > 0. \tag{8.12}$$

Example 8.5. Consider the problem

$$
\begin{aligned}
\max z = \quad & 4x_1 + 3x_2 \\
\text{s.t.} \quad & 1.5x_1 + x_2 \le 16 \\
& x_1 + x_2 \le 12 \\
& x_2 \le 10 \\
& x_j \ge 0.
\end{aligned}
$$

Figure 8.1 shows the feasible region, the gradient **c** of the objective function, and the line where z is constant at several extreme points. Improving directions are those that point above the line of constant z. At the extreme point $(2, 10)$, there are improving feasible directions because the z line enters the feasible region. At $(0, 0)$, c is an improving feasible direction. However, at $(8, 4)$, the z line does not enter the interior of the feasible region, and "sweeping" the line makes it immediately leave the feasible region as z is increased.

Now we give a second geometric explanation of why $(8, 4)$ is optimal. At this point, constraints 1 and 2 are active, with coefficient matrix

$$\mathbf{A} = \begin{bmatrix} 1.5 & 1 \\ 1 & 1 \end{bmatrix}.$$

These constraints are represented in Figure 8.2 by their normal vectors, which are the rows of **A**. Compare this to $(2, 10)$, where constraints 2 and 3 are active, with coefficient matrix

$$\mathbf{A} = \begin{bmatrix} 1 & 1 \\ 0 & 1 \end{bmatrix}.$$

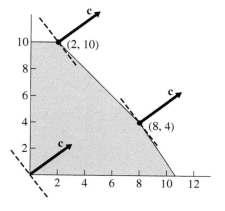

Figure 8.1 Gradient vectors and lines of constant z.

Figure 8.2 Gradient vectors and active constraint normal vectors.

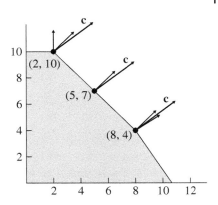

At $(8, 4)$, **c** is between the constraint normal vectors, while at $(2, 10)$ (and all other extreme points), it is not. Indeed, this is equivalent to what was seen in Figure 8.1: the line of constant z, which is orthogonal to **c**, is between the active constraint lines at $(8, 4)$, but not at the other extreme points. A similar statement can be made at other feasible solutions. At $(5, 7)$, only constraint 2 is active and its normal vector does not lie in the same direction as **c**, so it is not optimal.

These observations extend to higher dimensions if we interpret lying "between" a set of vectors as being a nonnegative linear combination of them. A feasible solution is optimal if and only if **c** is a nonnegative linear combination of the normal vectors to its active constraints. Remembering that the normal vectors are the rows of **A**, this linear combination can be written $\mathbf{y}^T\mathbf{A} = \mathbf{c}^T$. Algebraically, this second optimality condition is the existence of a solution to the system

$$\mathbf{A}^T\mathbf{y} = \mathbf{c}, \quad \mathbf{y} \geq \mathbf{0}. \tag{8.13}$$

Call (8.12) System 1 and (8.13) System 2. Table 8.3 summarizes the situation: either System 1 or System 2 has a solution, but not both. Farkas' lemma makes this statement about systems of linear inequalities, without reference to optimization.

Lemma 8.1 *(Farkas' lemma):* *For any $m \times n$ matrix* **A** *and vector* **c** *in $\mathbb{R}^n$, exactly one of the following alternatives holds: (i) System 1.* $\mathbf{Ax} \leq \mathbf{0}$ *and* $\mathbf{c}^T\mathbf{x} > 0$ *has a solution* **x**. *(ii) System 2.* $\mathbf{A}^T\mathbf{y} = \mathbf{c}$ *and* $\mathbf{y} \geq \mathbf{0}$ *has a solution* **y**.

Table 8.3 When Systems 1 and 2 have solutions.

Optimal solution	Improving feasible direction	Solution to System 1	Solution to System 2
Yes	No	No	Yes
No	Yes	Yes	No

Proof: Suppose System 2 has a solution $\mathbf{y}$. If $\mathbf{Ax} \leq \mathbf{0}$, then $\mathbf{c}^T\mathbf{x} = (\mathbf{A}^T\mathbf{y})^T\mathbf{x} = \mathbf{y}^T\mathbf{Ax} \leq \mathbf{0}$, so the first alternative cannot hold.

Now suppose System 2 does not have a solution. Consider the following primal–dual pair.

$$\begin{array}{ll} \max & \mathbf{c}^T\mathbf{x} \\ \text{s.t.} & \mathbf{Ax} \leq \mathbf{0} \end{array} \qquad \begin{array}{ll} \min & \mathbf{0}^T\mathbf{y} \\ \text{s.t.} & \mathbf{A}^T\mathbf{y} = \mathbf{c} \\ & \mathbf{y} \geq \mathbf{0}. \end{array}$$

The minimization problem is infeasible, which implies by strong duality that the maximization problem is either infeasible or unbounded. Since, for example, $\mathbf{x} = \mathbf{0}$ is feasible for the maximization problem, it is unbounded. Thus, it must have a feasible solution $\mathbf{x}$ with $\mathbf{Ax} \leq \mathbf{0}$ and objective function value $\mathbf{c}^T\mathbf{x} > 0$ satisfying System 1. □

We related Farkas' lemma to optimality of a linear program that only has inequality constraints. There are other versions of the lemma that give optimality conditions for problems in canonical and other forms. Another interesting feature of Farkas' lemma is its relation to the concept of a separating hyperplane. The line where z is constant through $(8, 4)$ in Figure 8.1 is an example of a *supporting hyperplane*. If we move slightly up and to the right to, say, $(8.01, 4.01)$ then we can shift the line slightly so that it separates the point from the feasible region but touches neither; it is called a *separating hyperplane*. For any convex set and point not in the set, the Separating Hyperplane Theorem guarantees that there is a separating hyperplane. Farkas' lemma can be proven as a consequence of the Separating Hyperplane Theorem. Further, the Strong Duality Theorem can be proven using Farkas' lemma; see, for example, Luenberger and Ye (2016). This approach resolves the issue of degeneracy, which was not addressed in the proof of strong duality given in this chapter.

Problems

For Exercises 8.1–8.4, find the dual of the linear program.

1      $\begin{array}{ll} \max & 5x_1 + x_2 + 4x_3 \\ \text{s.t.} & 6x_1 + 5x_2 + 3x_3 \leq 11 \\ & 3x_1 + 2x_2 + 9x_3 \leq 4 \\ & x_1, x_2, x_3 \geq 0. \end{array}$

2      $\begin{array}{ll} \max & 7x_1 + 3x_2 - 5x_3 \\ \text{s.t.} & 4x_1 \qquad\quad - 6x_3 \leq 12 \\ & 2x_1 + 9x_2 + x_3 = 8 \\ & 5x_1 - 3x_2 - 2x_3 \geq 6 \\ & x_1, x_2, x_3 \geq 0. \end{array}$

3
$$\min \quad 6x_1 + 11x_2$$
$$\text{s.t.} \quad -3x_1 - 2x_2 \leq 19$$
$$15x_1 + 7x_2 \geq 25$$
$$x_1 \qquad \geq 8$$
$$x_1 \geq 0, \; x_2 \text{ u.r.s.}$$

4
$$\min \quad 14x_1 - 6x_2 + 3x_3$$
$$\text{s.t.} \quad -3x_1 + x_2 + 4x_3 \leq 12$$
$$6x_1 + 7x_2 - x_3 \geq 17$$
$$2x_1 \qquad + x_3 = 10$$
$$x_1 \geq 0, x_2 \leq 0, x_3 \text{ u.r.s.}$$

5 Consider the linear program

$$\max \quad 2x_1 + 4x_2$$
$$\text{s.t.} \quad x_1 + 2x_2 \leq 3$$
$$x_1 - x_2 \leq 1$$
$$x_1, x_2 \geq 0.$$

(a) Find the dual linear program.
(b) Plot the feasible region of both problems.
(c) Find any degenerate basic feasible solutions for each problem.
(d) Find all optimal basic feasible solutions for both problems. Verify complementary slackness between these optimal solutions.

6 For the primal problem max $\mathbf{c}^T \mathbf{x}$ s.t. $\mathbf{Ax} \leq \mathbf{b}$, $\mathbf{x}$ u.r.s.,
(a) State the dual problem.
(b) Give a shorter proof of weak duality than Theorem 8.2's general proof.

7 Consider a linear program in standard form. Show that if its coefficient matrix $\mathbf{A} \geq \mathbf{0}$ and every column has at least one positive element, then the dual feasible region is unbounded.

8 Consider the linear program

$$\max \quad x_1 + 2x_2 - 6x_3$$
$$\text{s.t.} \quad x_1 + 4x_2 + x_3 \leq 5$$
$$5x_1 + 5x_2 - 2x_3 \leq 10$$
$$x_1, x_2, x_3 \geq 0.$$

Using complimentary slackness, show that $x = (1, 1, 0)$ is an optimal solution of the linear program. Also find an optimal solution of the dual problem.

9   Consider the linear program

$$\begin{array}{rl} \max & 4x_1 - x_2 + 3x_3 \\ \text{s.t.} & 5x_1 - x_2 + 2x_3 \le 4 \\ & -2x_1 + x_2 + x_3 \le 3 \\ & x_1, x_2, x_3 \ge 0. \end{array}$$

Using complimentary slackness, show that $y = (\frac{4}{3}, \frac{1}{3})$ is an optimal solution to the dual of the linear program. Also find an optimal solution of the primal problem.

10   Find the Lagrangian dual (8.10) of Exercise 8.5 and show that it is the same as the dual linear program.

11   For the linear program min $c^T x$ s.t. $Ax \ge b$, $x \ge 0$
(a) Find a formula like (8.10) for the Lagrangian dual.
(b) Find the Lagrangian dual of (8.3) by substituting it into the formula found in (a). Show that it is the same as (8.1).

9

Simplex Method

This chapter presents the simplex method for solving linear programs. It was the first such algorithm, pioneered by George Dantzig in 1947. Enhanced versions of it are widely used today. It is still one of the best linear programming algorithm available and can be used to solve very large problems. The general idea was described in Chapter 7. Starting at an extreme point, or basic feasible solution (bfs), move along an edge of the feasible region to an adjacent extreme point that is better. Continue until there is no better adjacent extreme point. Then, assuming the problem has an optimal solution, the current extreme point is optimal. Chapter 7 provides the background needed to perform these steps and understand why the algorithm is correct.

The efficiency of the algorithm depends on the number of iterations needed to reach an optimal solution and the amount of computation (and storage) required in an iteration. Choosing and moving to an adjacent extreme point can be done more efficiently with equality constraints, so the algorithm works with problems in canonical form. Serendipitously, the many variables that are added when converting to canonical form generally do *not* increase the number of iterations. These observations make canonical form appealing. Here is an outline of the algorithm.

Simplex Method

Phase I. Find a bfs and basis or show that the problem is infeasible.
Phase II.
1. For the current bfs, check whether there is an improving simplex direction.
2. If not, stop. The current solution is optimal. If there is, choose one.
3. Determine how far one can move in that direction and stay in the feasible region. If there is no limit, stop. The linear program is unbounded.

(Continued)

Linear and Convex Optimization: A Mathematical Approach, First Edition. Michael H. Veatch.
© 2021 John Wiley & Sons, Inc. Published 2021 by John Wiley & Sons, Inc.
Companion website: www.wiley.com/go/veatch/convexandlinearoptimization

(Continued)

4. Move to the new solution found in Step 3. The basis changes by one variable. If the bfs was degenerate, the new solution may be the same point (no move). Otherwise, it will be an adjacent bfs. Go to Step 1.

For some problems it is easy to find a bfs to start from, so we wait until Section 9.3 to describe Phase I. The heart of the algorithm is Phase II, introduced in Section 9.1. The question of correctness of the algorithm is taken up in Section 9.2. For problems without degeneracy, it is easy to show that the simplex method is correct. Modifications are described to ensure that it also terminates for degenerate problems. Section 9.3 shows that the same (Phase II) algorithm can be applied to a modified problem in Phase I to find an initial bfs to the original problem.

The simplex method takes remarkably few iterations on almost all problems of practical interest. Section 9.4 describes this situation, then discusses variants and implementations that can improve its speed. One of these implementations, called the revised simplex method, makes a fundamental improvement to the complexity of an iteration. Readers who are not interested in computational complexity may wish to skip this section.

9.1   Simplex Method From a Known Feasible Solution

For a linear program in canonical form,

$$\begin{aligned}
\max \quad & \mathbf{c}^T\mathbf{x} \\
\text{s.t.} \quad & \mathbf{A}\mathbf{x} = \mathbf{b} \\
& \mathbf{x} \geq \mathbf{0},
\end{aligned}$$

(9.1)

this section describes the Phase II simplex method to solve it. We use the following example to demonstrate the algorithm.

Example 9.1. Consider the following linear program:

$$\begin{aligned}
\max \quad & 4x + 3y \\
\text{s.t.} \quad & 1.5x + y \leq 16 \\
& x + y \leq 12 \\
& y \leq 10 \\
& x, y \geq 0,
\end{aligned}$$

which was solved graphically in Chapter 1. Converting to canonical form,

$$\begin{aligned}
\max \quad & 4x + 3y \\
\text{s.t.} \quad & 1.5x + y + s_1 && = 16 \\
& x + y && + s_2 && = 12 \\
& y && + s_3 = 10 \\
& x, y, s_1, s_2, s_3 \geq 0.
\end{aligned}$$

When using variable names other than the standard notation $\mathbf{x} = (x_1, \ldots, x_n)$ we will use the variable name as its index. For example, the column of the coefficient matrix for the variable y in Example 9.1 is $\mathbf{a}_y = (1, 1, 1)$, not $\mathbf{a}_2$. We also use the Section 7.3 notation for basic solutions in canonical form.

9.1.1 Finding an Improving Direction

For a bfs with basis matrix $\mathbf{B}$, Section 7.4 defined the simplex direction $\mathbf{d}^{(k)}$ for the nonbasic variable x_k as $d_k^{(k)} = 1$, $d_j^{(k)} = 0, j \in \mathcal{N} \backslash \{k\}$ and $\mathbf{d}_B^{(k)} = -\mathbf{B}^{-1}\mathbf{a}_k$. These directions include all edges emanating from $\mathbf{x}$. However, if $\mathbf{x}$ is degenerate, there may be simplex directions that have no edge and are not feasible directions. A direction $\mathbf{d}$ is improving if $\mathbf{c}^T\mathbf{d} > 0$. The simplex direction for x_k is improving if

$$c_k - \mathbf{c}_B^T\mathbf{B}^{-1}\mathbf{a}_k > 0.$$

This quantity will be used repeatedly (we already used it Chapter 8) and has a nice interpretation, given in Section 10.2, as a *reduced cost*. The term "cost" describes the objective function coefficient of a minimization problem. In maximization problems, where the objective function coefficients are better described as prices, it is customary to still call them costs.

The **reduced cost** of the variable x_k is

$$\bar{c}_k = c_k - \mathbf{c}_B^T\mathbf{B}^{-1}\mathbf{a}_k.$$

Every reduced cost uses the same vector $\mathbf{y}^T = \mathbf{c}_B^T\mathbf{B}^{-1}$, so it is convenient to compute reduced costs as

$$\bar{\mathbf{c}}^T = \mathbf{c}^T - \mathbf{y}^T\mathbf{A}.$$

Chapter 8 gives another meaning to $\mathbf{y}$: the values of the dual variables corresponding to the bfs. In the simplex method, these are called *simplex multipliers*.

Reduced costs of basic variables are zero. To see this, recall that, for a basic variable, $\mathbf{a}_k$ is a column of $\mathbf{B}$, so the basic reduced cost vector is

$$\bar{\mathbf{c}}_B^T = \mathbf{c}_B^T - \mathbf{c}_B^T\mathbf{B}^{-1}\mathbf{B} = \mathbf{c}_B^T - \mathbf{c}_B^T = \mathbf{0}.$$

Thus, we focus on the nonbasic reduced costs

$$\overline{\mathbf{c}}_N^T = \mathbf{c}_N^T - \mathbf{y}^T \mathbf{N}.$$

The simplex direction for x_k is improving if its reduced cost $\overline{c}_k > 0$. The bfs is optimal if $\overline{\mathbf{c}}_N \leq \mathbf{0}$, i.e. there are no positive reduced costs.

If there is more than one positive reduced cost, we will use the following rule to choose one.

Dantzig Rule for Choosing a Reduced Cost Choose a variable whose reduced cost is most positive.

The rule used will not affect the correctness of the algorithm, but will influence its speed. Other rules are discussed in Section 9.4.

9.1.2 Determining the Step Size

To move to an adjacent bfs, we take the largest feasible step. As in Section 5.2, from a current feasible solution $\mathbf{x}$ a step size λ in the direction $\mathbf{d}$ moves to the solution $\mathbf{x} + \lambda \mathbf{d}$. If all constraints are still satisfied, the new solution is feasible. In canonical form, the simplex directions preserve the equality constraints, so feasibility is simply nonnegativity. Thus, in canonical form, the largest feasible step size is the largest λ for which $\mathbf{x} + \lambda \mathbf{d} \geq \mathbf{0}$. The simplex method uses the maximum step size. This is desirable both because the step is always an improving direction and, starting from a bfs, the maximum step size moves to an adjacent bfs. The following example illustrates the step size calculation.

Example 9.2 (Step size). Suppose for some canonical form linear program $(x, y, z, s_1, s_2, s_3) = (0, 6, 0, 0, 16, 9)$ is a bfs with basis (y, s_2, s_3).

(a) Suppose $\mathbf{d}^{(x)} = (1, -3, 0, 0, -2, 2)$ is a simplex direction at this bfs. A feasible step size λ must satisfy

$$
\begin{bmatrix} x \\ y \\ z \\ s_1 \\ s_2 \\ s_3 \end{bmatrix} = \begin{bmatrix} 0 \\ 6 \\ 0 \\ 0 \\ 16 \\ 9 \end{bmatrix} + \lambda \begin{bmatrix} 1 \\ -3 \\ 0 \\ 0 \\ -2 \\ 2 \end{bmatrix} \geq 0 \qquad
\begin{aligned}
&\lambda \geq 0 \\
&\lambda \leq \tfrac{6}{3} \\
&\\
&\lambda \leq \tfrac{16}{2} \\
&\lambda \geq -\tfrac{9}{2}
\end{aligned}.
$$

The largest λ that satisfies all of these inequalities is $\lambda = 2$, giving a new solution

$$
\begin{bmatrix} x \\ y \\ z \\ s_1 \\ s_2 \\ s_3 \end{bmatrix} = \begin{bmatrix} 0 \\ 6 \\ 0 \\ 0 \\ 16 \\ 9 \end{bmatrix} + 2 \begin{bmatrix} 1 \\ -3 \\ 0 \\ 0 \\ -2 \\ 2 \end{bmatrix} = \begin{bmatrix} 2 \\ 0 \\ 0 \\ 0 \\ 12 \\ 13 \end{bmatrix}.
$$

This is a bfs with basis (x, s_2, s_3).

(b) Suppose $\mathbf{d}^{(z)} = (0, 3, 1, 0, 0, 4)$ is another simplex direction at the original bfs. Because $\mathbf{d}^{(z)} \geq 0, \mathbf{x} + \lambda \mathbf{d}^{(z)} \geq 0$ for all $\lambda \geq 0$ and there is no maximum step size. The feasible region is unbounded in this direction.

In part (a), because there are negative components of $\mathbf{d}^{(x)}$, the step size is finite. Notice that only the negative components need be considered, and that they are all basic. In part (b), $\mathbf{d}^{(z)} \geq 0$ and there is no limit to the step size. The simplex method only uses improving directions, so $\mathbf{d}^{(z)}$ would be an unbounded improving direction, showing that the linear program is unbounded. The following formula computes the step size for a simplex direction $\mathbf{d}$.

Maximum Step Size

(a) If $\mathbf{d} \geq \mathbf{0}$, the linear program is unbounded.

(b) If some component of $\mathbf{d}$ is negative, then the maximum step size is
$\lambda = \min \left\{ \frac{x_j}{-d_j} : d_j < 0, j \in B \right\}$.

Using this notation for part (a) of Example 9.2,

$$
\lambda = \min \left\{ \frac{6}{-(-3)}, \frac{16}{-(-2)} \right\} = 2.
$$

Only negative entries of $\mathbf{d}$ are considered; they give positive ratios. The first ratio is for the variable y and the second for s_2.

9.1.3 Updating the Basis

The nonbasic variable whose simplex direction is chosen will typically become positive. Thus, it becomes a basic variable. Call this the *entering variable*, x_e. When we use the largest possible step size, at least one basic variable equals zero at the new solution. In particular, any variable that achieves the minimum ratio in the step size calculation decreases to zero. We choose one of these to become a non-basic variable. Call this the *leaving variable*, x_l. At each iteration, we update the basis and compute the new solution $\mathbf{x} + \lambda\mathbf{d}$. Actually, since only the variables in the new basis can be nonzero, only these x_j need be computed. In Example 9.2, the variable x is the entering variable and y is the leaving variable.

This process identifies a new set of m variables, but how do we know the new solution is a bfs? Notice that the other variables, including x_l, are set to zero. The unique solution of $\mathbf{Ax} = \mathbf{b}$ for the remaining variables is then found, implying that the columns of $\mathbf{A}$ for these variables are linearly independent. By construction $\mathbf{x} \geq \mathbf{0}$. Thus, these variables are a basis and the new solution is a bfs. It is possible that $\lambda = 0$ and the new value of x_e is zero because of degeneracy, but we will still designate it a basic variable.

9.1.4 Simplex Method

Putting these pieces together, we can state the simplex method. We assume an initial bfs is available, so we are describing Phase II of the overall algorithm.

Algorithm 9.1 (Simplex Method)

Step 0. Initialization. Identify a bfs $\mathbf{x}$ and its basis $\mathcal{B}$.

Step 1. Optimality check. Compute the simplex multipliers $\mathbf{y}^T = \mathbf{c}_B^T \mathbf{B}^{-1}$ and the reduced costs $\overline{\mathbf{c}}_N^T = \mathbf{c}_N^T - \mathbf{y}^T \mathbf{N}$. If $\overline{\mathbf{c}}_N \leq \mathbf{0}$, stop. The current solution is optimal.

Step 2. Compute simplex direction. Otherwise, choose a variable x_e with the most positive reduced cost as the entering variable. Compute its simplex direction $\mathbf{d}_B^{(e)} = -\mathbf{B}^{-1}\mathbf{a}_e$. Let $\mathbf{d} = \mathbf{d}_B^{(e)}$.

Step 3. Compute step size. If $\mathbf{d} \geq \mathbf{0}$, stop. The linear program is unbounded. Otherwise, the step size is $\lambda = \min\left\{\frac{x_j}{-d_j} : d_j < 0\right\}$ and the leaving variable x_l is one that achieves the minimum ratio.

Step 4. Update solution and basis. The new solution $\mathbf{x}^{\text{new}}$ is $\mathbf{x}_B^{\text{new}} = \mathbf{x}_B + \lambda\mathbf{d}$, which makes $x_l^{\text{new}} = 0$, and $x_e^{\text{new}} = \lambda$. Replace x_l by x_e in the basis. Go to Step 1.

Steps 1 and 2 require solving systems of linear equations. This can be done by finding $\mathbf{B}^{-1}$ or by other methods. Specifying the basis determines $\mathbf{B}, \mathbf{N}, \mathbf{c}_B$, and $\mathbf{c}_N$. Although they are used in the formulas and we write them out in the following examples, a computer implementation might not actually store them, since the columns of $\mathbf{B}$ and $\mathbf{N}$ are in $\mathbf{A}$ and the components of $\mathbf{c}_B$ and $\mathbf{c}_N$ are in $\mathbf{c}$. Also, in the examples in the following text we construct $\mathbf{d}^{(e)}$, not just $\mathbf{d}_B^{(e)}$, and use it to compute $\mathbf{x}^{\text{new}}$. These few extra computations make the examples easier to follow.

Example 9.3. We apply the algorithm to Example 9.1, which has

$$\mathbf{A} = \begin{bmatrix} 1.5 & 1 & 1 & 0 & 0 \\ 1 & 1 & 0 & 1 & 0 \\ 0 & 1 & 0 & 0 & 1 \end{bmatrix}, \quad \mathbf{b} = \begin{bmatrix} 16 \\ 12 \\ 10 \end{bmatrix},$$

$$\mathbf{c} = (4, 3, 0, 0, 0,), \quad \mathbf{x} = (x, y, s_1, s_2, s_3).$$

Initialization. Because the problem was converted from standard form with $\mathbf{b} \geq \mathbf{0}$, an obvious initial bfs is $\mathbf{x} = (0, 0, 16, 12, 10)$ with basis (s_1, s_2, s_3), where the original variables are set to zero and the slack variables are set to the right-hand sides. For this basis, $\mathbf{B} = [\mathbf{a}_{s_1} | \mathbf{a}_{s_2} | \mathbf{a}_{s_3}]$ is the identity matrix and

$$\mathbf{N} = [\mathbf{a}_x | \mathbf{a}_y] = \begin{bmatrix} 1.5 & 1 \\ 1 & 1 \\ 0 & 1 \end{bmatrix}.$$

Notice that the order of the basic variables and of the nonbasic variables (x, y) must be specified to construct $\mathbf{B}$, $\mathbf{N}$, $\mathbf{c}_B$, and $\mathbf{c}_N$ and interpret $\bar{\mathbf{c}}_N$ and $\mathbf{d}_B^{(e)}$.

Iteration 1. First we compute

$$\mathbf{y}^T = \begin{bmatrix} 0 & 0 & 0 \end{bmatrix} \mathbf{B}^{-1} = \begin{bmatrix} 0 & 0 & 0 \end{bmatrix},$$

$$\bar{\mathbf{c}}_N^T = \begin{bmatrix} 4 & 3 \end{bmatrix} - \begin{bmatrix} 0 & 0 & 0 \end{bmatrix} \begin{bmatrix} 1.5 & 1 \\ 1 & 1 \\ 0 & 1 \end{bmatrix} = \begin{bmatrix} 4 & 3 \end{bmatrix}.$$

Since $\bar{c}_x = 4$ is largest, choose x as the entering variable. Its simplex direction is

$$\mathbf{d}_B^{(x)} = -\mathbf{B}^{-1} \begin{bmatrix} 1.5 \\ 1 \\ 0 \end{bmatrix} = \begin{bmatrix} -1.5 \\ -1 \\ 0 \end{bmatrix},$$

or $\mathbf{d}^{(x)} = (1, 0, -1.5, -1, 0)$. The step size is

$$\lambda = \min \left\{ \frac{16}{-(-1.5)}, \frac{12}{-(-1)} \right\} = \min \left\{ \frac{32}{3}, 12 \right\} = \frac{32}{3}.$$

The first ratio, for s_1, achieves the minimum, so s_1 is the leaving variable. The new basis is (x, s_2, s_3) with bfs

$$\begin{bmatrix} x \\ y \\ s_1 \\ s_2 \\ s_3 \end{bmatrix} = \begin{bmatrix} 0 \\ 0 \\ 16 \\ 12 \\ 10 \end{bmatrix} + \frac{32}{3} \begin{bmatrix} 1 \\ 0 \\ -1.5 \\ -1 \\ 0 \end{bmatrix} = \begin{bmatrix} \frac{32}{3} \\ 0 \\ 0 \\ \frac{4}{3} \\ 10 \end{bmatrix}.$$

The nonbasic variables (y, s_1) are zero as required. For the new basis,

$$
\mathbf{B} = \begin{bmatrix} 1.5 & 0 & 0 \\ 1 & 1 & 0 \\ 0 & 0 & 1 \end{bmatrix}, \quad \mathbf{N} = \begin{bmatrix} 1 & 1 \\ 1 & 0 \\ 1 & 0 \end{bmatrix}.
$$

Iteration 2. We begin by computing

$$
\mathbf{B}^{-1} = \begin{bmatrix} \frac{2}{3} & 0 & 0 \\ -\frac{2}{3} & 1 & 0 \\ 0 & 0 & 1 \end{bmatrix}, \quad \mathbf{y}^T = \begin{bmatrix} 4 & 0 & 0 \end{bmatrix} \begin{bmatrix} \frac{2}{3} & 0 & 0 \\ -\frac{2}{3} & 1 & 0 \\ 0 & 0 & 1 \end{bmatrix} = \begin{bmatrix} \frac{8}{3} & 0 & 0 \end{bmatrix},
$$

$$
\mathbf{\bar{c}}_N^T = \begin{bmatrix} 3 & 0 \end{bmatrix} - \begin{bmatrix} \frac{8}{3} & 0 & 0 \end{bmatrix} \begin{bmatrix} 1 & 1 \\ 1 & 0 \\ 1 & 0 \end{bmatrix} = \begin{bmatrix} \frac{1}{3} & -\frac{8}{3} \end{bmatrix}.
$$

There is one positive reduced cost, $\bar{c}_y = \frac{1}{3}$, so y is the entering variable. The simplex direction is

$$
\mathbf{d}_B^{(y)} = -\begin{bmatrix} \frac{2}{3} & 0 & 0 \\ -\frac{2}{3} & 1 & 0 \\ 0 & 0 & 1 \end{bmatrix} \begin{bmatrix} 1 \\ 1 \\ 1 \end{bmatrix} = \begin{bmatrix} -\frac{2}{3} \\ -\frac{1}{3} \\ -1 \end{bmatrix},
$$

or $\mathbf{d}^{(y)} = (-\frac{2}{3}, 1, 0, -\frac{1}{3}, -1)$. The step size is

$$
\lambda = \min \left\{ \frac{32/3}{-(-2/3)}, \frac{4/3}{-(-1/3)}, \frac{10}{-(-1)} \right\} = \min\{16, 4, 10\} = 4.
$$

The second ratio, for s_2, achieves the minimum, so s_2 is the leaving variable. The new basis is (x, y, s_3) with bfs

$$
\begin{bmatrix} x \\ y \\ s_1 \\ s_2 \\ s_3 \end{bmatrix} = \begin{bmatrix} \frac{32}{3} \\ 0 \\ 0 \\ \frac{4}{3} \\ 10 \end{bmatrix} + 4 \begin{bmatrix} -\frac{2}{3} \\ 1 \\ 0 \\ -\frac{1}{3} \\ -1 \end{bmatrix} = \begin{bmatrix} 8 \\ 4 \\ 0 \\ 0 \\ 6 \end{bmatrix}.
$$

Updating, the nonbasic variables are (s_1, s_2) with

$$
\mathbf{B} = \begin{bmatrix} 1.5 & 1 & 0 \\ 1 & 1 & 0 \\ 0 & 1 & 1 \end{bmatrix}, \quad \mathbf{N} = \begin{bmatrix} 1 & 0 \\ 0 & 1 \\ 0 & 0 \end{bmatrix}.
$$

Iteration 3. Again we compute

$$\mathbf{B}^{-1} = \begin{bmatrix} 2 & -2 & 0 \\ -2 & 3 & 0 \\ 2 & -3 & 1 \end{bmatrix}, \quad \mathbf{y}^T = \begin{bmatrix} 4 & 3 & 0 \end{bmatrix} \begin{bmatrix} 2 & -2 & 0 \\ -2 & 3 & 0 \\ 2 & -3 & 1 \end{bmatrix} = \begin{bmatrix} 2 & 1 & 0 \end{bmatrix},$$

$$\bar{\mathbf{c}}_N^T = \begin{bmatrix} 0 & 0 \end{bmatrix} - \begin{bmatrix} 2 & 1 & 0 \end{bmatrix} \begin{bmatrix} 1 & 0 \\ 0 & 1 \\ 0 & 0 \end{bmatrix} = \begin{bmatrix} -2 & -1 \end{bmatrix}.$$

Since neither reduced cost is positive, there is no improving simplex direction and the current solution $\mathbf{x} = (8, 4, 0, 0, 6)$ is optimal.

We can visualize the progress of the simplex method on this example in Figure 7.1. The initial bfs is at the origin. The first iteration moves to $(x, y) = (\frac{32}{3}, 0)$ and the second to $(8, 4)$. The next example illustrates how the simplex method can find that a linear program is unbounded.

Example 9.4. Consider the following linear program:

$$\begin{array}{rrrl} \max & -2x + & 6y & \\ \text{s.t.} & 5x - & 20y \le & 80 \\ & 2x - & 2y \le & 36 \\ & -3x + & 2y \le & 24 \\ & & x, y \ge & 0. \end{array}$$

Initialization. Slack variables are added as in the previous example. The initial bfs is $(x, y, s_1, s_2, s_3) = (0, 0, 80, 36, 24)$ with basis (s_1, s_2, s_3), $\mathbf{B}$ is the identity matrix, and

$$\mathbf{N} = \begin{bmatrix} 5 & -20 \\ 2 & -2 \\ -3 & 2 \end{bmatrix}.$$

Iteration 1. As in Example 9.3, in the first iteration $\mathbf{y} = (0, 0, 0)$ and $\bar{\mathbf{c}}_N = \mathbf{c}_N = (-2, 6)$. The only positive reduced cost is $\bar{c}_y = 6$, so y is the entering variable. The simplex direction is

$$\mathbf{d}_B^{(y)} = -\mathbf{B}^{-1} \begin{bmatrix} -20 \\ -2 \\ 2 \end{bmatrix} = \begin{bmatrix} 20 \\ 2 \\ -2 \end{bmatrix},$$

or $\mathbf{d}^{(y)} = (0, 1, 20, 2, -2)$. The step size is

$$\lambda = \frac{24}{-(-2)} = 12.$$

The only ratio is for s_3, so s_3 is the leaving variable. The new basis is (y, s_1, s_2) with bfs

$$
\begin{bmatrix} x \\ y \\ s_1 \\ s_2 \\ s_3 \end{bmatrix} = \begin{bmatrix} 0 \\ 0 \\ 80 \\ 36 \\ 24 \end{bmatrix} + 12 \begin{bmatrix} 0 \\ 1 \\ 20 \\ 2 \\ -2 \end{bmatrix} = \begin{bmatrix} 0 \\ 12 \\ 320 \\ 60 \\ 0 \end{bmatrix}.
$$

Updating, the nonbasic variables are (x, s_3) with

$$
\mathbf{B} = \begin{bmatrix} -20 & 1 & 0 \\ -2 & 0 & 1 \\ 2 & 0 & 0 \end{bmatrix}, \quad \mathbf{N} = \begin{bmatrix} 5 & 0 \\ 2 & 0 \\ -3 & 1 \end{bmatrix}.
$$

Iteration 2. First we compute

$$
\mathbf{B}^{-1} = \begin{bmatrix} 0 & 0 & 0.5 \\ 1 & 0 & 10 \\ 0 & 1 & 1 \end{bmatrix}, \quad \mathbf{y}^T = \begin{bmatrix} 6 & 0 & 0 \end{bmatrix} \begin{bmatrix} 0 & 0 & 0.5 \\ 1 & 0 & 10 \\ 0 & 1 & 1 \end{bmatrix} = \begin{bmatrix} 0 & 0 & 3 \end{bmatrix},
$$

$$
\bar{\mathbf{c}}_N^T = \begin{bmatrix} -2 & 0 \end{bmatrix} - \begin{bmatrix} 0 & 0 & 3 \end{bmatrix} \begin{bmatrix} 5 & 0 \\ 2 & 0 \\ -3 & 1 \end{bmatrix} = \begin{bmatrix} 7 & -3 \end{bmatrix}.
$$

Here the positive reduced cost is $\bar{c}_x = 7$, so x is the entering variable. The simplex direction is

$$
\mathbf{d}_B^{(x)} = - \begin{bmatrix} 0 & 0 & 0.5 \\ 1 & 0 & 10 \\ 0 & 1 & 1 \end{bmatrix} \begin{bmatrix} 5 \\ 2 \\ -3 \end{bmatrix} = \begin{bmatrix} 1.5 \\ 25 \\ 1 \end{bmatrix} \geq 0,
$$

showing that the problem is unbounded. We have found the improving unbounded direction $\mathbf{d} = (d_x, d_y) = (1, 1.5)$.

9.1.5 A Comment on Names

The simplex method is named after a *simplex*, which can be defined as a k-dimensional bounded polyhedron with as few extreme points as possible, namely, $k + 1$. A triangle is a two-dimensional simplex, a tetrahedron is a three-dimensional simplex. The feasible region of a canonical form linear program with just one constraint is an $(n - 1)$ - dimensional simplex (if bounded). However, the name comes from Dantzig (1963) and what is called column geometry. The convex hull of the basic columns is a simplex in this $(m + 1)$-dimensional geometry and is rotated, or *pivoted*, when one of its extreme points (basic columns) is replaced in an iteration of the simplex method. Thus, an iteration is called a *pivot*.

9.2 Degeneracy and Correctness

The simplex method presented in Section 9.1 is not guaranteed to be correct, that is, not guaranteed to either terminate at an optimal solution or show that the problem is unbounded in a finite number of iterations. The difficulty that arises involves degeneracy. This section presents a simple proof that the algorithm is correct when there is no degeneracy. Degeneracy is then discussed and a modification of the algorithm presented that guarantees it is correct.

Theorem 9.1: *For a given linear program, assume every bfs is nondegenerate. Then the simplex method terminates after a finite number of iterations. At termination, either the current bfs is optimal or an improving unbounded direction has been found, showing that the problem is unbounded.*

Proof: The fact that each iteration moves to a bfs is established in Section 9.1. Because they are nondegenerate, the step size is $\lambda > 0$ in the improving direction $\mathbf{d}^{(e)}$. Therefore, the objective function value increases at each iteration, and no bfs can be visited twice. There are a finite number of bfs's, providing a limit on the number of iterations before the algorithm terminates. If it terminates at Step 1, then the reduced costs $\overline{\mathbf{c}}_N^T$ are nonnegative, and by Theorem 7.5 the bfs is optimal. If it terminates at Step 3, then the simplex direction $\mathbf{d}^{(e)} \geq 0$ is an unbounded direction and it is improving, since $\overline{c}_e > 0$. □

Because it finds an optimal bfs, the simplex method provides an alternative proof of the Fundamental Theorem of Linear Programming (Theorem 7.3). The assumption that a problem has a bfs is met by starting in canonical form. However, extending the proof to include degeneracy adds complications.

Recall that a degenerate bfs has more than n active constraints and in canonical form has some basic variables equal to zero. When computing the step size in Step 3, if one of these variables has a negative component in the simplex direction $\mathbf{d}^{(e)}$, then $\lambda = 0$ and the next bfs is the same as the current one, with a different basis. We could change the rule for choosing an entering variable to try to avoid step sizes of zero, but they cannot always be avoided. In most cases (and nearly all practical problems), the algorithm can handle degeneracy. It continues to update the basis at this degenerate bfs until it finds a basis that has all reduced costs nonnegative, or a basis for which the entering variable has a nonzero step size, leading to a different bfs. However, it is possible that a sequence of basis changes at the same bfs will return to a basis that was already used. Once this happens, the algorithm has failed. It will continue to cycle through this sequence of bases indefinitely, remaining at the same bfs. Although rare, cycling can occur.

The following example from Beale (1955) illustrates cycling. It is the smallest size linear program that can cycle at a nonoptimal bfs.

Example 9.5. Consider the following linear program:

$$\max \quad 0.75x_4 - 20x_5 + 0.5x_6 - 6x_7$$
$$\text{s.t.} \quad x_1 \quad + 0.25x_4 - 8x_5 - x_6 + 9x_7 = 0$$
$$x_2 \quad + 0.5x_4 - 12x_5 - 0.5x_6 + 3x_7 = 0$$
$$x_3 \quad + x_6 = 1$$
$$x_j \geq 0.$$

Starting at the degenerate bfs $\mathbf{x} = (0, 0, 1, 0, 0, 0, 0)$ with basis (x_1, x_2, x_3), the progress of the algorithm is shown in Table 9.1. Blanks indicate entries that must be zero, i.e. reduced costs of basic variables or nonbasic variables in the simplex direction that are not entering the basis. The entering and leaving variables have bold reduced cost and simplex direction entries, respectively. Whenever there is more than one negative entry in the simplex direction, there is a tie for the leaving variable, so we must specify how to break ties in the algorithm: when more than one variable achieves the minimum ratio in the step size calculation, we will choose the one with smallest index. For example, at iteration 0, x_1 and x_2 are tied, so x_1 leaves and the basis at iteration 1 is (x_2, x_3, x_4). At iteration 6, the algorithm has cycled back to the basis from iteration 0.

Fortunately, it is easy to prevent cycling. Although in principle one could do this by remembering prior bases that have been used at the current degenerate bfs, there are much simpler rules. The easiest rule to implement is the smallest subscript pivoting rule, also known as Bland's rule (Bland, 1977). It changes the rule for choosing an entering variable and tells us how to break ties for the leaving variable.

Smallest Subscript Pivoting Rule

1. Choose the variable with smallest index among those with positive reduced cost as the entering variable.
2. Of the variables that are tied in the test for the leaving variable, choose the one with smallest index.

With this rule, the simplex algorithm is correct.

Theorem 9.2: *For a linear program with a given initial bfs, the simplex method using the smallest subscript pivoting rule terminates after a finite number of iterations.*

Table 9.1 Reduced costs and simplex directions for Example 9.5.

Iteration		x_1	x_2	x_3	x_4	x_5	x_6	x_7
0	$\bar{c}$				0.75	−20	0.5	−6
	$\mathbf{d}^{(e)}$	−0.25	−0.5	0	1			
1	$\bar{c}$	−3				4	3.5	−33
	$\mathbf{d}^{(e)}$		−4	32	0	1		
2	$\bar{c}$	−1	−1				2	−18
	$\mathbf{d}^{(e)}$			−1	−8	−0.375	1	
3	$\bar{c}$	2	−3		0.75			3
	$\mathbf{d}^{(e)}$			−10.5		−0.1875	10.5	1
4	$\bar{c}$	1	−1		1	−16		
	$\mathbf{d}^{(e)}$	1		2			−2	−1/3
5	$\bar{c}$		2		1.75	−0.5	−44	
	$\mathbf{d}^{(e)}$	3	1	0				−1/3
6	$\bar{c}$				0.75	−20	0.5	−6

Proof: Proofs that these, and other, anti-cycling rules work are given in Chvátal (1983). □

The reason that more complex anti-cycling rules are used in solvers is that this rule alters the way the entering variable is chosen, potentially increasing the number of iterations.

Knowing that the simplex method terminates also allows us to strengthen the optimality condition. Theorem 7.5 showed that the optimality condition $\bar{\mathbf{c}}_N \leq \mathbf{0}$ is sufficient but not necessary for degenerate problems, i.e. it might not hold at an optimal bfs. When the simplex method terminates at an optimal solution, it finds a basis for which $\bar{\mathbf{c}}_N \leq \mathbf{0}$. A basis that is feasible and satisfies this optimality condition is called *optimal*. Thus, we have shown that any canonical form linear program with an optimal solution has an optimal basis. If the bfs for this basis is degenerate, the other bases corresponding to it may not be optimal.

9.3   Finding an Initial Feasible Solution

Phase II of the simplex algorithm starts with an initial bfs. This section describes Phase I, which is used when an initial bfs is not available. It either finds a bfs or shows that the problem is infeasible. Because Example 9.3 was converted from standard form, there is a slack variable for each constraint. In this case, as long as $\mathbf{b} \geq \mathbf{0}$, setting $(\mathbf{x}, \mathbf{s}) = (\mathbf{0}, \mathbf{b})$ is a bfs with basis $(s_1, \ldots, s_m)$. If a bfs is not readily available, we can find one by solving a linear program related to the constraints.

We assume $\mathbf{b} \geq \mathbf{0}$; this can be achieved by multiplying constraints by -1 as needed. In order to construct a bfs, introduce an artificial variable y_i for each functional constraint and consider the auxiliary problem

$$\min y_1 + y_2 + \cdots + y_m$$
$$\text{s.t. } \mathbf{Ax} + \mathbf{y} = \mathbf{b} \tag{9.2}$$
$$\mathbf{x} \geq \mathbf{0}, \ \mathbf{y} \geq \mathbf{0}.$$

Note that $\mathbf{y}$ appears in the constraints in the way that slack variables would, so we can use $(\mathbf{x}, \mathbf{y}) = (\mathbf{0}, \mathbf{b})$ as a bfs with basis $(y_1, \ldots, y_m)$. However, $\mathbf{y}$ measures infeasibility: a feasible solution to (9.2) is only feasible in the original problem if $\mathbf{y} = \mathbf{0}$. That is why the objective is to minimize the sum of the y_i. Now, (9.2) has an optimal solution and the optimal value cannot be negative. If it has an optimal value of zero, then an optimal solution has $\mathbf{y} = \mathbf{0}$ and the corresponding $\mathbf{x}$ is feasible for the original problem. However, if the optimal value is positive, then we conclude that the original problem is infeasible. Thus, solving the auxiliary problem either finds a solution to the original problem or shows that it is infeasible.

A final observation is that we only need to add artificial variables to constraints that do not have slack variables. More generally, if there is a column j with ith component $a_{ij} > 0$ and zeros elsewhere, then x_j can be the variable associated with row i in the initial basis and $x_j = b_i / a_{ij}$ in the initial bfs.

9.3.1   Artificial Variables in the Basis

Suppose we solve the auxiliary problem using the simplex method and its optimal value is zero. Then we have found not just a feasible solution but a bfs $\mathbf{x}$ for the original problem. However, if the optimal bfs for the auxiliary problem was degenerate, its basis could include some artificial variables y_i. To start Phase II, we need to remove them from the basis, replacing them with some of the x_j. All of the variables swapped in and out of the basis have values of zero. Not just any nonbasic x_j can be chosen; to form a basis, the basic columns of $\mathbf{A}$ must be linearly independent. A systematic method to do this is given in (Bertsimas and Tsitsiklis, 1997, Section 3.5). Another option is to leave the basic artificial variables in the Phase II problem. Because their objective function coefficients are zero, they will not affect the optimal solution.

9.3.2 Two-Phase Simplex Method

With these additions, we have a complete algorithm for solving linear programs.

Algorithm 9.2 (Two-Phase Simplex Method)

For any maximization problem, formulate the auxiliary problem:

1. Eliminate negative right-hand sides by multiplying these constraints by -1.
2. Introduce slack or surplus variables to give "=" constraints.
3. For each constraint i that does not have an initial basic variable, add an artificial variable y_i and include it in the initial basis. In the initial bfs, $y_i = b_i$.
4. The objective is $\min \sum_i y_i$.

Phase I
Apply the simplex method to the auxiliary problem.

Case 1. If the optimal value is positive (negative for the canonical form objective function max $-\sum_i y_i$), stop. The original problem is infeasible.
Case 2. If the optimal value is zero and no artificial variables are in the basis, then a basis and bfs for the original problem have been found. Go to Phase II.
Case 3. If the objective function is zero and artificial variables are in the basis, find a basis for this (degenerate) bfs with no artificial variables. Go to Phase II.

Phase II
Apply the simplex method to the original problem in canonical form (after Steps 1 and 2), starting with the final bfs and basis from Phase I.

Example 9.6. Consider the linear program

$$
\begin{aligned}
\max \quad & 2x_1 + 5x_2 + 3x_3 \\
\text{s.t.} \quad & x_1 - 2x_2 + x_3 \geq 20 \\
& 2x_1 + 4x_2 + x_3 = 50 \\
& x_j \geq 0.
\end{aligned}
$$

The auxiliary problem is

$$
\begin{aligned}
\min \quad & y_1 + y_2 \\
\text{s.t.} \quad & x_1 - 2x_2 + x_3 - s_1 + y_1 = 20 \\
& 2x_1 + 4x_2 + x_3 \qquad\quad + y_2 = 50 \\
& x_1, x_2, x_3, s_1, y_1, y_2 \geq 0.
\end{aligned}
$$

Because of the direction of the first constraint, we subtracted the surplus variable s_1, then still needed to add an artificial variable to it. To apply the simplex method, convert to the maximization problem $\max - (y_1 + y_2)$.

Phase I. We solve the auxiliary (maximization) problem. The initial basis is (y_1, y_2) with bfs $\mathbf{x} = (0, 0, 0, 0, 20, 50)$,

$$\mathbf{B} = \begin{bmatrix} 1 & 0 \\ 0 & 1 \end{bmatrix}, \quad \mathbf{N} = \begin{bmatrix} 1 & -2 & 1 & -1 \\ 2 & 4 & 1 & 0 \end{bmatrix}.$$

Iteration 1. First we compute

$$\mathbf{y}^T = \begin{bmatrix} -1 & -1 \end{bmatrix} \mathbf{B}^{-1} = \begin{bmatrix} -1 & -1 \end{bmatrix},$$

$$\bar{\mathbf{c}}_N^T = \begin{bmatrix} 0 & 0 & 0 & 0 \end{bmatrix} - \begin{bmatrix} -1 & -1 \end{bmatrix} \begin{bmatrix} 1 & -2 & 1 & -1 \\ 2 & 4 & 1 & 0 \end{bmatrix} = \begin{bmatrix} 3 & 2 & 2 & -1 \end{bmatrix}.$$

The largest reduced cost is $\bar{c}_{x_1} = 3$, so x_1 is the entering variable. The simplex direction is

$$\mathbf{d}_B^{(x_1)} = -\mathbf{B}^{-1} \begin{bmatrix} 1 \\ 2 \end{bmatrix} = \begin{bmatrix} -1 \\ -2 \end{bmatrix},$$

or $\mathbf{d}^{(x_1)} = (1, 0, 0, 0, -1, -2)$. The step size is

$$\lambda = \min \left\{ \frac{20}{-(-1)}, \frac{50}{-(-2)} \right\} = \min\{20, 25\} = 20.$$

The first ratio, for y_1, achieves the minimum, so y_1 is the leaving variable. The new basis is (x_1, y_2) with bfs

$$\begin{bmatrix} x_1 \\ x_2 \\ x_3 \\ s_1 \\ y_1 \\ y_2 \end{bmatrix} = \begin{bmatrix} 0 \\ 0 \\ 0 \\ 0 \\ 20 \\ 50 \end{bmatrix} + 20 \begin{bmatrix} 1 \\ 0 \\ 0 \\ 0 \\ -1 \\ -2 \end{bmatrix} = \begin{bmatrix} 20 \\ 0 \\ 0 \\ 0 \\ 0 \\ 10 \end{bmatrix}.$$

The nonbasic variables are (x_2, x_3, s_1, y_1) with

$$\mathbf{B} = \begin{bmatrix} 1 & 0 \\ 2 & 1 \end{bmatrix}, \quad \mathbf{N} = \begin{bmatrix} -2 & 1 & -1 & 1 \\ 4 & 1 & 0 & 0 \end{bmatrix}.$$

Iteration 2. We compute

$$\mathbf{B}^{-1} = \begin{bmatrix} 1 & 0 \\ -2 & 1 \end{bmatrix}, \quad \mathbf{y}^T = \begin{bmatrix} 0 & -1 \end{bmatrix} \begin{bmatrix} 1 & 0 \\ -2 & 1 \end{bmatrix} = \begin{bmatrix} 2 & -1 \end{bmatrix},$$

$$\bar{\mathbf{c}}_N^T = \begin{bmatrix} 0 & 0 & 0 & -1 \end{bmatrix} - \begin{bmatrix} 2 & -1 \end{bmatrix} \begin{bmatrix} -2 & 1 & -1 & 1 \\ 4 & 1 & 0 & 0 \end{bmatrix} = \begin{bmatrix} 8 & -1 & 2 & -3 \end{bmatrix}.$$

The largest reduced cost is $\bar{c}_{x_2} = 8$, so x_2 is the entering variable. The simplex direction is

$$\mathbf{d}_B^{(x_2)} = -\begin{bmatrix} 1 & 0 \\ -2 & 1 \end{bmatrix}\begin{bmatrix} -2 \\ 4 \end{bmatrix} = \begin{bmatrix} 2 \\ -8 \end{bmatrix},$$

or $\mathbf{d}^{(x_2)} = (2, 1, 0, 0, 0, -8)$. The step size is

$$\lambda = \frac{10}{-(-8)} = 1.25$$

and y_2 is the leaving variable. The new basis is (x_1, x_2) with bfs

$$\begin{bmatrix} x_1 \\ x_2 \\ x_3 \\ s_1 \\ y_1 \\ y_2 \end{bmatrix} = \begin{bmatrix} 20 \\ 0 \\ 0 \\ 0 \\ 0 \\ 10 \end{bmatrix} + 1.25\begin{bmatrix} 2 \\ 1 \\ 0 \\ 0 \\ 0 \\ -8 \end{bmatrix} = \begin{bmatrix} 22.5 \\ 1.25 \\ 0 \\ 0 \\ 0 \\ 0 \end{bmatrix}.$$

Since the artificial variables are zero and nonbasic, we know that this bfs is optimal and we move to Phase II (Case 2).

Phase II: The corresponding bfs for the original problem, in canonical form, is $(x_1, x_2, x_3, s_1) = (22.5, 1.25, 0, 0)$ with basis (x_1, x_2) and

$$\mathbf{B} = \begin{bmatrix} 1 & -2 \\ 2 & 4 \end{bmatrix}, \quad \mathbf{N} = \begin{bmatrix} 1 & -1 \\ 1 & 0 \end{bmatrix}.$$

Iteration 1. We compute

$$\mathbf{B}^{-1} = \frac{1}{8}\begin{bmatrix} 4 & 2 \\ -2 & 1 \end{bmatrix}, \quad \mathbf{y}^T = \begin{bmatrix} 2 & 5 \end{bmatrix}\frac{1}{8}\begin{bmatrix} 4 & 2 \\ -2 & 1 \end{bmatrix} = \frac{1}{8}\begin{bmatrix} -2 & 9 \end{bmatrix},$$

$$\bar{\mathbf{c}}_N^T = \begin{bmatrix} 3 & 0 \end{bmatrix} - \frac{1}{8}\begin{bmatrix} -2 & 9 \end{bmatrix}\begin{bmatrix} 1 & -1 \\ 1 & 0 \end{bmatrix} = \begin{bmatrix} \frac{17}{8} & -\frac{2}{8} \end{bmatrix}.$$

The positive reduced cost is $\bar{c}_{x_3} = 17/8$, so x_3 is the entering variable. The simplex direction is

$$\mathbf{d}_B^{(x_3)} = -\frac{1}{8}\begin{bmatrix} 4 & 2 \\ -2 & 1 \end{bmatrix}\begin{bmatrix} 1 \\ 1 \end{bmatrix} = \begin{bmatrix} -\frac{3}{4} \\ \frac{1}{8} \end{bmatrix},$$

or $\mathbf{d}^{(x_3)} = (-3/4, 1/8, 1, 0)$. The step size is

$$\lambda = \frac{22.5}{-(-3/4)} = 30$$

and x_1 is the leaving variable. The new basis is (x_2, x_3) with bfs

$$
\begin{bmatrix} x_1 \\ x_2 \\ x_3 \\ s_1 \end{bmatrix} = \begin{bmatrix} 22.5 \\ 1.25 \\ 0 \\ 0 \end{bmatrix} + 30 \begin{bmatrix} -3/4 \\ 1/8 \\ 1 \\ 0 \end{bmatrix} = \begin{bmatrix} 0 \\ 5 \\ 30 \\ 0 \end{bmatrix}.
$$

The nonbasic variables are (x_1, s_1) with

$$
\mathbf{B} = \begin{bmatrix} -2 & 1 \\ 4 & 1 \end{bmatrix}, \quad \mathbf{N} = \begin{bmatrix} 1 & -1 \\ 2 & 0 \end{bmatrix}.
$$

Iteration 2. We compute

$$
\mathbf{B}^{-1} = \frac{1}{6} \begin{bmatrix} -1 & 1 \\ 4 & 2 \end{bmatrix}, \quad \mathbf{y}^T = \begin{bmatrix} 5 & 3 \end{bmatrix} \frac{1}{6} \begin{bmatrix} -1 & 1 \\ 4 & 2 \end{bmatrix} = \frac{1}{6} \begin{bmatrix} 7 & 11 \end{bmatrix},
$$

$$
\bar{\mathbf{c}}_N^T = \begin{bmatrix} 2 & 0 \end{bmatrix} - \frac{1}{6} \begin{bmatrix} 7 & 11 \end{bmatrix} \begin{bmatrix} 1 & -1 \\ 2 & 0 \end{bmatrix} = \frac{1}{6} \begin{bmatrix} -17 & 7 \end{bmatrix}.
$$

The entering variable is s_1, with simplex direction

$$
\mathbf{d}_B^{(s_1)} = -\frac{1}{6} \begin{bmatrix} -1 & 1 \\ 4 & 2 \end{bmatrix} \begin{bmatrix} -1 \\ 0 \end{bmatrix} = \begin{bmatrix} -\frac{1}{6} \\ \frac{2}{3} \end{bmatrix},
$$

or $\mathbf{d}^{(s_1)} = (0, -1/6, 2/3, 1)$. The step size is

$$
\lambda = \frac{5}{-(-1/6)} = 30
$$

and x_2 is the leaving variable. The new basis is (x_3, s_1) with bfs

$$
\begin{bmatrix} x_1 \\ x_2 \\ x_3 \\ s_1 \end{bmatrix} = \begin{bmatrix} 0 \\ 5 \\ 30 \\ 0 \end{bmatrix} + 30 \begin{bmatrix} 0 \\ -1/6 \\ 2/3 \\ 1 \end{bmatrix} = \begin{bmatrix} 0 \\ 0 \\ 50 \\ 30 \end{bmatrix}.
$$

The nonbasic variables are (x_1, x_2) with

$$
\mathbf{B} = \begin{bmatrix} 1 & -1 \\ 1 & 0 \end{bmatrix}, \quad \mathbf{N} = \begin{bmatrix} 1 & -2 \\ 2 & 4 \end{bmatrix}.
$$

Iteration 3. We compute

$$
\mathbf{B}^{-1} = \begin{bmatrix} 0 & 1 \\ -1 & 1 \end{bmatrix}, \quad \mathbf{y}^T = \begin{bmatrix} 3 & 0 \end{bmatrix} \begin{bmatrix} 0 & 1 \\ -1 & 1 \end{bmatrix} = \begin{bmatrix} 0 & 3 \end{bmatrix},
$$

$$
\bar{\mathbf{c}}_N^T = \begin{bmatrix} 2 & 5 \end{bmatrix} - \begin{bmatrix} 0 & 3 \end{bmatrix} \begin{bmatrix} 1 & -2 \\ 2 & 4 \end{bmatrix} = \begin{bmatrix} -4 & -7 \end{bmatrix},
$$

showing that the current bfs is optimal. The optimal solution to the original problem is $\mathbf{x} = (0, 0, 50)$.

The next example illustrates the two-phase method on an infeasible problem.

Example 9.7. Consider the following linear program:

$$
\begin{array}{ll}
\max & 4x_1 + 3x_2 \\
\text{s.t.} & 1.5x_1 + x_2 \le 16 \\
& x_1 + x_2 \le 12 \\
& 3x_1 + 2x_2 = 33 \\
& x_j \ge 0.
\end{array}
$$

After adding slack variables, only one artificial variable is needed. The auxiliary problem is

$$
\begin{array}{ll}
\min & y_3 \\
\text{s.t.} & 1.5x_1 + x_2 + s_1 = 16 \\
& x_1 + x_2 + s_2 = 12 \\
& 3x_1 + 2x_2 + y_3 = 33 \\
& x_1, x_2, s_1, s_2, y_3 \ge 0.
\end{array}
$$

Initialization. We begin Phase I on the (maximization) auxiliary problem with bfs $\mathbf{x} = (0, 0, 0, 16, 12, 33)$, basis (s_1, s_2, y_3), and

$$
\mathbf{B} = \begin{bmatrix} 1 & 0 & 0 \\ 0 & 1 & 0 \\ 0 & 0 & 1 \end{bmatrix}, \quad \mathbf{N} = \begin{bmatrix} 1.5 & 1 \\ 1 & 1 \\ 3 & 2 \end{bmatrix}.
$$

Iteration 1. We compute

$$
\mathbf{y}^T = \begin{bmatrix} 0 & 0 & -1 \end{bmatrix} \mathbf{B}^{-1} = \begin{bmatrix} 0 & 0 & -1 \end{bmatrix},
$$

$$
\bar{\mathbf{c}}_N^T = \begin{bmatrix} 0 & 0 \end{bmatrix} - \begin{bmatrix} 0 & 0 & -1 \end{bmatrix} \begin{bmatrix} 1.5 & 1 \\ 1 & 1 \\ 3 & 2 \end{bmatrix} = \begin{bmatrix} 3 & 2 \end{bmatrix}.
$$

The entering variable is x_1 with reduced cost $\bar{c}_{x_1} = 3$. The simplex direction is

$$
\mathbf{d}_B^{(x_1)} = -\mathbf{B}^{-1} \begin{bmatrix} 1.5 \\ 1 \\ 3 \end{bmatrix} = \begin{bmatrix} -1.5 \\ -1 \\ -3 \end{bmatrix},
$$

or $\mathbf{d}^{(x_1)} = (1, 0, -1.5, -1, -3)$. The step size is

$$
\lambda = \min \left\{ \frac{16}{-(-1.5)}, \frac{12}{-(-1)}, \frac{33}{-3} \right\} = \min \left\{ \frac{32}{3}, 12, 11 \right\} = \frac{32}{3}.
$$

The first ratio, for s_1, achieves the minimum, so s_1 is the leaving variable. The new basis is (x_1, s_2, y_3) with bfs

$$
\begin{bmatrix} x_1 \\ x_2 \\ s_1 \\ s_2 \\ y_3 \end{bmatrix} = \begin{bmatrix} 0 \\ 0 \\ 16 \\ 12 \\ 33 \end{bmatrix} + \frac{32}{3} \begin{bmatrix} 1 \\ 0 \\ -1.5 \\ -1 \\ -3 \end{bmatrix} = \begin{bmatrix} 32/3 \\ 0 \\ 0 \\ 4/3 \\ 1 \end{bmatrix}.
$$

Updating,

$$
\mathbf{B} = \begin{bmatrix} 1.5 & 0 & 0 \\ 1 & 1 & 0 \\ 3 & 0 & 1 \end{bmatrix}, \quad \mathbf{N} = \begin{bmatrix} 1 & 1 \\ 1 & 0 \\ 2 & 0 \end{bmatrix}.
$$

Iteration 2. We compute

$$
\mathbf{B}^{-1} = \begin{bmatrix} \frac{2}{3} & 0 & 0 \\ -\frac{2}{3} & 1 & 0 \\ -2 & 0 & 1 \end{bmatrix}, \quad \mathbf{y}^T = \begin{bmatrix} 0 & 0 & -1 \end{bmatrix} \begin{bmatrix} \frac{2}{3} & 0 & 0 \\ -\frac{2}{3} & 1 & 0 \\ -2 & 0 & 1 \end{bmatrix} = \begin{bmatrix} 2 & 0 & -1 \end{bmatrix},
$$

$$
\bar{\mathbf{c}}_N^T = \begin{bmatrix} 0 & 0 \end{bmatrix} - \begin{bmatrix} 2 & 0 & -1 \end{bmatrix} \begin{bmatrix} 1 & 1 \\ 1 & 0 \\ 2 & 0 \end{bmatrix} = \begin{bmatrix} 0 & -2 \end{bmatrix}.
$$

Because $\bar{\mathbf{c}}_N^T \leq \mathbf{0}$, the current solution is optimal for the auxiliary problem, with $y_3 = 1$. We conclude that the original problem is infeasible (Case 1).

9.4 Computational Strategies and Speed

The simplex method as described in Section 9.1 is sometimes called the naive implementation. It allowed us to crisply describe the algorithm. Other implementations are much more efficient. In this section, we present some of the most important ideas for speeding up the algorithm. These improvements and others have had a huge impact on the size of linear programs that can be solved. In fact, over a 16-year period, Bixby (2012) notes that improvements in linear programming algorithms (software) outpaced the improvements in speed of computers (hardware), though the pace of improvement has now slowed. We begin this section by discussing the number of iterations required by the simplex method. Next, two more efficient implementations, the revised simplex method and the tableau method, are described. We close the section with a discussion of other improvements.

9.4.1 Number of Iterations Required

Extensive experience with using the simplex method on examples from various fields indicates that the method can be expected to reach an optimal solution in about m to $3m/2$ iterations, where m is the number of functional constraints, roughly independent of the number of variables n. This is surprisingly few iterations. One implication of this experience is that, for a problem with more constraints than variables ($m > n$), the simplex method is likely to be faster on the dual problem, where the role of m and n are reversed. Also, converting inequality constraints and unrestricted variables to canonical form only increases the number of variables, so the estimated number of iterations is the same, regardless of whether a problem of a given size is in canonical form.

It is also surprising that the typical number of iterations does not depend on the dimension n. The number of extreme points of the feasible region can increase exponentially with m or n. For example, the unit cube, $0 \leq x_j \leq 1, j = 1, \dots, n$ has 2^n extreme points, formed by setting each variable to 0 or 1. When the simplex method terminates, it has followed a path starting at some extreme point $\mathbf{x}^0$ and moved to adjacent extreme points until it reaches one, say, $\mathbf{x}^*$, that is optimal. Each move along this path must strictly increase the objective function. How many extreme points could lie on such a path? It is easy to construct examples where these paths have exponentially many extreme points. A perturbed version of the unit cube is described in (Bertsimas and Tsitsiklis, 1997, Section 3.7) where a path includes all 2^n extreme points. Although the Dantzig rule for choosing an entering variable doesn't follow this path, some rule does. Thus, if we use a bad rule and start at the worst extreme point, the simplex method takes $2^n - 1$ iterations. Even worse, there are examples given in Klee and Minty (1972) where the simplex method takes $2^n - 1$ iterations using the Dantzig rule. The same is true of other popular rules. Thus, the worst-case performance of the simplex method (using any of these rules) is an exponential number of iterations. It is not known whether for *every* rule there are examples that take an exponential number of steps, but there are reasons to think this is the case. When we use the simplex method on a large practical problem, then, we expect it to take around m iterations but cannot guarantee that it will not take an enormous (exponential) number of iterations.

Another way of looking at the number of iterations required is to consider the path with the *smallest* number of extreme points, rather than the largest. Let $d(\mathbf{x}^0, \mathbf{x}^*)$ be the minimum number of moves to adjacent extreme points needed to get from $\mathbf{x}^0$ to $\mathbf{x}^*$. Then $d(\mathbf{x}^0, \mathbf{x}^*)$ is a lower bound on the number of iterations of the simplex method. Of course, this bound may not be achievable because we can only move to extreme points with better objective function values. Even an improving path with the smallest number of extreme points doesn't seem likely to be chosen by the rule for the entering variable. But the bound may be close

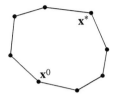

Figure 9.1 A polygon with $m = 8$ sides has diameter 4.

enough to give some insight into the small average number of iterations observed. Given a feasible region, we don't know which extreme points the algorithm will start and stop at, so we consider the *diameter* of the polyhedron, defined as the maximum of $d(\mathbf{x}^0, \mathbf{x}^*)$ over any extreme points $\mathbf{x}^0$ and $\mathbf{x}^*$. For example, the polygon in Figure 9.1 has diameter 4.

If the largest diameter for a certain size problem could be no more than, say, $3m/2$, that would help explain why the average number of iterations observed was around $3m/2$, or at least linear in m. The Hirsch conjecture is that the diameter of a bounded polyhedron in R^n defined by m inequality constraints is no more than $m - n$. In Figure 9.1, $m - n = 8 - 2 = 6 \geq 4$ as conjectured. If we consider standard form polyhedra, then the number of constraints (including nonnegativity) is $m + n$, so the Hirsch conjecture is that the diameter is no more than $(m + n) - n = m$. While this would be very helpful in explaining the performance of the simplex method, very little is known about maximum diameters (Kim and Santos, 2010). The conjecture is true in two or three dimensions. A cube, for example, has $m - n = 6 - 3 = 3$ and diameter 3, equaling the Hirsch bound. In high dimension there are examples that just barely violate it, but no reasonable upper bound on the diameter is known–no polynomial bound, much less a linear bound.

9.4.2 Revised Simplex Method

The naive simplex method requires solving two related $m \times m$ linear systems at each iteration; this can be done by finding $\mathbf{B}^{-1}$ or some other method. Either way, the number of operations required is $O(m^3)$. A faster way of doing an iteration makes use of the fact only one column of $\mathbf{B}$ changes at each iteration. We will update $\mathbf{B}^{-1}$ using row operations (see Section A.5). If the jth variable enters the basis and the lth basic variable $B(l)$ leaves the basis at an iteration, the basis matrix changes from

$$\mathbf{B} = [\mathbf{a}_{B(1)} \dots \mathbf{a}_{B(m)}]$$

to

$$\overline{B} = [\mathbf{a}_{B(1)} \dots \mathbf{a}_{B(l-1)} \ \mathbf{a}_j \ \mathbf{a}_{B(l+1)} \dots \mathbf{a}_{B(m)}].$$

Recall that one way to find the inverse of a matrix is to apply a sequence of row operations to the matrix, augmented by the identity matrix $\mathbf{I}$. The row operations that transform $\mathbf{B}$ into $\mathbf{I}$ transform $\mathbf{I}$ into $\mathbf{B}^{-1}$. We write this as

$$[\mathbf{B} \mid \mathbf{I}] \to [\mathbf{I} \mid \mathbf{B}^{-1}].$$

Now, because $\overline{\mathbf{B}}$ differs from $\mathbf{B}$ only in column l, $\mathbf{B}^{-1}\overline{\mathbf{B}}$ differs from $\mathbf{I}$ only in column l, and the lth column is $\mathbf{d} = \mathbf{B}^{-1}\mathbf{a}_j$:

$$\mathbf{B}^{-1}\overline{\mathbf{B}} = \begin{bmatrix} 1 & & d_1 & & \\ & \ddots & \vdots & & \\ & & d_l & & \\ & & \vdots & \ddots & \\ & & d_m & & 1 \end{bmatrix}. \tag{9.3}$$

One way to transform $\overline{\mathbf{B}}$ into the identity is to first perform the row operations that transform $\mathbf{B}$ into the identity. Recalling that row operations are equivalent to premultiplying by a certain invertible matrix, the effect of these row operations is

$$[\overline{\mathbf{B}} \mid \mathbf{I}] \to [\mathbf{B}^{-1}\overline{\mathbf{B}} \mid \mathbf{B}^{-1}].$$

From there, we apply the row operations that transform (9.3) into the identity:

$$[\mathbf{B}^{-1}\overline{\mathbf{B}} \mid \mathbf{B}^{-1}] \to [\mathbf{I} \mid \overline{\mathbf{B}}^{-1}].$$

Notice that the row operations that transform $\mathbf{B}^{-1}\overline{\mathbf{B}}$ into the identity also transform $\mathbf{B}^{-1}$ into $\overline{\mathbf{B}}^{-1}$; they are the row operations needed to update $\mathbf{B}^{-1}$.

For example, if the inverse basis matrix is

$$\mathbf{B}^{-1} = \begin{bmatrix} 2 & 4 & 0 \\ -2 & 3 & 0 \\ 4 & -2 & 6 \end{bmatrix},$$

the entering variable's (basic) simplex direction is $\mathbf{d} = (-3, 1, 2)$ and the third basic variable leaves the basis ($l = 3$), we apply row operations as follows.

$$\begin{bmatrix} 2 & 4 & 0 & -3 \\ -2 & 3 & 0 & 1 \\ 4 & -2 & 6 & 2 \end{bmatrix} \to \begin{bmatrix} 8 & -2 & 18 & 0 \\ -4 & 4 & -3 & 0 \\ 2 & -1 & 3 & 1 \end{bmatrix}. \tag{9.4}$$

The first three columns of the resulting matrix contain $\overline{\mathbf{B}}^{-1}$. The row operations were to add 1.5 times row 3 to row 1, subtract 0.5 times row 3 from row 2, and then multiply row 3 by 0.5. We must use these row operations; other row operations could achieve the desired right-most column of (9.4) but they would alter the other columns in (9.3).

Now we can describe the revised simplex method. We refer to the leaving variable by its position l in the basis, corresponding to row l in the constraints, so that we know what row operations to perform. Let $\mathbf{e}_i$ be a vector with a one in the ith component and zeros elsewhere.

Algorithm 9.3 (Revised Simplex Method)

Step 0. Initialization. Identify a bfs $\mathbf{x}$ and its basis $\mathcal{B}$. Compute $\mathbf{B}^{-1}$.

Step 1. Optimality check. Compute the simplex multipliers $\mathbf{y}^T = \mathbf{c}_B^T \mathbf{B}^{-1}$ and the reduced costs $\overline{\mathbf{c}}_N^T = \mathbf{c}_N^T - \mathbf{y}^T \mathbf{N}$. If $\overline{\mathbf{c}}_N \leq \mathbf{0}$, stop. The current solution is optimal.

Step 2. Compute simplex direction. Otherwise, choose a variable x_e with the most positive reduced cost as the entering variable. Compute its simplex direction $\mathbf{d}_B^{(e)} = -\mathbf{B}^{-1}\mathbf{a}_e$. Let $\mathbf{d} = \mathbf{d}_B^{(e)}$.

Step 3. Compute step size. If $\mathbf{d} \geq \mathbf{0}$, stop. The linear program is unbounded. Otherwise, the step size is $\lambda = \min\left\{ \frac{x_j}{-d_j} : d_j < 0 \right\}$ and the leaving variable is one that achieves the minimum ratio. Suppose the lth basic variable leaves the basis.

Step 4. Update the solution and basis. The new solution $\mathbf{x}^{\text{new}}$ is $\mathbf{x}_B^{\text{new}} = \mathbf{x}_B + \lambda\mathbf{d}$ and $x_e^{\text{new}} = \lambda$. Replace the leaving variable (with index $B(l)$) by x_e in the basis.

Step 5. Update $\mathbf{B}^{-1}$. Form the augmented matrix $[\mathbf{B}^{-1} \,|\, \mathbf{d}]$. Use the row operations described earlier to convert $\mathbf{d}$ to $\mathbf{e}_l$. The left part of the resulting matrix is the new $\mathbf{B}^{-1}$. Go to Step 1.

9.4.3 Simplex Tableaus

The revised simplex method performs row operations on the matrix $\mathbf{B}^{-1}$. Another way of keeping track of all the information needed in the simplex method is to perform row operations on an augmented matrix known as the *simplex tableau*. If we start with a standard form problem with the objective function $z = \mathbf{c}^T\mathbf{x}$, after adding slack variables we have the equations

$$z - \mathbf{c}^T\mathbf{x} = 0,$$

$$\mathbf{Ax} + \mathbf{Is} = \mathbf{b}.$$

The simplex tableau for these equations is

$$
\begin{array}{cccc}
z & \mathbf{x} & \mathbf{s} & \text{rhs}
\end{array}
$$
$$
\left[\begin{array}{ccc|c}
1 & -\mathbf{c}^T & 0 & 0 \\
0 & \mathbf{A} & \mathbf{I} & \mathbf{b}
\end{array}\right].
$$

We refer to the objective function row as row 0; the constraints are in rows 1 to m. Because their columns form an identity matrix, the variables $(s_1, \ldots, s_m)$ are a basis with $\mathbf{B} = \mathbf{I}$ and their values for this bfs can be read off the right-hand side as $\mathbf{s} = \mathbf{b}$. Also, because there are zeros in row 0 for the basic variables, row 0 of the right-hand side contains the value of z, which is 0 for this basis.

For other bases, the same information can be found by performing row operations on the tableau that change $\mathbf{B}$ to $\mathbf{I}$ and eliminate the basic variables from row 0 (the objective function). These row operations are equivalent to premultiplying by a certain matrix. In terms of equations, we get

$$z + (\mathbf{c}_B^T \mathbf{B}^{-1} \mathbf{A} - \mathbf{c}^T)\mathbf{x} + \mathbf{c}_B^T \mathbf{B}^{-1} \mathbf{s} = \mathbf{c}_B^T \mathbf{B}^{-1} \mathbf{b},$$

$$\mathbf{B}^{-1} \mathbf{A} \mathbf{x} + \mathbf{B}^{-1} \mathbf{s} = \mathbf{B}^{-1} \mathbf{b}.$$

As we have seen before, the basic components of $\mathbf{c}_B^T \mathbf{B}^{-1} \mathbf{A} - \mathbf{c}^T$ are zero. The tableau for this basis (dropping the z column, which is unchanged) is

$$
\begin{array}{ccc}
\mathbf{x} & \mathbf{s} & \text{rhs}
\end{array}
$$
$$
\begin{bmatrix}
\mathbf{c}_B^T \mathbf{B}^{-1} \mathbf{A} - \mathbf{c}^T & \mathbf{c}_B^T \mathbf{B}^{-1} & \Big| & \mathbf{c}_B^T \mathbf{B}^{-1} \mathbf{b} \\
\mathbf{B}^{-1} \mathbf{A} & \mathbf{B}^{-1} & \Big| & \mathbf{B}^{-1} \mathbf{b}
\end{bmatrix}.
$$

In addition to the objective function value and bfs, it contains the reduced costs and simplex multipliers (or dual variables):

$$
\begin{array}{ccc}
\mathbf{x} & \mathbf{s} & \text{rhs}
\end{array}
$$
$$
\begin{bmatrix}
-\bar{\mathbf{c}}^T & \mathbf{y}^T & \Big| & z \\
\mathbf{B}^{-1} \mathbf{A} & \mathbf{B}^{-1} & \Big| & \mathbf{x}_B
\end{bmatrix}.
$$

In order to read the bfs off a tableau we must know the order of the basic variables, i.e. which variable is associated with row 1, row 2, etc. Although we might be able to find this out from the fact that the column for the ith basic variable is the unit vector $\mathbf{e}_i$, it is customary to label each row with its basic variable. Finally, if we did not add slack variables, or more generally if there is no identity submatrix in the initial tableau, then the tableau does not contain $\mathbf{B}^{-1}$ or $\mathbf{y}$.

The simplex method can be performed using a tableau as follows.

Algorithm 9.4 (Tableau Simplex Method)

Step 0. Initialization. Start with a simplex tableau in row reduced form: the basic columns form an identity matrix in the constraint rows and have zeros in row 0. A basic variable is assigned to each constraint row.

Step 1. Entering variable. If there are no negative entries in row 0 (positive reduced costs), stop. The current solution is optimal. Otherwise, choose the column j with the most positive row 0 as the pivot column. Let $\bar{\mathbf{a}}$ be this column (excluding row 0).

Step 2. Step size and leaving variable. If $\bar{\mathbf{a}} \leq 0$, stop. The linear program is unbounded. Otherwise, the step size is $\lambda = \min \left\{ \frac{x_{B(i)}}{\bar{a}_i} : \bar{a}_i > 0 \right\}$ and the pivot row l is one of the i that achieves the minimum ratio. Here $x_{B(i)}$ is the right-hand side of row i. The leaving variable is $x_{B(l)}$, i.e., the variable associated with row l.

Step 3. Pivot. Perform a pivot on row l column j: Add a multiple of row l to each other row so that their entries in column j become zero. Multiply row l by a constant so that row l of column j becomes one. Update the basic variable associated with row l from $x_{B(l)}$ to x_j (the variable of the pivot column). Go to Step 1.

Example 9.8. Example 9.1 was solved in Example 9.3 using the naive simplex method. Now we rework it using tableaus. The initial tableau is shown first. The pivot column is x because -4 is the most negative entry in row 0. The ratios are $16/1.5 = 32/3$ in row 1 and $12/1 = 12$ in row 2. Row 1 has the smallest ratio, so it is the pivot row. Performing the pivot gives the second tableau. The tableau includes the objective function value ($z = 128/3$) and $\mathbf{B}^{-1}$ (rows 1–3 of the s columns). Now the pivot column is y and the ratios are $(32/2)/(2/3) = 16$ in row 1, $(4/3)/(1/3) = 4$ in row 2, and $10/1 = 10$ in row 3. Thus, we pivot on row 2, giving the third tableau. There are no negatives in row 0, so this tableau is optimal with $z^* = 44$ and $\mathbf{x} = (8, 4, 0, 0, 6)$.

x	y	s_1	s_2	s_3	rhs	
-4	-3	0	0	0	0	
1.5	1	1	0	0	16	s_1
1	1	0	1	0	12	s_2
0	1	0	0	1	10	s_3

x	y	s_1	s_2	s_3	rhs	
0	$-1/3$	8/3	0	0	128/3	
1	2/3	2/3	0	0	32/3	x
0	**1/3**	$-2/3$	1	0	4/3	s_2
0	1	0	0	1	10	s_3

x	y	s_1	s_2	s_3	rhs	
0	0	3	2	0	44	
1	0	2	-2	0	8	x
0	1	-2	3	0	4	y
0	0	2	-3	1	6	s_3

The main difference in the computations done by these two implementations is that the tableau method does row operations on the full $\mathbf{A}$ matrix, while the revised simplex method does row operations just on $\mathbf{B}^{-1}$. The number of computations required is $O(mn)$ to update the tableau and only $O(m^2)$ to update $\mathbf{B}^{-1}$. This can be a major advantage, as m is often much less than n. However, if the revised simplex uses the Dantzig rule, it also must compute all $n - m$ reduced costs, each involving multiplying two m-vectors, so the total number of operations is $O(mn)$ as in the tableau method. Other rules for choosing an entering variable stop as soon as one positive reduced cost is found, which can make the revised simplex method much faster. There is also a difference in the memory required. In most large-scale problems, the matrix $\mathbf{A}$ is very sparse, so only its nonzero entries are stored, not all mn entries. The revised simplex method also stores $\mathbf{B}^{-1}$ (as we have presented it; there are techniques that store less), which has m^2 entries while the tableau method stores the tableau, with about mn entries.

9.4.4 Other Implementation Issues

Finally, we mention some additional techniques commonly used to address roundoff error and improve efficiency of the simplex method. Because the revised simplex method updates the matrix $\mathbf{B}^{-1}$ at each iteration, roundoff errors will accumulate and accuracy will degrade over many iterations. To compensate, it is customary to recompute $\mathbf{B}^{-1}$ from scratch, called *reinversion*, after some number of iterations. Various techniques from numerical linear algebra are used to do the reinversion. Also, because what is really needed is to solve the two linear systems involving $\mathbf{B}$, the matrix $\mathbf{B}^{-1}$ does not have to be computed explicitly. At a reinversion, the LU decomposition method may be used to express $\mathbf{B} = \mathbf{LU}$, where $\mathbf{L}$ and $\mathbf{U}$ are (sparse) lower and upper triangular matrices, respectively. Then each linear system involving $\mathbf{B}$ may be solved by quickly solving two triangular systems.

When updating $\mathbf{B}^{-1}$, it also not necessary to compute it explicitly. Given the old $\mathbf{B}^{-1}$, we can store the row operations needed to update rather than computing the new $\mathbf{B}^{-1}$. These can be stored for several iterations, not just the current one, allowing a new inverse basis matrix to be expressed as the old $\mathbf{B}^{-1}$ and a sequence of row operations for multiple iterations.

Now we turn to ideas that affect the number of iterations. There are better methods of choosing an initial basis. These heuristic procedures, called a "crash" in Bixby (1992), try to find a basis that is close to feasible by having few artificial variables, but also contains many variables that may end up in the optimal basis. Using as few slack variables as possible is one strategy. These procedures also try to choose a basis that will allow faster iterations due to the basis matrix being closer to triangular. If a similar problem has been solved, the optimal basis can be modified and used as a "warm start."

The choice of an entering variable, known as *pricing*, is also important. The Dantzig rule requires computing each variable's reduced cost, which is computationally expensive. An alternative approach is to first compute reduced costs for a subset of the variables. If none of them are positive, then another subset is chosen until either an entering variable is found or the current solution is determined to be optimal. Such a rule saves computation but may not find as good an entering variable. Other useful rules are Devex (Harris, 1973) and steepest edge (Forrest and Goldfarb, 1992); (Goldfarb and Reid, 1977) pricing. Both of these approximately compute the angle that the simplex direction makes with the gradient (the direction of most rapid increase in the objective) and chooses the one with the smallest angle as the entering variable.

Problems

Solve the linear programs in Exercises 9.1–9.4 using the simplex method (Algorithm 9.1). Use the slack variables as the initial basis.

1
$$\max \ 3x_1 + 4x_2$$
$$\text{s.t.} \quad x_1 + 2x_2 \leq 9$$
$$\quad 4x_1 + 2x_2 \leq 24$$
$$x_j \geq 0.$$

2
$$\max \ 2x + 3y$$
$$\text{s.t.} \ -2x + y \leq 4$$
$$x - 2y \leq 2$$
$$x + y \leq 10$$
$$x, y \geq 0.$$

3
$$\max \ 2x_1 \qquad + x_3$$
$$\text{s.t.} \quad 6x_1 + 2x_2 + 2x_3 \leq 12$$
$$2x_1 - 2x_2 + 4x_3 \leq 5$$
$$x_j \geq 0.$$

4
$$\max \ 4x_1 + 2x_2 + 4x_3$$
$$\text{s.t.} \quad 4x_1 + x_2 + 2x_3 \leq 24$$
$$2x_1 + 5x_2 + 3x_3 \leq 30$$
$$2x_2 + x_3 \leq 16$$
$$x_j \geq 0.$$

5 (a) Solve the following linear program using the simplex method.

(b) How can you tell that the optimal solution is degenerate?

$$\begin{array}{ll} \max & 2x + 3y \\ \text{s.t.} & -x + 2y \le 12 \\ & 3x + 2y \le 12 \\ & x, y \ge 0. \end{array}$$

6 (a) Solve the following linear program using the simplex method.

(b) How does the simplex method indicate that there are multiple optimal solutions?

(c) Solve the linear program graphically. Find all optimal bfs's.

$$\begin{array}{ll} \max & 10x_1 + 5x_2 \\ \text{s.t.} & 3x_1 + x_2 \le 27 \\ & 2x_1 + x_2 \le 21 \\ & x_2 \le 15 \\ & x_j \ge 0. \end{array}$$

7 Use the simplex method to show that the linear program is unbounded.

$$\begin{array}{ll} \max & 4x_1 + 6x_2 + x_3 + x_4 \\ \text{s.t.} & -2x_1 + x_2 + 4x_3 + 4x_4 \le 4 \\ & x_1 - 2x_2 + 2x_3 + 2x_4 \le 2 \\ & x_j \ge 0. \end{array}$$

For Exercises 9.8–9.11, use Phase I of the simplex method (Algorithm 9.1) to find a feasible solution or show that the linear program is infeasible. For Exercise 9.11, when there is a tie for the leaving variable choose the artificial variable.

8
$$\begin{array}{ll} \min & 2x_1 + 5x_2 + x_3 \\ \text{s.t.} & 6x_1 + 3x_2 + 9x_3 = 120 \\ & 3x_1 + 3x_2 + 5x_3 \ge 80 \\ & x_j \ge 0. \end{array}$$

9
$$\begin{array}{ll} \max & 10x_1 + 5x_2 \\ \text{s.t.} & 2x_1 + x_2 \le 4 \\ & x_1 - 3x_2 \ge 3 \\ & x_j \ge 0. \end{array}$$

10
$$\max \quad 4x_1 + x_2 - 2x_3$$
$$\text{s.t.} \quad 2x_1 - 3x_2 - 4x_3 \geq 40$$
$$7x_1 + 2x_2 - 2x_3 \leq 70$$
$$4x_1 + 5x_2 - 2x_3 \geq 30$$
$$x_j \geq 0.$$

11
$$\max \quad 2x_1 + 3x_2 + x_3$$
$$\text{s.t.} \quad 2x_1 + x_2 + x_3 = 4$$
$$3x_1 + 3x_2 + 6x_3 = 6$$
$$x_j \geq 0.$$

12 Solve the following linear program using the two phase simplex method.
$$\max \quad 2x_1 + 7x_2$$
$$\text{s.t.} \quad x_1 \quad\quad \leq 12$$
$$x_2 \leq 18$$
$$1.5x_1 + x_2 = 27$$
$$x_j \geq 0.$$

13 Solve the following linear program using the two phase simplex method.
$$\max \quad 4x_1 + 10x_2 + 6x_3$$
$$\text{s.t.} \quad 2x_1 - 4x_2 + 2x_3 \geq 60$$
$$4x_1 + 8x_2 + 2x_3 = 150$$
$$x_j \geq 0.$$

14 Solve Exercise 9.1 using the tableau simplex method (Algorithm 9.4).

15 Solve Exercise 9.2 using the tableau simplex method (Algorithm 9.4).

16 Solve Exercise 9.12 using the tableau simplex method (Algorithm 9.4).

10

Sensitivity Analysis

We have been treating linear programs as if the data for a problem is exactly known. For real-world problems, some of the values are likely to be inaccurate. Even when we are confident about the numbers that go into a model, we often want to know about another scenario where the inputs are changed.

What happens when one tries to implement the optimal solution to a linear program when the input values were incorrect? If the constraints were inaccurate, the solution could be infeasible, forcing the solution to be adjusted or perhaps not implemented at all. Methods to avoid violating uncertain constraints are known as *robust optimization* (Ben-Tal et al., 2009). However, adding robustness generally changes the form of the problem; a robust linear program is no longer linear. We take a less ambitious approach and ask how the optimal solution and value change if the input values change. Similarly, if the objective coefficients are uncertain, optimizing their nominal value in a linear program may not be desirable. If the objective is profit, maximizing a mean or nominal value might make sense, while if the objective is the risk in a portfolio, minimizing its largest value (as can be done using robust optimization) might be preferable. Again, our approach is less ambitious: we will investigate how the optimal value changes and the range of objective function values over which the same solution is optimal. Although one could approach these questions by repeatedly changing the inputs to a linear program and solving the perturbed version, more insight is gained using sensitivity analysis techniques.

We begin by investigating these sensitivities graphically. Then in Section 10.2 the matrix equations we have developed are used to systematically compute the sensitivities. A connection between sensitivity analysis and duality theory (Chapter 8) is also made. Section 10.3 explores this connection further, interpreting sensitivities to the constraint right-hand sides as prices to be set by the dual. An interpretation of strong duality in terms of these prices is also given.

Linear and Convex Optimization: A Mathematical Approach, First Edition. Michael H. Veatch.
© 2021 John Wiley & Sons, Inc. Published 2021 by John Wiley & Sons, Inc.
Companion website: www.wiley.com/go/veatch/convexandlinearoptimization

10.1 Graphical Sensitivity Analysis

Many of the ideas in sensitivity analysis can be understood graphically using a problem with two variables. In this section we explore how the optimal solution and value of a linear program change with the following changes to the problem:

- Adding or removing a constraint.
- Changing one of the constraint right-hand sides.
- Changing one of the objective function coefficients.

First we introduce some general language for interpreting linear programs and their sensitivities. For the standard form of a linear program

$$z^* = \max \; \mathbf{c}^T\mathbf{x}$$
$$\text{s.t. } \mathbf{Ax} \le \mathbf{b}$$
$$\mathbf{x} \ge \mathbf{0},$$

we will use the terminology in Table 10.1. For other forms, we will generally stick to this terminology. Although "=" and "≥" constraints may be requirements rather than resources, we will use the generic term "resource." For a minimization problem, the objective function coefficients c_j will still be called costs or prices. For example, in a product mix problem, an activity is a product and its level is the quantity produced, while in a blending problem the activity is an ingredient, and in an employee scheduling problem the activity is the employees of a certain type assigned to a certain shift. In a production problem, the resources might be labor time, machine time, or raw material. In an employee scheduling problem, the "resources" are the staffing requirements.

Table 10.1 General terminology for a linear program.

Quantity	Interpretation
Decision variable	Activity
x_j	Level of activity j
c_j	Cost or price of activity j
z	Total profit or revenue
Constraint	Resource
b_i	Availability of resource i
a_{ij}	Consumption of resource i by each unit of activity j

Now consider the linear program that we solved graphically in Chapter 1:

$$z^* = \max \quad 4x + 3y$$
$$\text{s.t.} \quad 1.5x + y \le 16$$
$$x + y \le 12 \tag{10.1}$$
$$y \le 10$$
$$x, y \ge 0.$$

Suppose we add the constraint $x \le 5$. Adding a constraint can only make the feasible region smaller, so the optimal value can only get worse; since this is a maximization problem, it can only decrease. The original problem is called a relaxation of the new problem:

Given an optimization problem (P) $\max_{x \in S} f(\mathbf{x})$, the problem (P′) $\max_{x \in S'} f(\mathbf{x})$ is a **relaxation** of (P) if $S \subseteq S'$. The same definition applies to minimization problems.

Because every feasible solution of (P) is also a feasible solution of (P′), the optimal value of (P′) must be the same or better than the optimal value of (P). Thus, for a maximization (minimization) problem, the relaxation can have a larger (smaller) optimal value. When a constraint is added, the original problem is a relaxation of the modified problem. Returning to (10.1), the optimal solution is $(x, y) = (8, 4)$ with value $z^* = 44$. When we add the constraint $x \le 5$ this solution is no longer feasible. The new optimal solution is $(5, 7)$ with value $z^* = 42$; see Figure 10.1.

However, if instead we add the constraint $y \le 11$, the optimal solution is the same as the original problem. Furthermore, this constraint does not change the feasible region, so it would not change the optimal solution for any objective function. Such a constraint is redundant. Now consider removing the second constraint from (10.1), creating a relaxation. The new optimal solution is $(4, 10)$ with value $z^* = 46$. This follows the principle that the relaxation can only have a better optimal value.

10.1.1 Changes in the Right-hand Side

Next consider changes in the amount of a resource available. For (10.1), suppose that we have 15 units of the first resource instead of 16. The modified constraint is parallel to the original, but shifted to reduce the feasible region as shown in Figure 10.2. The new optimal solution is still at the point where constraints 1 and 2 intersect, which is now $(6, 6)$ with value $z^* = 42$. Notice that reducing b_1 by 1 (from 16 to 15) decreased z^* by 2. In general, if we change b_1 by Δ and the optimal

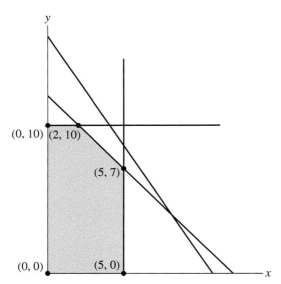

Figure 10.1 Feasible region for (10.1) with constraint $x \leq 5$.

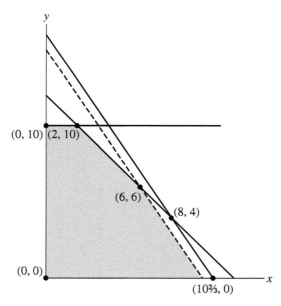

Figure 10.2 Feasible region for (10.1) with $b_1 = 15$.

solution is still where constraints 1 and 2 intersect, then the active constraints

$$1.5x + y = 16 + \Delta$$
$$x + y = 12$$

have solution $x = 8 + 2\Delta$, $y = 4 - 2\Delta$, and $z^* = 4(8 + 2\Delta) + 3(4 - 2\Delta) = 44 + 2\Delta$. This tells us that z^* increases by 2 for each unit increase in b_1. A similar analysis finds that the sensitivity to resource 2 is 1. Constraint 3, on the other hand, is not active at the optimal solution. Thus, small changes in b_3 cannot change the optimal solution and there is zero sensitivity to resource 3.

The sensitivities computed earlier assume that constraints 1 and 2 are still active at the optimal solution of the perturbed problem. As b_2 decreases, constraint 2 shifts down and the optimal point moves along constraint 1 from $(8, 4)$ until it reaches the constraint $y \geq 0$ at $(10\frac{2}{3}, 0)$. At this point $b_2 = 10\frac{2}{3}$ and $z^* = 42\frac{2}{3}$. When b_2 decreases further, the optimal solution is at the intersection of constraint 2 and $y = 0$, until it reaches the origin when $b_2 = 0$. If instead b_2 is increased, the optimal solution moves along constraint 1 to $(4, 10)$ at $b_2 = 14$ and $z^* = 46$. Increasing b_2 further has no effect on the optimal solution. The dependence of z^* on b_2 is graphed in Figure 10.3. The graph is piecewise linear and concave, with turns occurring when another constraint becomes active and the optimal basis changes. This behavior will always be seen for maximization problems that have an optimal solution. In economic terms, availability of the resource has decreasing economies of scale because the graph is concave. Also, the problem becomes infeasible beyond some minimal value of b_2 (in this example, below 0) and has a horizontal asymptote as b_2 increases. This behavior occurs because, even without constraint 2, the feasible region is bounded.

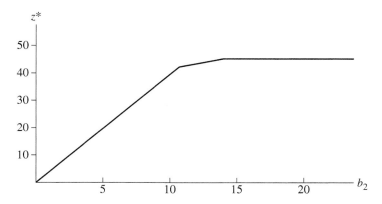

Figure 10.3 Optimal value as a function of b_2.

Sensitivity to Constraint Right-hand Sides

1. The sensitivity to a change in the right-hand side remains constant as long as the same basis is optimal.
2. If the change in the right-hand side relaxes the constraint, the optimal solution can only improve.
3. If a constraint is not active at the optimal solution, the sensitivity to a change in its right-hand side is 0.

10.1.2 Changes in the Objective Function

An objective function coefficient, or price, can change within some range without changing the optimal solution to a linear program. In (10.1), the point $(8, 4)$ is optimal because the slope of the line where z is constant lies between the slopes of constraints 1 and 2, which are active at this point. Now allow c_x, the price of the first activity, to vary. The point remains optimal if

$$-1.5 \le -\frac{c_x}{3} \le -1, \text{ or}$$

$$3 \le c_x \le 4.5.$$

When c_x moves outside of this range, an adjacent extreme point becomes optimal. For $0 \le c_x \le 3$, the slope is between the slopes of constraints 2 and 3, so $(2, 10)$ is optimal, while for $c_x \ge 3$, the slope is steeper than constraint 3 and $(10\frac{2}{3}, 0)$ is optimal.

10.2 Shadow Prices and Reduced Costs

The sensitivity computations done in Section 10.1 can be generalized using the matrix equations for basic solutions. We assume that the problem is in canonical form and that we have an optimal basis with basis matrix B. Then

$$z = c_B^T B^{-1} b, \quad x_B = B^{-1} b.$$

When we start with a general linear program and convert to canonical form, there is a straightforward way to translate the results of this section back to the original constraints and variables. If we solved the problem with the simplex algorithm, then we already computed $(y^*)^T = c_B^T B^{-1}$. From the proof of the Strong Duality Theorem, we know that these are the optimal values of the dual variables. If we perturb the data $\mathbf{A}, \mathbf{b}, \mathbf{c}$ the solution $\mathbf{x}^*$ remains optimal if:

1. *Feasibility.* $x_B^* = B^{-1} b \ge 0$ (nonnegative solution).
2. *Optimality.* $\overline{\mathbf{c}}_N^T = c_N^T - c_B^T B^{-1} N \le 0$ (nonpositive reduced costs).

10.2.1 Changes in the Right-hand Side

Consider changing the resources available from $\mathbf{b}$ to $\mathbf{b}^{\text{new}}$. If the same basis remains optimal, then by the Strong Duality Theorem

$$z^*_{\text{new}} = (\mathbf{y}^*)^T \mathbf{b}^{\text{new}} = \sum_{i=1}^{m} y_i^* b_i^{\text{new}}.$$

Notice that y_i^* is the rate of change of z^* with respect to b_i. This rate of change can be interpreted as the marginal profit derived from resource i; for this reason it is called a price.

The **shadow price** of a constraint is the rate of change of the optimal value with respect to the right-hand side of the constraint.

The shadow prices, then, are the optimal values of the dual variables. In the nondegenerate case, z^* is a differentiable function of b_i, and we can also call the shadow price y_i^* the partial derivative $\frac{\partial z^*}{\partial b_i}$. Having solved the original problem, we know z^*, so if b_i changes by Δ it is convenient to compute the new optimal value as

$$z^*_{\text{new}} = z^* + y_i^* \Delta. \tag{10.2}$$

Now, since changing $\mathbf{b}$ does not affect the optimality condition, the basis remains optimal if the feasibility condition is met. Writing $\mathbf{b}^{\text{new}} = \mathbf{b} + \Delta \mathbf{e}_i$, the condition is

$$\mathbf{B}^{-1}(\mathbf{b} + \Delta \mathbf{e}_i) = \mathbf{x}_B + \Delta \mathbf{B}^{-1} \mathbf{e}_i \geq \mathbf{0}.$$

Here $\mathbf{B}^{-1}\mathbf{e}_i$ is the ith column of $\mathbf{B}^{-1}$. If we let $\mathbf{B}^{-1} = [\beta_{ij}]$, the condition can be written

$$\begin{bmatrix} x_{B(1)} \\ \vdots \\ x_{B(m)} \end{bmatrix} + \Delta \begin{bmatrix} \beta_{1i} \\ \vdots \\ \beta_{mi} \end{bmatrix} \geq \mathbf{0}. \tag{10.3}$$

These constraints determine an interval for Δ in which the basis remains feasible and the shadow price applies. The interval for Δ is easily converted to an interval for b_i^{new} by adding b_i to the end points.

Example 10.1. For the problem (10.1), converted to canonical form, the optimal solution is $\mathbf{x}^* = (8, 4, 0, 0, 6)$ with basis (x, y, s_3) and dual solution $\mathbf{y}^* = (2, 1, 0)$. To find the allowable range for b_2, we use the second column of

$$\mathbf{B}^{-1} = \begin{bmatrix} 2 & -2 & 0 \\ -2 & 3 & 0 \\ 2 & -3 & 1 \end{bmatrix}.$$

Condition (10.3) is

$$
\begin{bmatrix} 8 \\ 4 \\ 6 \end{bmatrix} + \Delta \begin{bmatrix} -2 \\ 3 \\ -3 \end{bmatrix} \geq 0,
$$

giving the interval $\Delta \in [-\frac{4}{3}, 2]$ or $b_2 \in [10\frac{2}{3}, 14]$. This interval agrees with the graphical analysis in Section 10.1. Similarly, to find the allowable range for b_1, we use the first column of $\mathbf{B}^{-1}$, yielding the condition

$$
\begin{bmatrix} 8 \\ 4 \\ 6 \end{bmatrix} + \Delta \begin{bmatrix} 2 \\ -2 \\ 2 \end{bmatrix} \geq 0
$$

and the interval $\Delta \in [-3, 2]$ or $b_1 \in [13, 18]$. Because constraint 3 is inactive, its slack variable is basic, there are no negative entries in column 3 of $\mathbf{B}^{-1}$, and the allowable range is infinite: $b_3 \in [4, \infty)$.

Checking whether the basis remains optimal is not as simple if multiple components of $\mathbf{b}$ change. One approach is to directly check $\mathbf{B}^{-1}\mathbf{b} \geq \mathbf{0}$. Another approach is called the 100% rule. If b_i is allowed to increase by Δ and it is increased by, say, 0.30Δ, the percent of the allowable range used is 30%. Similarly, if it decreased 30% of the way to the lower endpoint of its allowable interval, 30% is used. If the total of these percentages over all constraints is no more than 100%, then the basis must remain optimal. The 100% rule is very conservative: often when the total of the percentages exceeds 100 the basis is still optimal.

10.2.2 Sign of Shadow Prices

For inequality constraints, the sign of a shadow price can be inferred from its definition. First consider a "$\leq$" constraint. Increasing the right-hand side relaxes the constraint, so the perturbed problem is a relaxation of the original problem and the optimal value can only improve. Thus, for a maximization problem, the optimal value can only increase and the shadow price is nonnegative. Conversely, for a "$\geq$" constraint increasing the right-hand side tightens the constraint, so the original problem is a relaxation of the perturbed problem and the optimal value can only get worse. Hence, for a maximization problem the shadow price is nonpositive. Also, if a constraint is inactive at an optimal solution, perturbing the constraint has no effect on the optimal solution, so the shadow price is zero. The situation is summarized in Table 10.2.

Table 10.2 Sign of shadow prices.

Constraint	$\leq$	$\geq$	$=$	Inactive
Max LP	≥ 0	≤ 0	Unrestricted	0
Min LP	≤ 0	≥ 0	Unrestricted	0

10.2.3 Changes in the Objective Function

Next consider changing the objective function coefficient of a variable by Δ. If the basis remains optimal, the solution $\mathbf{x}^*$ does not change but, if the variable is basic, the optimal value changes. The more interesting question is whether the basis remains optimal. Feasibility is not affected, so it remains optimal if the optimality condition holds.

Case 1. Change the coefficient of nonbasic variable x_j to $c_j + \Delta$. The only optimality condition that changes is the inequality for x_j, which becomes

$$c_j + \Delta - \mathbf{c}_B^T \mathbf{B}^{-1} \mathbf{a}_j = \bar{c}_j + \Delta \leq 0. \tag{10.4}$$

In other words, the basis remains optimal if the reduced cost of this nonbasic variable remains negative or zero. Indeed, another interpretation of a reduced cost is the amount by which the cost of an activity must change in order for it to be optimal to perform that activity, i.e. for the variable to become basic. Another way to put this is that if we use a suboptimal solution that includes an activity that is not in the optimal basis, the optimal value gets worse by the reduced cost for every unit of that activity. We can see this by rearranging the canonical form constraint

$$\mathbf{B}\mathbf{x}_B + \mathbf{N}\mathbf{x}_N = \mathbf{b}$$

as

$$\mathbf{x}_B = \mathbf{B}^{-1}\mathbf{b} - \mathbf{B}^{-1}\mathbf{N}\mathbf{x}_N$$

and substituting it in the objective function:

$$z = \mathbf{c}_B^T(\mathbf{B}^{-1}\mathbf{b} - \mathbf{B}^{-1}\mathbf{N}\mathbf{x}_N) + \mathbf{c}_N^T\mathbf{x}_N = \mathbf{c}_B^T\mathbf{B}^{-1}\mathbf{b} + \bar{\mathbf{c}}_N^T\mathbf{x}_N.$$

Notice that the sensitivity to the nonbasic variable x_j is $\bar{c}_j$.

Case 2. Change the coefficient of a basic variable. Let x_j be the kth basic variable (the basic variable associated with the kth constraint in the optimal basis).

Then $\mathbf{c}_B$ changes to $\mathbf{c}_B + \Delta\mathbf{e}_k$, and the optimality condition is

$$\mathbf{c}_N^T - (\mathbf{c}_B^T + \Delta\mathbf{e}_k^T)\mathbf{B}^{-1}\mathbf{N} = \bar{\mathbf{c}}_N^T - \Delta\mathbf{e}_k^T\mathbf{B}^{-1}\mathbf{N} \leq \mathbf{0}.$$

Here $\mathbf{e}_k^T\mathbf{B}^{-1}\mathbf{N}$ is the kth row of $\mathbf{B}^{-1}\mathbf{N}$, which we denote $\mathbf{q}^k$. Then the condition is

$$\bar{\mathbf{c}}_N^T - \Delta\mathbf{q}^k \leq \mathbf{0}. \tag{10.5}$$

In this case, the basis remains optimal if the reduced cost of all nonbasic variables remain negative or zero. Again these inequalities, one for each nonbasic variable, determine an interval for Δ in which the basis remains feasible. Adding c_j to the end points gives the desired interval for c_j^{new}.

These calculations required computing one row of $\mathbf{B}^{-1}\mathbf{N}$, which can be done for very large problems, even if $\mathbf{B}^{-1}$ is not found directly. In the tableau implementation described in Section 9.4, $\mathbf{B}^{-1}\mathbf{N}$ is computed in the tableau. If more than one objective coefficient is changed, one could recompute the reduced costs to check the optimality condition.

Example 10.2. Having solved problem (10.1), we know the reduced costs $\bar{\mathbf{c}}_N^T = [-2 - 1]$. To find the allowable range of c_x, because x is the first variable in the basis, we use the first row of

$$\mathbf{B}^{-1}\mathbf{N} = \begin{bmatrix} 2 & -2 & 0 \\ -2 & 3 & 0 \\ 2 & -3 & 1 \end{bmatrix} \begin{bmatrix} 1 & 0 \\ 0 & 1 \\ 0 & 0 \end{bmatrix} = \begin{bmatrix} 2 & -2 \\ -2 & 3 \\ 2 & -3 \end{bmatrix},$$

namely, $\mathbf{q}^1 = [2 - 2]$. The optimality condition is

$$\begin{bmatrix} -2 & -1 \end{bmatrix} - \Delta \begin{bmatrix} 2 & -2 \end{bmatrix} \leq \mathbf{0},$$

giving the interval $\Delta \in [-1, 0.5]$ or $c_x \in [3, 4.5]$. This interval agrees with the graphical analysis in Section 10.1. Similarly, since y is the second variable in the basis, to find the allowable range of c_y we use the second row of $\mathbf{B}^{-1}\mathbf{N}$, so (10.5) is

$$\begin{bmatrix} -2 & -1 \end{bmatrix} - \Delta \begin{bmatrix} -2 & 3 \end{bmatrix} \leq \mathbf{0},$$

giving the interval $\Delta \in [-\frac{1}{3}, 1]$ or $c_y \in [2\frac{2}{3}, 4]$. Because the nonbasic variables are slack variables, their allowable ranges are not of practical interest. Still, if we use (10.4) for s_1, then $\bar{c}_{s_1} + \Delta = -2 + \Delta \leq 0$, giving the interval $c_{s_1} \in (-\infty, 2]$.

Notice that in this example the reduced costs of the slack variables are just the negative of the shadow prices. This is generally true (see Exercise 10.3):

- At a nondegenerate optimal solution, the reduced cost of a slack variable is the negative of the shadow price of the corresponding constraint.

- At a nondegenerate optimal solution, the reduced cost of a surplus variable is the shadow price of the corresponding constraint.

Because these reduced costs contain no additional information, they are not reported by optimization software. A similar sensitivity analysis can be done when an entry in **A** changes.

10.2.4 Sensitivity Analysis Examples

Two examples will be used to illustrate the application of sensitivity analysis.

Example 10.3 (Kan Jam sensitivity). The Kan Jam production problem was solved in Example 2.1. We repeat it here for convenience. They produce the Ultimate Disc Game, Splash Pool Disc Game, and sell the disc individually. The material requirements and availability are shown in Table 10.3, along with the retail price on Amazon and the profit for each product. Demand for the new pool version is limited; they are forecasting a demand of 2000 for the month at this price. How many of each product should be produced to maximize profit for the month?

Letting U, P, and D denote the number of Ultimate, Pool, and separate discs produced, we formulate the following linear program:

$$
\begin{aligned}
\max\ & 10U + 12P + 2D \\
\text{s.t.}\ & 48U + 40P && \leq 240\,000 \ \text{(Plastic)} \\
& U + P + D && \leq 8000 \ \text{(Flying discs)} \\
& 10U + 10P + 2D && \leq 75\,000 \ \text{(Packaging)} \\
& P && \leq 2000 \ \text{(Demand for Pool)} \\
& U, P, D && \geq 0.
\end{aligned}
$$

The optimal profit is $62 666.67, with optimal production levels of 3333, 2000, and 2667 for Ultimate, Pool, and separate discs. The sensitivity report from MPL is

Table 10.3 Data for Kan Jam production.

Material	Ultimate	Pool	Disc	Available
Plastic (oz)	48	40	0	240 000
Flying discs	1	1	1	8000
Packaging (ft^2)	10	10	2	75 000
Selling price	$40	$38	$9	–
Profit margin	$10	$12	$2	–

shown. Because all three products are produced, they are basic variables and their reduced costs are zero. The unit profit for Pool can be decreased by $3.33 without changing the optimal production levels.

Variable	Activity	Reduced Cost	Lower Range	Upper Range
U	3333.3333	0.0000	2.0000	14.0000
P	2000.0000	0.0000	8.6667	1E+020
D	2666.6667	0.0000	0.0000	10.0000

Constraint	Slack	Shadow Price	Lower Range	Upper Range
Plastic	0.0000	0.1667	80000.0000	338000.0000
Flying_discs	0.0000	2.0000	5333.3333	16166.6667
Packaging	16333.3333	0.0000	58666.6667	1E+020
Demand_Pool	0.0000	3.3333	0.0000	6000.0000

The "activity" for a variable is its optimal value. All of the plastic, flying discs, and demand for Pool were used; these constraints are active at the optimal solution. However, the packaging constraint is not active: only 58 667 ft^2 of packaging was used. The shadow price for plastic is $\frac{1}{6}$, meaning that each additional ounce of plastic increases profit by about $0.17. For example, if they could obtain an additional 3000 oz of plastic then, since the new amount available of 243 000 is below the allowable upper range of 338 000, the new optimal profit can be computed using (10.2) as

$$z^*_{new} = 62\ 666.67 + \frac{1}{6}(3000) = \$63\ 166.67.$$

We can also compute the new optimal profit from just the shadow prices and constraint right-hand sides using strong duality:

$$\frac{1}{6}(243\ 000) + 2(8000) + 0(75\ 000) + \frac{10}{3}(2000) = \$63\ 166.67.$$

Since the normal cost of plastic was already factored into the profit margins, additional plastic purchased at a premium of less than the shadow price ($0.17 per oz) can be used to increase profit. Note that the shadow price of packaging is zero. This must be the case because the packaging constraint is inactive at the optimal solution. Only 58 667 ft^2 of packaging was used, so additional packaging without other additional resources has no value.

If we change the profit margin for Pool from $12 to $8.50, the change is more than the allowable decrease and we expect the basis to change. The resulting sensitivity report is shown. The Pool product is not produced. Its reduced cost of − $0.17

means that its profit margin would have to increase by $0.17 before it would be profitable to produce.

Variable	Activity	Reduced Cost	Lower Range	Upper Range
U	5000.0000	0.0000	9.8000	1E+020
P	0.0000	-0.1667	-1E+020	8.6667
D	3000.0000	0.0000	1.0000	10.0000

Example 10.4 (Online ads sensitivity). A Custom Tees advertising problem was solved in Example 2.3. Now consider an expanded version. They are planning an online advertising campaign with two web companies. The first sells banner and popup ad space on their web pages; Custom Tees can select the parts of web pages where their ad appears. The second is a social media company that sells ads that appear on their visitor's newsfeed; Custom Tees can select which categories of visitors will see their ads, based on various characteristics. These choices allow them to show their ads to visitors who are more likely to be interested in clicking on them and making a purchase. Both companies charge when a visitor to the site clicks on an ad (called engagement) rather than when they visit a page containing the ad (called an impression). Customers in the 18–25 age range are particularly valuable to them, so they have set goals of 500 000 clicks by visitors in this age range and 600 000 clicks from older visitors. After making their placement choices and running test ads, Custom Tees was able to estimate the percent of clicks in the 18–25 age range for each type of ad.

Another concern is that with this large of an ad campaign, repeat visitors to the web sites are likely to see many Custom Tees ads. Although they only pay when the visitor clicks, some of these clicks will be by people who have already clicked and been directed to the Custom Tees site. These repeats are not as valuable as new visitors, so they have also set goals for unique visitors at 50% of the total click goals in each age group. Forecasting unique visitors as a function of total visitors is more challenging. They have chosen functions with two linear pieces, one when the number of clicks is below a "market saturation" point and one above. Table 10.4 lists this data and the prices the companies charge per click.

To formulate these decisions as a linear program, we use two variables for each type of ad to model the nonlinearity at the saturation point. Let x_i be the number of clicks on type i ads below the saturation point and s_i the number above the saturation point (1000s), where $i = 1$ (banner), 2 (popup), or 3 (newsfeed). The

Table 10.4 Data for Custom Tees ads.

	Banner	Popup	Newsfeed
Cost ($ per 1000 clicks)	75	100	120
Clicks from age 18 –25	40%	30%	70%
Clicks from age 26 and over	60%	70%	30%
Unique clicks	40%	75%	90%
Unique clicks after saturation	25%	40%	30%
Saturation point (1000 clicks)	450	450	300

linear program to minimize cost is

$$\min \quad 75(x_1 + s_1) + 100(x_2 + s_2) + 120(x_3 + s_3)$$

$$\text{s.t.} \quad 0.4(x_1 + s_1) + 0.3(x_2 + s_2) + 0.7(x_3 + s_3) \geq 500$$

$$0.4(0.4x_1 + 0.25s_1) + 0.3(0.75x_2 + 0.4s_2) + 0.7(0.9x_3 + 0.3s_3) \geq 250$$

$$0.6(x_1 + s_1) + 0.7(x_2 + s_2) + 0.3(x_3 + s_3) \geq 600$$

$$0.6(0.4x_1 + 0.25s_1) + 0.7(0.75x_2 + 0.4s_2) + 0.3(0.9x_3 + 0.3s_3) \geq 300$$

$$0 \leq x_1 \leq 450, \ 0 \leq x_2 \leq 450, \ 0 \leq x_3 \leq 300, \ s_i \geq 0.$$

The sensitivity report from MPL is shown. Total cost of the ads is $98 295.18. Only the banner ads exceed the saturation level. Because it is a minimization problem the reduced costs can only be positive, the shadow prices of "≥" constraints positive, and the shadow prices of "≤" constraints negative. The constraint labeled "Clicks Banner" is the saturation upper bound of 450 000 clicks. Its shadow price means that each 1000 increase in this saturation point would decrease the optimal cost $8.67. It is a decrease because more of the clicks would be unique clicks, so fewer clicks are required to meet the unique clicks constraint – which is active for age 26 and over.

```
Variable    Activity   Reduced Cost  Lower Range  Upper Range

Banner      450.0000     0.0000        -1E+020      83.6747
Popup       169.8795     0.0000        60.0000     157.6471
Newsfeed    256.6265     0.0000        44.4643     132.0455
Sat_Banner  223.4940     0.0000        68.0628     105.0000
Sat_Popup     0.0000    23.6145        76.3855      1E+020
Sat_News      0.0000    17.3494       102.6506      1E+020


Constraint      Slack   Shadow Price  Lower Range  Upper Range

Clicks_18_25   0.0000    127.4096       449.5189    525.7143
Unique_18_25 -44.2470      0.0000        -1E+020    294.2470
```

Clicks_26_over	0.0000	15.9639	567.2727	673.8191
Unique_26_over	0.0000	96.3855	239.1298	354.0000
Clicks_Banner	0.0000	-8.6747	0.0000	764.4068
Clicks_Popup	280.1205	0.0000	169.8795	1E+020
Clicks_News	43.3735	0.0000	256.6265	1E+020

Suppose Custom Tees decreased the required age 18–25 clicks by 10%, or 50 000. Keeping the unique clicks requirement at 50% of total clicks, it would decrease by 25 000. The new right-hand side for the age 18–25 clicks constraint is 450 000. Since that is within the allowable lower range, and the unique clicks constraint is allowed to decrease by any amount, the shadow price can be used. The new optimal cost is

$$z^*_{new} = \$98\ 295.18 - 127.4096(50) = \$91\ 924.70.$$

If instead they increased the required age 26 and over clicks goal by 30 000 and the required age 26 and over unique clicks by 15 000, these are both less than half of the allowable increase, so by the 100% rule the shadow prices can be used. The new optimal cost is

$$z^*_{new} = \$98\ 295.18 + 15.9639(30) + 96.3855(15) = \$100\ 219.88.$$

The reduced costs of the saturated clicks are not of practical interest; however, if a fourth type of ad was added to the model but was not in the optimal solution, its reduced cost would be the amount its price would need to decrease in order for it to be used in an optimal solution.

10.2.5 Degeneracy

Suppose the following constraint is added to problem (10.1):

$$4x + 5y \leq 52.$$

This constraint passes through the optimal point $(8, 4)$ but is redundant the feasible region does not change (see Figure 10.2). The optimal solution is the same, but is now degenerate with three active constraints. The sensitivity report from MPL has nonzero shadow prices for constraints 1 and 4. From this we infer that it corresponds to the basis (x, y, s_2, s_3), where s_i is the slack variable for constraint i. It is degenerate, with $s_2 = 0$. Note that the shadow prices for constraints 1, 2, and 4 only apply in one direction. For constraint 1, the allowable increase is zero (the upper range and the current right-hand side are both 16). Any increase makes constraint 1 inactive, changing the basis to include its slack variable s_1. For constraint 2, the allowable decrease is zero because any decrease makes constraint 4 inactive, changing the basis to include s_4. Other optimal bases would have different shadow prices.

Constraint	Slack	Shadow Price	Lower Range	Upper Range
Constraint1	0.0000	2.2857	10.7500	16.0000
Constraint2	0.0000	0.0000	12.000	1E+020
Constraint3	6.0000	0.0000	4.0000	1E+020
Constraint4	0.0000	0.1429	42.6667	52.0000

In general, degeneracy can have the following effects on sensitivity results.

- The stated shadow price may only apply in one direction. In this case the right-hand side range has an allowable increase or decrease of zero and the shadow price, defined as a derivative, does not exist. The reported value of the dual variable is the sensitivity in one direction.
- The dual solution, reported as the shadow prices, may not be unique. In the example earlier, s_2 is basic, so its reduced cost and the associated dual value must be zero. However, there is another optimal basis with s_4 basic instead of s_2, where $y_4^* = 0$. Thus, the shadow prices reported may depend on which optimal basis is found.
- Changing an objective function coefficient by its reduced cost may not be sufficient for there to be an optimal solution where that variable has a positive value. The coefficient may have to be changed further for this to happen. Thus, the reported allowable range of the coefficient may be too conservative.

10.2.6 Parametric Programming

This section has focused on the range in which the current basis remains optimal, sometimes called *local* sensitivity analysis. To explore the sensitivity to a single cost c_j or resource availability b_i beyond this range, one can use a method called parametric programming; see, for example, (Bertsimas and Tsitsiklis, 1997, Section 5.2, 5.4). The information obtained is similar to Figure 10.3, which shows the optimal value of the perturbed problem for all values of the input that is varied.

As we saw in Section 10.1, the optimal value of a maximization problem is a concave function of a resource availability b_i. This result extends to when more than one resource is varied: the optimal value of a maximization (minimization) problem is a concave (convex) function of **b** over those values for which the problem is feasible and has an optimal solution. Similarly, assuming it is feasible, the optimal value of a maximization (minimization) problem is a convex (concave) function of **c** over those values for which the problem has an optimal solution. Note that the concavity of the optimal value function is reversed for **b** and **c**. One way of understanding the reversal is to use the Fundamental Theorem of Linear Programming to write z^* as a maximum over the bfs's $\mathbf{x}^1, \mathbf{x}^2, \dots, \mathbf{x}^N$, which do not depend on **c**.

For a maximization problem, then,

$$z^*(\mathbf{c}) = \max_{i=1,\ldots,N} \mathbf{c}^T \mathbf{x}^i.$$

This is the maximum of linear functions of $\mathbf{c}$ and therefore convex. However, by strong duality, when the primal problem has an optimal solution, so does the dual problem. Let $\mathbf{y}^1, \mathbf{y}^2, \ldots, \mathbf{y}^M$ be the bfs's of the dual problem, which do not depend on $\mathbf{b}$. Since the dual problem is a minimization,

$$z^*(\mathbf{b}) = \min_{i=1,\ldots,M} \mathbf{b}^T \mathbf{y}^i.$$

As a function of $\mathbf{b}$, z^* is the minimum of linear functions and therefore concave.

10.3 Economic Interpretation of the Dual

Now we elaborate on interpreting the optimal value of a dual variable as the marginal profit, or shadow price, of a resource. Consider again the strong duality relationship

$$\mathbf{c}^T \mathbf{x}^* = \mathbf{b}^T \mathbf{y}^*. \tag{10.6}$$

In the activity "market," c_j is the price of activity j and x_j is its level or quantity, so $\mathbf{c}^T \mathbf{x}$ is the total profit for all activities. In the resource "market," y_i is the implicit price of resource i and b_i is the quantity, so $\mathbf{b}^T \mathbf{y}$ is the total profit. By (10.6), the optimal activity level $\mathbf{x}^*$ and resource prices $\mathbf{y}^*$ give the same total profit; the two pricing methods agree when prices are chosen optimally.

Interestingly, y_i^* appears in (10.6) as an *average* cost (or profit), i.e. the total value of b_i units of resource i is $b_i y_i^*$, while in the definition of a shadow price, y_i^* is a *marginal* cost. Thus, the average cost of a resource equals its marginal cost. It may seem contradictory that Figure 10.3 shows decreasing economies of scale for the availability of a resource and yet average cost equals marginal cost. Indeed, if there is only one resource, the only way for average and marginal cost to be equal is for marginal cost to be constant, making $z^*(b_1) = b_1 y_1^*$ linear (no economies of scale). What makes Figure 10.3 nonlinear is that one b_i is being changed at a time. After changing all m resource availabilities from 0 to b_i, the total effect on z^* is the same as if the shadow prices applied for all $\mathbf{b}$.

To state this formally, consider a linear program with right-hand side $\lambda \mathbf{b}$. Here λ is a scalar parameter.

Theorem 10.1: *For the parametric linear program with right-hand side $\lambda \mathbf{b}$ and optimal value $z^*(\lambda)$, if $z^*(1)$ is finite, then $z^*(\lambda) = \lambda z^*(1)$ for all $\lambda \geq 0$.*

Proof: Since $z^*(1)$ is finite, this problem has an optimal solution $\mathbf{x}^*$. We will show that $\lambda\mathbf{x}^*$ is an optimal solution for all $\lambda \geq 0$, so that $z^*(\lambda) = \mathbf{c}^T\lambda\mathbf{x}^* = \lambda z^*(1)$. Rearrange the constraints to have the form $\mathbf{Ax} \leq \lambda\mathbf{b}$ with solution set S_λ. First, note that $\lambda\mathbf{x}^*$ is feasible because $\mathbf{Ax}^* \leq \mathbf{b}$ implies $\mathbf{A}(\lambda\mathbf{x}^*) \leq \lambda\mathbf{b}$ for $\lambda \geq 0$. If there is no feasible direction in S_λ at $\lambda\mathbf{x}^*$, then S_λ contains only $\lambda\mathbf{x}^*$ and we are done. Otherwise, let $\mathbf{d}$ be a feasible direction. Then $\mathbf{A}(\lambda\mathbf{x}^* + \varepsilon\mathbf{d}) \leq \lambda\mathbf{b}$ for some $\varepsilon > 0$, which we rewrite as $\mathbf{A}(\mathbf{x}^* + \lambda^{-1}\varepsilon\mathbf{d}) \leq \mathbf{b}$, showing that $\mathbf{d}$ is a feasible direction in S_1 at $\mathbf{x}^*$. From Section 5.2, because $\mathbf{x}^*$ is optimal there is no improving feasible direction in S_1 at $\mathbf{x}^*$, so $\mathbf{d}$ is not improving. Hence, there is no improving feasible direction in S_λ at $\lambda\mathbf{x}^*$ and this solution is optimal. □

Theorem 10.1 is illustrated in Figure 10.4. The feasible region scales by the parameter λ and the optimal solutions lie on the dashed line with optimal values proportional to λ. The sensitivity to λ is a linear combination of the shadow prices at $\lambda = 1$, namely, $\frac{\partial z^*}{\partial \lambda} = \mathbf{b}^T\mathbf{y}^*$. Unlike shadow prices, which only apply in a certain range, this sensitivity applies for any $\lambda \in [0, \infty)$.

Returning to the notion of two markets in which prices are set, we can interpret a linear program and its dual as the mechanisms for correctly setting the prices. In Example 10.3, Wild Sports decides on production levels to maximize profit. Suppose a second firm wants to buy all of the resources identified in this problem and must decide on prices to offer Wild Sports. Let y_1 be the price for plastic ($ per oz), y_2 for discs ($ per disc), y_3 for packaging ($ per ft^2), and y_4 for potential Pool customers ($ per item). Since producing a Universal is worth $10, the resources used to produce it must price out at more than $10 for Wild Sports to be willing to sell them, so

$$48y_1 + y_2 + 10y_3 \geq 10.$$

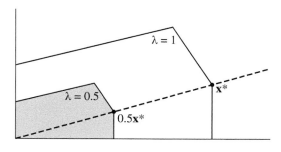

Figure 10.4 Feasible regions with right-hand sides $\lambda\mathbf{b}$.

Writing similar constraints for the other products, the second firm wants to minimize the cost of buying all the resources subject to these constraints:

$$
\begin{aligned}
\min \quad & 240\,000y_1 + 8000y_2 + 75\,000y_3 + 2000y_4 \\
\text{s.t.} \quad & 48y_1 + y_2 + 10y_3 \geq 10 \quad \text{(Unlimited)} \\
& 40y_1 + y_2 + 10y_3 + y_4 \geq 12 \quad \text{(Pool)} \\
& y_2 + 2y_3 \geq 2 \quad \text{(Disc)} \\
& \; y_i \geq 0.
\end{aligned}
$$

This is the dual of the Wild Sports problem. When the resource prices are set by solving the dual, the cost for the second firm to buy out the resources is equal to the profit that could be made using these resources. It would be rational for Wild Sport to sell at these prices, but no lower.

It is important to note that the shadow price of a resource only applies when it is used with this combination of resources. For example, the shadow price of packaging is zero because it is not all needed; the second firm pays nothing for the packaging. The second firm pays for the potential Pool customers because they are in limited supply. There are sufficient Unlimited and Disc customers (the demand for these items were not even listed as constraints) so the second firm doesn't pay for them. Similar economic interpretations apply to the duals of most linear programs.

We saw in Section 10.1 that changing the right-hand side of a constraint that is inactive does not affect the optimal solution. In economic terms, the shadow price of this resource is zero. We have seen this relationship before. From Theorem 8.4, the complementary slackness conditions can be written

$$
x_j^*(c_j - \mathbf{a}_j^T \mathbf{y}^*) = 0, \; j = 1, \ldots, m,
$$
$$
y_i^*(b_i - \mathbf{A}_i \mathbf{x}^*) = 0, \; i = 1, \ldots, n.
$$

The second condition shows that either a primal constraint is active or its shadow price y_i^* is zero. The first condition says the same thing about the dual problem.

Problems

1   The linear program

$$
\begin{aligned}
z^* = \max \quad & 3x + 4y \\
\text{s.t.} \quad & 2x + 2y \leq 200 \\
& x + 3y \leq 150 \\
& x, y \geq 0
\end{aligned}
$$

has optimal solution $(75, 25)$.

(a) Find the shadow price of each constraint using matrix calculations.

    (b) Graph the feasible region and use the graphical technique of Section 10.1 to find the shadow price of the second constraint.

    (c) Use (10.3) to find the range of b_2 (the right-hand side of the second constraint) for which the basis remains optimal.

    (d) Repeat (c) using the graphical technique of Section 10.1.

    (e) Find the shadow price of the second constraint where b_2 is (i) below the range in (c) and (ii) above the range in (c). Graph z^* as a function of b_2. What is the slope of each linear segment?

2 For the linear program in Exercise 10.1, find the range of c_1 (the objective function coefficient of x) for (75, 25) to remain optimal using

    (a) The graphical technique of Section 10.1.

    (b) Equation (10.5).

3 For a standard form problem, let s_i be the slack variable for constraint i.

    (a) If the optimal basis consists of the slack variables, show that the shadow prices are zero.

    (b) If s_i is nonbasic, show that its reduced cost is the negative of the shadow price of constraint i.

4 Refer to the sensitivity report for Example 10.3.

    (a) Suppose the demand for the Pool game increased from 2000 to 3500. Find the new optimal profit.

    (b) Suppose 2000 more discs could be manufactured, but the cost would be $1.25 more than the normal cost. Should Wild Sports produce them? What would the change in profit be if they were produced?

    (c) If the selling price of the Ultimate game decreased $5, would the optimal solution change? Explain.

5 Refer to the sensitivity report for Example 10.4.

    (a) Suppose the required number of age 18–25 clicks increased by 10 000 and the required age 18–25 unique clicks increased by 5000. Find the new optimal cost.

    (b) How much can the cost of Popup ads decrease before the optimal solution changes?

    (c) How much can the cost of Newsfeed ads increase before the optimal solution changes?

6 For the TV advertising problem in Exercise 2.8, MPL sensitivity results are shown. We allow the number of ads be non-integral, since these quantities could be the average number of ads over a season. The objective function that is minimized is the cost of ads (in $1000s); the optimal cost is 966.4207, or $966 420.70.

(a) Suppose the goal for women 18–35 is increased from 40 to 45 million impressions. Find the new optimal cost.
(b) Suppose the goals for men 18–35 and men 36–55 are both decreased from 30 to 27 million. Find the new optimal cost.
(c) This Is Us ads cost $140 000. How much would their cost have to change before they could be part of an optimal solution?
(d) The Bachelor ads cost $50 000. How much would their cost have to increase before the optimal solution changes?

Variable	Activity	Reduced Cost	Lower Range	Upper Range
This_Is_Us	0.0000	3.6531	136.3469	1E+020
Greys_Anat	4.0590	0.0000	74.9490	123.1034
NFL	1.9188	0.0000	40.7143	127.3881
Bachelor	8.2288	0.0000	28.0702	51.3543

Constraint	Slack	Shadow Price	Lower Range	Upper Range
M18_35	0.0000	9.2251	23.8409	48.9655
M35_55	0.0000	12.2878	17.6190	35.3708
W18_35	-6.6273	0.0000	-1E+020	46.6273
W36_55	0.0000	6.4207	43.5489	73.6364

7 The linear program (10.1) has optimal value $z^* = 44$. Without solving another linear program, find the optimal value if
(a) The right-hand side is changed to $\mathbf{b} = (8, 6, 5)$.
(b) The right-hand side is changed to $\mathbf{b} = \lambda(16, 12, 10)$ for $\lambda \geq 0$.

11

Algorithmic Applications of Duality

The simplex method was the first efficient linear programming algorithm. Variants of it are still the fastest known algorithm on some large problems. However, other algorithms are often faster on sparse problems. There are also specialized algorithms for some important types of linear programs. We present four algorithms, all of which use ideas from duality theory. Two of them can be applied to general linear programs: the dual simplex method (Section 11.1) and an interior point method called primal-dual path following (Section 11.3), which also illustrates some of the techniques used to solve nonlinear programs. These methods and the (primal) simplex are the commonly used in solvers.

This interior point algorithm not only performs well in practice, it is provably efficient in a certain sense. We noted in Section 9.4 that the worst-case number of iterations required by the simplex algorithm is exponential in the number of variables n, also making the worst-case number of arithmetic operations exponential. Bounds on the number of iterations required by interior point algorithms are stated differently because they converge to an optimal solution, rather than reaching it exactly. The bound mentioned in Section 11.3 is $O(\sqrt{n})$, so its dependence on n is much better than the worst case bound for the simplex or dual simplex method. There are several other interior point methods for linear programs; see Den Hertog (2012) for a review. Implementation details of interior point methods are discussed in Lustig et al. (1994) and Andersen and Andersen (2000).

For special types of linear programs, variants of the simplex algorithm can be used that are much faster because they take advantage of the special structure. Section 11.2 describes the network simplex method, which applies to the network flow problems of Section 2.6. Other variants apply to other problem structures. A method called *delayed column generation* can be applied if the maximum reduced cost can be found without computing all of the reduced costs. Columns, or variables, are ignored unless they enter the basis. A classic problem where this applies is called the cutting stock problem, which can have an exponentially large number of variables. Another special structure is when the constraint matrix $\mathbf{A}$ is nearly

Linear and Convex Optimization: A Mathematical Approach, First Edition. Michael H. Veatch.
© 2021 John Wiley & Sons, Inc. Published 2021 by John Wiley & Sons, Inc.
Companion website: www.wiley.com/go/veatch/convexandlinearoptimization

block-diagonal. Notice that if **A** really is block diagonal then the constraints can be written

$$\begin{bmatrix} \mathbf{A}_1 & \mathbf{0} \\ \mathbf{0} & \mathbf{A}_2 \end{bmatrix} \begin{bmatrix} \mathbf{x}_1 \\ \mathbf{x}_2 \end{bmatrix} \leq \begin{bmatrix} \mathbf{b}_1 \\ \mathbf{b}_2 \end{bmatrix},$$

$$\mathbf{x}_1, \mathbf{x}_2 \geq \mathbf{0}.$$

We have no reason to solve this linear program! Instead, we can solve two separate problems, one involving $\mathbf{x}_1$ and the other $\mathbf{x}_2$. If there are a small number of linking constraints containing $\mathbf{x}_1$ and $\mathbf{x}_2$, then a method called *Dantzig–Wolfe decomposition* can be used. These methods can be found in any advanced text on linear programming.

11.1 Dual Simplex Method

The simplex method starts with a feasible solution and at each iteration maintains feasibility, working toward optimality by selecting a variable whose optimality condition is violated. We saw in Section 8.2 that, for basic solutions, the primal optimality condition is the same as the dual feasibility condition. Thus, the simplex method works toward dual feasibility. Such methods are called *primal* algorithms. If instead we apply the simplex method to the dual, we are maintaining dual feasibility and working toward primal feasibility (which is equivalent to dual optimality). From the perspective of the primal problem, this is called a *dual* algorithm. This section presents a different dual algorithm, called the *dual simplex method*. It is often faster than applying the regular simplex method to the dual.

To explain the dual simplex method, we first extend the idea of complementary slackness. Theorem 8.4 asserts that complimentary slackness holds between an optimal primal and optimal dual solution. Now assume that the primal is in canonical form

$$\begin{aligned} \max \quad & \mathbf{c}^T \mathbf{x} \\ \text{s.t.} \quad & \mathbf{A}\mathbf{x} = \mathbf{b} \\ & \mathbf{x} \geq \mathbf{0} \end{aligned} \tag{11.1}$$

with m linearly independent rows and n variables, so that we can apply the simplex method. The dual is

$$\begin{aligned} \min \quad & \mathbf{b}^T \mathbf{y} \\ \text{s.t.} \quad & \mathbf{A}^T \mathbf{y} \geq \mathbf{c}. \end{aligned} \tag{11.2}$$

From the proof of strong duality, if **B** is an optimal basis matrix, then

$$\mathbf{y}^T = \mathbf{c}_B^T \mathbf{B}^{-1} \tag{11.3}$$

is an optimal dual solution. Now consider other basis matrices. Because (11.3) satisfies the m constraints $\mathbf{B}^T\mathbf{y} = \mathbf{c}_B$, which by definition are linearly independent, there are m linearly independent active constraints and it is a basic solution of the dual. Thus, every basic solution $\mathbf{x}_B = \mathbf{B}^{-1}\mathbf{b}$, $\mathbf{x}_N = \mathbf{0}$ of the primal corresponds to a basic solution of the dual. The objective function values of corresponding solution are equal:

$$\mathbf{c}^T\mathbf{x} = \mathbf{c}_B^T\mathbf{x}_B = \mathbf{c}_B^T\mathbf{c}\mathbf{B}^{-1}\mathbf{b} = \mathbf{y}^T\mathbf{b}. \tag{11.4}$$

Further, they satisfy the complimentary slackness conditions from Section 8.3:

$$x_j(\mathbf{y}^T\mathbf{a}_j - c_j) = 0, \quad j = 1, \ldots, n,$$
$$y_i(\mathbf{A}_i\mathbf{x} - b_i) = 0, \quad i = 1, \ldots, m.$$

For $j \in \mathcal{N}, x_j = 0$ and for $j \in \mathcal{B}, \mathbf{y}^T\mathbf{a}_j = c_j$ by (11.3), so the first condition holds. The second condition holds because $\mathbf{A}\mathbf{x} = \mathbf{B}\mathbf{x}_B = \mathbf{B}\mathbf{B}^{-1}\mathbf{b} = \mathbf{b}$. Also, substituting (11.3) into the dual constraints, feasibility for the dual is

$$\bar{\mathbf{c}}_N^T = \mathbf{c}_N^T - \mathbf{c}_B^T\mathbf{B}^{-1}\mathbf{N} = \mathbf{c}_N^T - \mathbf{y}^T\mathbf{N} \leq \mathbf{0},$$

which we recognize as the optimality condition for the primal. We show later that optimality for the dual is equivalent to feasibility for the primal, namely, $\mathbf{x}_B = \mathbf{B}^{-1}\mathbf{b} \geq \mathbf{0}$.

To summarize, every basic solution for the primal corresponds to a basic solution for the dual, with the relationships shown in Table 11.1. We make the additional observations:

- If both are feasible, then both optimality conditions are met. Satisfying feasibility and the optimality condition makes a solution optimal. Another way to say this is that primal feasibility, dual feasibility, and complementary slackness are a necessary and sufficient condition for optimality.
- Because the primal and dual objective function values are the same, by strong duality they cannot both be feasible unless both are optimal. However, it could be that neither is feasible.

Table 11.1 Dual relationships for corresponding basic solutions.

Primal	Dual
Feasibility	Optimality
Optimality	Feasibility
Basic variable	Active constraint
Nonbasic variable	Inactive constraint

- If there are degenerate basic solutions, then the correspondence may not be one-to-one. Multiple bases could correspond to the same primal basic solution and different dual basic solutions or vice-versa.

We will use the following example.

Example 11.1. Consider the linear program

$$
\begin{aligned}
\min \ & 2x_1 + 4x_2 + x_3 \\
\text{s.t.} \ & 2x_1 - 4x_2 + 2x_3 \geq 24 \\
& 2x_1 + x_2 - x_3 \geq 18 \\
& x_j \geq 0.
\end{aligned}
$$

Converting to canonical form gives

$$
\begin{aligned}
\max \ & -2x_1 - 4x_2 - x_3 \\
\text{s.t.} \ & 2x_1 - 4x_2 + 2x_3 - s_1 \quad\quad = 24 \\
& 2x_1 + x_2 - x_3 \quad\quad -s_2 = 18 \\
& x_j, s_i \geq 0.
\end{aligned}
$$

There is no obvious initial bfs for this problem. Notice, though, that if we choose the surplus variables as an initial basis, we only need to multiple the constraints by -1 to get the infeasible basic solution $(\mathbf{x}, \mathbf{s}) = (0, 0, 0, -24, -18)$:

$$
\begin{aligned}
\max \ & -2x_1 - 4x_2 - x_3 \\
\text{s.t.} \ & -2x_1 + 4x_2 - 2x_3 + s_1 \quad\quad = -24 \\
& -2x_1 - x_2 + x_3 \quad\quad + s_2 = -18 \\
& x_j, s_i \geq 0.
\end{aligned}
$$

Also, surplus variables have zero cost coefficients, so $\mathbf{c}_B = \mathbf{0}$. Thus, the optimality condition $\overline{\mathbf{c}}_N \leq \mathbf{0}$ reduces to $\mathbf{c}_N \leq \mathbf{0}$, which holds for this problem: $\mathbf{c}_N = (-2, -4, -1)$. The dual of the canonical form is

$$
\begin{aligned}
\min \ & 24y_1 + 18y_2 \\
\text{s.t.} \ & 2y_1 + 2y_2 \geq -2 \\
& -4y_1 + y_2 \geq -4 \\
& 2y_1 - y_2 \geq -1 \\
& -y_1 \geq 0 \\
& -y_2 \geq 0.
\end{aligned}
\tag{11.5}
$$

In this example, we started with a minimization problem with "$\geq$" constraints and $\mathbf{c}, \mathbf{b} \geq \mathbf{0}$. The basis containing the surplus variables has negative variables. The positive objective function coefficients lead to negative reduced costs. This basic solution is infeasible but satisfies the optimality condition. Thus, the corresponding basic solution for the dual, $\mathbf{y}^T = \mathbf{c}_B^T \mathbf{B}^{-1} = (0,0)$, is feasible but not optimal. The dual simplex method requires just such an initial solution. It is also useful for resolving a linear program after a constraint is added or a right-hand side is changed. Rather than solve the revised problem from scratch, we would like to use the old optimal solution as the initial solution for the revised problem. These changes can make the old solution infeasible, so that it cannot be used to initialize the simplex method. However, they do not affect dual feasibility, so the dual simplex algorithm can be started from the old solution.

We take the following approach to apply the simplex method to (11.2). As usual, we subtract surplus variables. Instead of partitioning into basic and nonbasic variables, use a basis for the primal problem to partition the *constraints* into basic and nonbasic:

$$
\begin{aligned}
\min \quad & \mathbf{b}^T \mathbf{y} \\
\text{s.t.} \quad & \mathbf{B}^T \mathbf{y} - \mathbf{w}_B = \mathbf{c}_B \\
& \mathbf{N}^T \mathbf{y} - \mathbf{w}_N = \mathbf{c}_N \\
& \mathbf{w}_B, \mathbf{w}_N \geq \mathbf{0}.
\end{aligned}
\tag{11.6}
$$

Note that the variables in $\mathbf{w}_B$ are ordered according to the primal basis, so $w_{B(k)}$ corresponds to $x_{B(k)}$, the kth component of $\mathbf{x}_B$. Using (11.3), $\mathbf{w}_B = \mathbf{0}$. The m nonbasic variables are those in $\mathbf{w}_B$; the remaining n variables in $\mathbf{y}$ and $\mathbf{w}_N$ are basic. Note that $\mathbf{y}$ is always basic. That is possible because $\mathbf{y}$ is unrestricted in sign. We will modify the simplex method to keep $\mathbf{y}$ in the basis. At this basic solution, feasibility for the dual problem (11.6) is again optimality for the primal problem: $\bar{\mathbf{c}}_N^T = -\mathbf{w}_N \leq \mathbf{0}$.

To write the simplex equations for (11.6), let $\hat{\mathbf{A}}$ be its coefficient matrix with the variables ordered $(\mathbf{y}, \mathbf{w}_B, \mathbf{w}_N)$, $\hat{\mathbf{b}}$ its right-hand side, and $\hat{\mathbf{c}}$ its objective function coefficients:

$$
\hat{\mathbf{c}} = \begin{bmatrix} \mathbf{b} \\ \mathbf{0} \\ \mathbf{0} \end{bmatrix}, \quad \hat{\mathbf{A}} = \begin{bmatrix} \mathbf{B}^T & -\mathbf{I} & \mathbf{0} \\ \mathbf{N}^T & \mathbf{0} & -\mathbf{I} \end{bmatrix}, \quad \hat{\mathbf{b}} = \begin{bmatrix} \mathbf{c}_B \\ \mathbf{c}_N \end{bmatrix}.
\tag{11.7}
$$

The naive simplex method computes the following.

1. The basis matrix and its inverse:

$$
\hat{\mathbf{B}} = \begin{bmatrix} \mathbf{B}^T & \mathbf{0} \\ \mathbf{N}^T & -\mathbf{I} \end{bmatrix}, \quad \hat{\mathbf{B}}^{-1} = \begin{bmatrix} \mathbf{B}^{-T} & \mathbf{0} \\ \mathbf{N}^T \mathbf{B}^{-T} & -\mathbf{I} \end{bmatrix}.
\tag{11.8}
$$

2. The reduced costs for the nonbasic variables. We omit the "hat" notation since it is clear these are for the dual:

$$\bar{\mathbf{c}}_{w_B}^T = \mathbf{0}^T - \begin{bmatrix} \mathbf{b}^T & \mathbf{0}^T \end{bmatrix} \begin{bmatrix} \mathbf{B}^{-T} & \mathbf{0} \\ \mathbf{N}^T\mathbf{B}^{-T} & -\mathbf{I} \end{bmatrix} \begin{bmatrix} -\mathbf{I} \\ \mathbf{0} \end{bmatrix}$$

$$= -\begin{bmatrix} \mathbf{b}^T & \mathbf{0}^T \end{bmatrix} \begin{bmatrix} -\mathbf{B}^{-T} \\ -\mathbf{N}^T\mathbf{B}^{-T} \end{bmatrix}$$

$$= \mathbf{b}^T\mathbf{B}^{-T}$$

$$= (\mathbf{B}^{-1}\mathbf{b})^T$$

$$= \mathbf{x}_B^T.$$

Thus, the reduced costs of the dual problem are the basic variables of the primal problem. This shows that the dual optimality condition, $\bar{\mathbf{c}}_{w_B}^T \geq \mathbf{0}$ (remember it is a minimization), is equivalent to the primal feasibility condition.

3. The basic components $(\mathbf{d}_y, \mathbf{d}_{w_N})$ of the simplex direction for the entering variable $w_{B(l)}$. This notation refers to the lth component of $\mathbf{w}_B$. The column of $\hat{\mathbf{A}}$ for $w_{B(l)}$ is $(-\mathbf{e}_l, \mathbf{0})$ and

$$\begin{bmatrix} \mathbf{d}_y \\ \mathbf{d}_{w_N} \end{bmatrix} = -\hat{\mathbf{B}}^{-1}\hat{\mathbf{a}}_{w_{B(l)}} = -\begin{bmatrix} \mathbf{B}^{-T} & \mathbf{0} \\ \mathbf{N}^T\mathbf{B}^{-T} & -\mathbf{I} \end{bmatrix} \begin{bmatrix} -\mathbf{e}_l \\ \mathbf{0} \end{bmatrix} = \begin{bmatrix} \mathbf{B}^{-T}\mathbf{e}_l \\ \mathbf{N}^T\mathbf{B}^{-T}\mathbf{e}_l \end{bmatrix}.$$

Because $\mathbf{y}$ is allowed to be negative and never leaves the basis, we only need $\mathbf{d}_{w_N} = \mathbf{N}^T\mathbf{B}^{-T}\mathbf{e}_l$ to find the leaving variable. This formula is sometimes written $\mathbf{d}_{w_N}^T = \mathbf{e}_l^T\mathbf{B}^{-1}\mathbf{N}$ to make clear that it is row l of $\mathbf{B}^{-1}\mathbf{N}$. We know that moving in the simplex direction increases the entering variable $w_{B(l)}$ in the dual (the problem we are using the simplex method on), but what does it do to the primal solution? The entering variable must have a negative reduced cost. As shown earlier, it equals the corresponding primal variable, so $x_{B(l)} < 0$. After updating the basis, this reduced cost must be zero, so $x_{B(l)} = 0$. Thus, an iteration always increases a primal basic variable to zero, changing it from infeasible to feasible and from basic to nonbasic. That is why we used the index l, which stands for "leaving": $x_{B(l)}$ leaves the primal basis and $w_{B(l)}$ enters the dual basis.

4. The ratio test is $\lambda = \min\left\{ \frac{\bar{c}_{N(j)}}{d_j} : d_j < 0, \quad j \in \mathcal{N} \right\}$, where $\mathbf{d} = \mathbf{d}_{w_N}$ and we have used $\bar{\mathbf{c}}_N = \mathbf{w}_N$. Using the ratio test preserves dual feasibility, so $\bar{\mathbf{c}}_N \leq \mathbf{0}$ for a minimization problem. Because $\bar{\mathbf{c}}_N \leq \mathbf{0}$, these ratios are nonnegative. Note that only the basic variables $\mathbf{w}_N$ are considered because $\mathbf{y}$ remains basic and that $\mathbf{w}_N$ corresponds to the nonbasic variables in the primal problem. If $\mathbf{d} > \mathbf{0}$, then it is an unbounded improving direction for the dual problem, showing that the primal problem is infeasible.

The aforementioned equations allow us to apply the simplex method, adapted to handle unrestricted in sign variables, to the dual (11.2) with all computations expressed in terms of the primal problem. Using $\mathbf{B}^{-1}$ from the primal problem, not the dual, is important because the primal problem often has more variables than constraints ($n > m$), so it has a smaller basis matrix: $\mathbf{B}$ is $m \times m$, while (11.6) has n basic variables and $\hat{\mathbf{B}}$ is $n \times n$. Thus, the dual simplex method is quite different computationally from applying the simplex method to the dual.

Here is a summary of the algorithm for a problem in canonical form.

Algorithm 11.1 (Dual Simplex Method)

Step 0. Initialization. Find a basis matrix $\mathbf{B}$ for which the corresponding basic solution of the dual problem is feasible, i.e. $\bar{\mathbf{c}}_N^T \leq \mathbf{0}$.

Step 1. Primal optimality check. Compute the primal solution $\mathbf{x}_B = \mathbf{B}^{-1}\mathbf{b}$. If $\mathbf{x}_B \geq \mathbf{0}$, stop. The current solution is optimal.

Step 2. Compute direction. Otherwise, choose a most negative basic variable $x_{B(l)}$ as the leaving variable. Compute the simplex direction for the corresponding dual nonbasic variable: $\mathbf{d}^T = \mathbf{e}_l^T \mathbf{B}^{-1} \mathbf{N}$. This vector contains a component for each basic surplus variable in the dual.

Step 3. Perform ratio test. If $\mathbf{d} \geq \mathbf{0}$, stop. The linear program is infeasible. Otherwise, compute the reduced costs $\bar{\mathbf{c}}_N^T = \mathbf{c}_N^T - \mathbf{c}_B^T \mathbf{B}^{-1} \mathbf{N}$. The entering variable x_e is one that achieves the minimum ratio in $\left\{ \frac{\bar{c}_{N(j)}}{d_j} : d_j < 0, \quad j \in \mathcal{N} \right\}$. Since $\bar{\mathbf{c}}_N \leq \mathbf{0}$, these are nonnegative.

Step 4. Update basis. Replace x_l by x_e in the basis. Go to Step 1.

The algorithm is stated with a "most negative" rule for choosing the leaving variable. Other rules could be used. The dual solution $\mathbf{y}^T = \mathbf{c}_B^T \mathbf{B}^{-1}$ need not be stored, but if the optimal dual solution is needed for sensitivity analysis it is easily available from the other computations. Most refinements of the simplex algorithm mentioned in Section 9.4 can be used in the dual simplex method, including the revised simplex and full tableau methods.

The argument that the dual simplex method is correct and terminates in a finite number of iterations is essentially that it visits improving bfs's of the dual. Step 2 chooses an entering dual variable with negative reduced cost, and therefore a decreasing simplex direction. By (11.4), the objective function values of the primal and dual are equal at corresponding basic solutions. Hence, moving in this direction also decreases the primal value, so it decreases (or remains the same) at each iteration. If we stop the algorithm before it reaches optimality, the current solution is an upper bound on the optimal value: the algorithm starts with a "superoptimal" solution and the objective function decreases to the optimal value.

If there is degeneracy, a modification is needed to prevent cycling and guarantee that the algorithm terminates in a finite number of iterations. The step size, or minimum ratio, can only be zero if one of the nonbasic reduced costs is zero. Since basic reduced costs are always zero, that means more than m reduced costs are zero. At a basic solution, a reduced cost of zero in the primal problem is equivalent to an active constraint in the inequality form (11.2) of the dual problem. Now, (11.2) has m variables, so the dual solution is degenerate when there are more than m active constraints. Thus, the step size can only be zero if the dual bfs at some iteration is degenerate. The simplex method can stay at a degenerate primal bfs for multiple iterations; now we see that the dual simplex method can stay at a degenerate dual bfs for multiple iterations. There are simple rules for tie-breaking when choosing the entering variable that prevent cycling; see, e.g. (Bertsimas and Tsitsiklis, 1997, Section 4.5).

Example 11.2. We will solve the problem in Example 11.1 using the dual simplex method on the canonical form.

Initialization. Using the initial basis (s_1, s_2), $\mathbf{B} = -\mathbf{I}$ and the solution is the negative of the right-hand side: $(\mathbf{x}, \mathbf{s}) = (0, 0, 0, -24, -18)$. For this basis, we have

$$\mathbf{N} = \begin{bmatrix} 2 & -4 & 2 \\ 2 & 1 & -1 \end{bmatrix}, \quad \bar{\mathbf{c}}_N^T = \mathbf{c}_N^T = \begin{bmatrix} -2 & -4 & -1 \end{bmatrix}.$$

Iteration 1. The leaving variable is s_1 because it is most negative so we set $l = 1$, denoting the first variable in the basis. Its simplex direction is

$$\mathbf{d}^T = \begin{bmatrix} 1 & 0 \end{bmatrix} (-\mathbf{I}) \begin{bmatrix} 2 & -4 & 2 \\ 2 & 1 & -1 \end{bmatrix} = \begin{bmatrix} -2 & 4 & -2 \end{bmatrix}.$$

The ratio test compares $\left\{ x_1 : \frac{-2}{-2}, x_3 : \frac{-1}{-2} \right\}$. The minimum ratio is for x_3, so it is the entering variable. The new basis is (x_3, s_2) with

$$\mathbf{B} = \begin{bmatrix} 2 & 0 \\ -1 & -1 \end{bmatrix}, \quad \mathbf{N} = \begin{bmatrix} 2 & -4 & -1 \\ 2 & 1 & 0 \end{bmatrix}.$$

Iteration 2. First we compute

$$\mathbf{B}^{-1} = \begin{bmatrix} \frac{1}{2} & 0 \\ -\frac{1}{2} & -1 \end{bmatrix}, \quad \mathbf{x}_B = \begin{bmatrix} \frac{1}{2} & 0 \\ -\frac{1}{2} & -1 \end{bmatrix} \begin{bmatrix} 24 \\ 18 \end{bmatrix} = \begin{bmatrix} 12 \\ -30 \end{bmatrix},$$

$$\mathbf{B}^{-1}\mathbf{N} = \begin{bmatrix} \frac{1}{2} & 0 \\ -\frac{1}{2} & -1 \end{bmatrix} \begin{bmatrix} 2 & -4 & -1 \\ 2 & 1 & 0 \end{bmatrix} = \begin{bmatrix} 1 & -2 & -\frac{1}{2} \\ -3 & 1 & \frac{1}{2} \end{bmatrix}.$$

Since $s_2 = -30 < 0$, the current solution is not optimal. The leaving variable is s_2 and it is the second variable in the basis, so its simplex direction is row 2 of $\mathbf{B}^{-1}\mathbf{N}$: $\mathbf{d} = (-3, 1, \frac{1}{2})$. Using $\mathbf{B}^{-1}\mathbf{N}$, the reduced costs are

$$\bar{\mathbf{c}}_N^T = \begin{bmatrix} -2 & -4 & 0 \end{bmatrix} - \begin{bmatrix} -1 & 0 \end{bmatrix} \begin{bmatrix} 1 & -2 & -\frac{1}{2} \\ -3 & 1 & \frac{1}{2} \end{bmatrix} = \begin{bmatrix} -1 & -6 & -\frac{1}{2} \end{bmatrix}.$$

The only ratio is $\frac{-1}{-3}$ for x_1, so it is the entering variable. The new basis is (x_1, x_3) with

$$\mathbf{B} = \begin{bmatrix} 2 & 2 \\ 2 & -1 \end{bmatrix}, \quad \mathbf{N} = \begin{bmatrix} -4 & -1 & 0 \\ 1 & 0 & -1 \end{bmatrix}.$$

Iteration 3. We compute

$$\mathbf{B}^{-1} = \begin{bmatrix} \frac{1}{6} & \frac{1}{3} \\ \frac{1}{3} & -\frac{1}{3} \end{bmatrix}, \quad \mathbf{x}_B = \begin{bmatrix} \frac{1}{6} & \frac{1}{3} \\ \frac{1}{3} & -\frac{1}{3} \end{bmatrix} \begin{bmatrix} 24 \\ 18 \end{bmatrix} = \begin{bmatrix} 10 \\ 2 \end{bmatrix},$$

which is nonnegative, so the solution $\mathbf{x} = (10, 0, 2)$ with value 22 (or -22 in canonical form) is optimal. The optimal dual solution is

$$\mathbf{y}^T = \begin{bmatrix} -2 & -1 \end{bmatrix} \begin{bmatrix} \frac{1}{6} & \frac{1}{3} \\ \frac{1}{3} & -\frac{1}{3} \end{bmatrix} = \begin{bmatrix} -\frac{2}{3} & -\frac{1}{3} \end{bmatrix}.$$

We did not need to construct the dual to apply the dual simplex method. As an aside, we can interpret the direction used in the algorithm as a dual simplex direction. Adding surplus variables to (11.5), the dual is

$$\begin{aligned} \min \quad & 24y_1 + 18y_2 \\ \text{s.t.} \quad & 2y_1 + 2y_2 - w_1 = -2 \\ & -4y_1 + y_2 - w_2 = -4 \\ & 2y_1 - y_2 - w_3 = -1 \\ & -y_1 - w_4 = 0 \\ & -y_2 - w_5 = 0 \\ & w_i \geq 0. \end{aligned}$$

In the first iteration, the direction was $\mathbf{d}_{w_N} = (-2, 4, -2)$. If we also compute

$$\mathbf{d}_y^T = \mathbf{e}_i^T \mathbf{B}^{-1} = \begin{bmatrix} 1 & 0 \end{bmatrix} (-\mathbf{I}) = \begin{bmatrix} -1 & 0 \end{bmatrix},$$

and use $d_{w_4} = 1$ (because w_4 is the entering dual variable) and $d_{w_5} = 0$ for the nonbasic dual variables, then $(\mathbf{d}_y, \mathbf{d}_w) = (-1, 0, -2, 4, -2, 1, 0)$. In the dual variable space, this direction moves from the current dual bfs $(\mathbf{y}, \mathbf{w}) = (0, 0, 2, 4, 1, 0, 0)$ to an adjacent bfs.

The next example shows how the dual simplex algorithm can be used to resolve a problem when a constraint is added.

Example 11.3. In Example 9.1, the optimal solution of the linear program

$$
\begin{array}{rl}
\max & 4x_1 + 3x_2 \\
\text{s.t. } & 1.5x_1 + x_2 + s_1 &= 16 \\
& x_1 + x_2 + s_2 &= 12 \\
& x_2 + s_3 = 10 \\
& x_j, s_i \geq 0
\end{array}
$$

was found to be $\mathbf{x}^* = (8, 4, 0, 0, 6)$ with dual optimal solution $\mathbf{y}^* = (2, 1, 0)$. Suppose the constraint $2x_1 + x_2 \leq 18$ is added. Note that $\mathbf{x}^*$ violates this constraint. Adding a slack variable, the new problem is

$$
\begin{array}{rl}
\max & 4x_1 + 3x_2 \\
\text{s.t. } & 1.5x_1 + x_2 + s_1 &= 16 \\
& x_1 + x_2 + s_2 &= 12 \\
& x_2 + s_3 = 10 \\
& 2x_1 + x_2 + s_4 = 18 \\
& x_j, s_i \geq 0.
\end{array}
$$

Augmenting the old solution $(\mathbf{x}^*, s_4) = (8, 4, 0, 0, 6, -2)$ is infeasible. However, $\mathbf{y}^*$ is still feasible for the dual, so the dual simplex method can be used to find a new optimal solution, starting with $\mathbf{x}^*$.

Similarly, if the right-hand side of constraint 1 or 2 is reduced, then $\mathbf{x}^*$ is no longer feasible but $\mathbf{y}^*$ is still feasible for the dual. In this case, one could use sensitivity analysis to check if the same basis was still optimal. If not, then the dual simplex method could be used.

11.2 Network Simplex Method

This section presents a streamlined version of the simplex method for the minimum cost network flow problem of Section 2.6. It was developed by Dantzig (1951). The simplex method is already efficient on this special class of linear program. However, for truly large applications and some real-time applications, it is important to have a more efficient algorithm. There are other efficient algorithms for network flow and related problems that we do not mention; see Ahuja et al. (1993).

We present the algorithm, called the network simplex method, by specializing the computations in the simplex method. The algorithm can also be developed as an iterative search method without referring to the simplex method. As in Section 2.6, consider a directed graph on the set $\mathcal{V}$ of nodes, or vertices, and the set $\mathcal{A}$ of arcs. Each node has an external supply b_i, which is positive at supply nodes and negative at demand nodes. Each arc (i,j) has a cost c_{ij} per unit of flow from node i to node j. Let x_{ij} be the amount of flow through arc (i,j). The minimum cost network flow problem is

$$\min \sum_{(i,j) \in \mathcal{A}} c_{ij} x_{ij}$$

$$\text{s.t.} \sum_{j:(i,j) \in \mathcal{A}} x_{ij} - \sum_{k:(k,i) \in \mathcal{A}} x_{ki} = b_i, \ i \in \mathcal{V} \tag{11.9}$$

$$x_{ij} \geq 0.$$

A solution **x** is called a flow; a feasible solution is a feasible flow.

We will need the following definitions regarding graphs. All of them ignore the direction of arcs; they are properties of *undirected* graphs. An arc (i,j) or (j,i) is called an *edge* between i and j in the undirected graph. A *path* from node i_1 to node i_n is a sequence of nodes $i_1, i_2, \dots, i_n$ connected by edges, i.e. $(i_k, i_{k-1}) \in \mathcal{V}$ or $(i_{k-1}, i_k) \in \mathcal{V}$. We do not allow nodes to be repeated. A *cycle* is a path that starts and ends at the same node, i.e. $i_1 = i_n$. A graph is *connected* if there is a path between every pair of nodes. A *tree* is a graph that is connected and has no cycles. We will form a tree by selecting some of the edges in a graph (and the nodes in these edges). If this tree contains all the nodes of the graph, it is a *spanning tree* of the graph. Figure 11.1 shows a spanning tree. Note that the tree has six nodes and five edges. If an edge is removed, it is not connected. If an edge is added, say, $(3,4)$, a single cycle $1, 4, 3, 1$ is created. In general, a tree has $|\mathcal{V}| - 1$ edges.

We make the following assumptions.

1. Total supply equals total demand, so that

$$\sum_{i \in \mathcal{V}} b_i = 0.$$

Problems meeting this assumption are called *balanced*. For the flow balance equations (11.9) to hold, a problem must be balanced. Section 2.6 explains how to convert a problem with excess supply into a balanced problem.
2. The graph is connected.
3. There are no reverse arcs: if $(i,j) \in \mathcal{A}$ then $(j,i) \notin \mathcal{A}$.

11.2.1 Node-arc Incidence Matrix

The special structure of a linear program that makes it a minimum cost network flow can be described if we write (11.9) in matrix form. Let $\mathcal{V} = \{1, \dots, m\}$

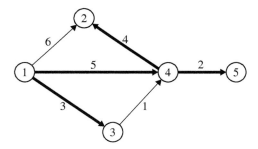

Figure 11.1 Heavy lines are a spanning tree of the directed graph.

and n be the number of arcs. The network in Figure 11.1 has the arcs $A = \{(1,2),(1,3),(1,4),(3,4),(4,2),(4,5)\}$ with $m = 5$ and $n = 6$. The flow balance equation (11.9) for this network is

$$
\begin{aligned}
x_{12} + x_{13} + x_{14} &= b_1 \\
-x_{12} \qquad\qquad\qquad\quad - x_{42} &= b_2 \\
- x_{13} \qquad + x_{34} &= b_3 \\
- x_{14} - x_{34} + x_{42} + x_{45} &= b_4 \\
- x_{45} &= b_5.
\end{aligned}
$$

The coefficient matrix is

$$
\mathbf{A} = \begin{bmatrix}
1 & 1 & 1 & 0 & 0 & 0 \\
-1 & 0 & 0 & 0 & -1 & 0 \\
0 & -1 & 0 & 1 & 0 & 0 \\
0 & 0 & -1 & -1 & 1 & 1 \\
0 & 0 & 0 & 0 & 0 & -1
\end{bmatrix}. \tag{11.10}
$$

There is a row for each node and a column for each arc, with the arcs in the order listed earlier. Each column has exactly two nonzero entries, $+1$ for the node where the arc starts and -1 for the node where the arc ends. Each row has $+1$ for each arc starting at that node and -1 for each arc ending at that node. We call such a matrix the *node-arc incidence matrix* of the network. In general, numbering the arcs $k = 1, \ldots, n$, it is an $m \times n$ matrix with elements

$$
a_{ik} = \begin{cases}
1, & \text{if } i \text{ is the start node of arc } k \\
-1, & \text{if } i \text{ is the end node of arc } k \\
0, & \text{otherwise.}
\end{cases}
$$

Also define n-vectors $\mathbf{x}$ and $\mathbf{c}$ containing the arc flows and unit costs and $\mathbf{b} = (b_1, \ldots, b_m)$ containing the external supply. Then (11.9) can be written

$$
\begin{aligned}
\min \quad & \mathbf{c}^T \mathbf{x} \\
\text{s.t.} \quad & \mathbf{A}\mathbf{x} = \mathbf{b} \\
& \mathbf{x} \geq \mathbf{0}.
\end{aligned} \tag{11.11}
$$

It is the structure of **A**, where each column has exactly two nonzero entries, $+1$ and -1, that describes this class of problem. Any linear program with this structure is called a minimum cost network flow problem.

11.2.2 Tree Solutions

The key idea behind the algorithm is that basic solutions correspond to flows that only use the arcs in a spanning tree. In Figure 11.1, the spanning tree consists of the arcs $(1, 3), (1, 4), (4, 2), (4, 5)$. How can we find the basic solution for this tree? The number of basic variables is the number of linearly independent constraints, or rows of **A**. If we add all of the rows of **A**, we get the zero vector, implying that they are not all linearly independent. It turns out that, for any connected graph, if any one row is eliminated, the remaining rows are linearly independent. For example, if we eliminate the last row of (11.10), the remaining rows

$$\tilde{\mathbf{A}} = \begin{bmatrix} 1 & 1 & 1 & 0 & 0 & 0 \\ -1 & 0 & 0 & 0 & -1 & 0 \\ 0 & -1 & 0 & 1 & 0 & 0 \\ 0 & 0 & -1 & -1 & 1 & 1 \end{bmatrix}$$

are linearly independent. In general, if $\tilde{\mathbf{A}}$ and $\tilde{\mathbf{b}}$ are formed by eliminating the last row of **A** and **b**, then we can use the linearly independent constraints

$$\tilde{\mathbf{A}}\mathbf{x} = \tilde{\mathbf{b}}.$$

Now if we choose the columns of $\tilde{\mathbf{A}}$ corresponding to the arcs in the spanning tree of Figure 11.1, we get

$$\mathbf{B} = \begin{bmatrix} 1 & 1 & 0 & 0 \\ 0 & 0 & -1 & 0 \\ -1 & 0 & 0 & 0 \\ 0 & -1 & 1 & 1 \end{bmatrix}.$$

These columns are linearly independent, so **B** is a basis matrix. The basic variables are the flows on the arcs of the spanning tree. Note that the number of arcs in the spanning tree and the number of linearly independent constraints are both $m - 1 = 4$. They match because we removed one constraint. In fact, these dimensions will always match, giving an $(m - 1) \times (m - 1)$ matrix **B**, because a spanning tree in a graph of m nodes has $m - 1$ edges.

Lemma 11.1: *The incidence matrix **A** of a connected directed graph with m nodes has the following properties.*

*(i) rank(**A**) = $m - 1$.*

(ii) An $(m-1) \times (m-1)$ submatrix **B** *has full rank if and only if the columns of* **B** *correspond to a spanning tree.*

Proof: For a proof, see, e.g. Bertsimas and Tsitsiklis (1997, Theorem 7.4). □

A result of the lemma is that the basic solutions of a network flow problem are the spanning tree solutions. To find a basic solution, choose a spanning tree, eliminate one row of **A** (we will eliminate the last row of **A**), and solve $\mathbf{Bx} = \tilde{\mathbf{b}}$.

An iteration of the simplex method starts with a bfs, which is spanning tree solution with $\mathbf{x} \geq \mathbf{0}$. It then adds a variable to the basis and removes another variable. Since the variables correspond to a spanning tree, adding a variable adds an arc and forms one cycle. Removing any other arc in that cycle forms a new spanning tree, giving a new basis. However, to preserve feasibility, we must remove an arc for which the new solution is nonnegative. There is an easy way to find an arc to remove.

For the network in Figure 11.1, suppose the external supplies are as shown in Figure 11.2. The solution for the first spanning tree is shown in Figure 11.3a. Suppose we add the arc $(1, 2)$, creating the cycle $1, 4, 2, 1$. The only way to preserve flow balance is to send an additional flow around this cycle. The direction of the entering arc determines the direction around the cycle, since otherwise its flow would become negative. Thus, the additional flow is in the direction $1, 2, 4, 1$. We send enough flow around the cycle to change one of the flows to zero. Only arcs in the reverse direction in the cycle will decrease; in this example, $(1, 4)$ and $(4, 2)$ are reverse arcs. The minimum of their flows is 10. Thus, we send 10 units around the cycle, adding 10 to the forward arc $(1, 2)$ and subtracting 10 from the reverse arcs. The new spanning tree and flow are shown in Figure 11.3b. A general formula for the amount of flow to send around the cycle is

$$\lambda = \min_{(k,l) \in \mathcal{R}} x_{kl}, \tag{11.12}$$

where $\mathcal{R}$ is the set of arcs in the reverse direction from (i, j) in the cycle.

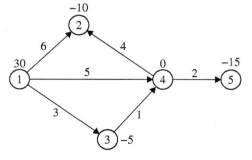

Figure 11.2 A network flow problem for Figure 11.1.

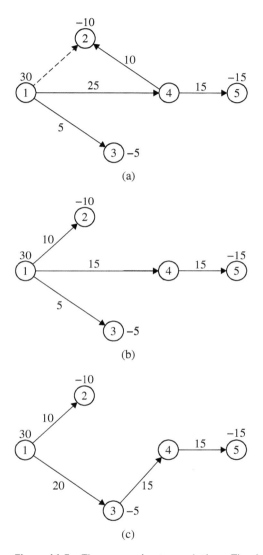

Figure 11.3 Three spanning tree solutions. The dotted line is an entering arc.

11.2.3 Network Simplex

The simplex method also requires the reduced costs of the nonbasic variables. We will use duality theory, which interprets a reduced cost as the slack in the corresponding dual constraint. The dual of (11.9) is

$$\max \sum_{i=1}^{m} b_i y_i$$

$$\text{s.t. } y_i - y_j \le c_{ij}, \ (i,j) \in \mathcal{A} \tag{11.13}$$

$$y_i \text{ u.r.s.}$$

The dual has a variable y_i for each node and a constraint for each arc. The two nonzero entries in each column of **A** become the two terms in each dual constraint, a positive term for the start node of the arc and a negative term for the end node. Given a basis, or spanning tree, $\mathcal{B}$ for the primal problem, we compute the corresponding dual solution as follows. The corresponding dual constraints are active:

$$y_i - y_j = c_{ij}, \ (i,j) \in \mathcal{B}. \tag{11.14}$$

However, there are only $|\mathcal{B}| = m - 1$ equations and there are m dual variables. We could create a dual with $m - 1$ dual variables by using the linearly independent primal constraints $\tilde{\mathbf{A}}\mathbf{x} = \tilde{\mathbf{b}}$ when forming the dual. However, it is equivalent to set one of the dual variables to a fixed value, say, $y_n = 0$, and solve for the rest – which is very easy. Given the dual solution, the reduced costs are

$$\bar{c}_{ij} = c_{ij} - (y_i - y_j). \tag{11.15}$$

Now we have all the elements of the simplex method.

Algorithm 11.2 (Simplex Method for Network Flow Problems)

Step 0. Initialization. Identify a bfs **x** and its basis $\mathcal{B}$, which is a spanning tree of the network.

Step 1. Dual solution. Compute the dual solution **y** by setting $y_n = 0$ and using (11.14).

Step 2. Optimality check and entering variable. Compute the reduced costs for all nonbasic arcs using (11.15). If all reduced costs are nonnegative, stop. The current solution is optimal. Otherwise, choose a nonbasic arc (i,j) with the most negative reduced cost as the entering variable.

Step 3. Update the flow. The entering arc (i,j) combines with the basic arcs to form one cycle. If all arcs on this cycle are oriented the same direction as (i,j), stop. The problem is unbounded. Otherwise, update **x** by sending a flow of λ, given by (11.12), around the cycle.

Step 4. Leaving variable. At least one basic variable in the cycle will have a new value of zero. Choose one of these as the leaving variable. Update the basis and go to Step 1.

Step 3 allows for a problem to be unbounded. However, a network flow problem is only unbounded if it contains a cycle with all arcs oriented in the same direction and the sum of the arc costs on the cycle is negative.

The algorithm requires an initial bfs, which will be a feasible spanning tree solution. For some types of network flow problems, including the transportation problem, it is easy to find an initial bfs. For problems where it is difficult to find a bfs, the problem can be modified so that it is easy to find one. Call arcs starting at a supply node and ending at a demand node *direct* arcs. We add auxiliary arcs from each supply node to each demand node (unless this arc already exists), so that the auxiliary graph has all possible direct arcs. The subgraph of direct arcs has the form of a transportation problem, so it is easy to find a bfs on this subgraph. Then we assign large costs to the auxiliary arcs and apply the algorithm above to the auxiliary graph. If the optimal solution has zero flow on all auxiliary arcs, it is an optimal solution to the original problem. However, if it has a nonzero auxiliary flow, the original problem is infeasible.

What could make a network flow problem infeasible? We have already assumed that supply is balanced with demand and the graph is connected. However, because it is a directed graph, not all supply points can send to a demand point. The example in Figure 11.4 is infeasible because node 4 cannot supply node 3. Also, if we add capacity constraints to the arcs, they could change a problem from feasible to infeasible.

Example 11.4. Consider the network and arc costs in Figure 11.1 with the external supplies $\mathbf{b} = (30, -10, -5, 0, -15)$ as shown in Figure 11.3. We use the initial bfs shown in Figure 11.3a. To find the dual solution, start with $y_5 = 0$. Then

$$y_4 = y_5 + c_{45} = 0 + 2 = 2,$$
$$y_2 = y_4 - c_{42} = 2 - 4 = -2,$$
$$y_1 = y_4 + c_{14} = 2 + 5 = 7,$$
$$y_3 = y_1 - c_{13} = 7 - 3 = 4.$$

Figure 11.4 An infeasible network flow problem.

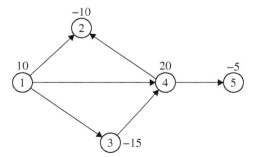

Note that only arcs in the spanning tree are used. The nonbasic reduced costs are

$$\bar{c}_{12} = c_{12} - (y_1 - y_2) = 6 - (7 - (-2)) = -3,$$
$$\bar{c}_{34} = c_{34} - (y_3 - y_4) = 1 - (4 - 2) = -1.$$

The entering arc is $(1, 2)$, shown by the dotted line in Figure 11.3a. The cycle $1, 2, 4, 1$ has arcs $(1, 4)$ and $(4, 2)$ in the reverse direction, so $\lambda = \min\{x_{14}, x_{42}\} = \min\{25, 10\} = 10$. Sending a flow of 10 around the cycle results in arc $(4, 2)$ having zero flow and leaving the basis, giving the new spanning tree and flow of Figure 11.3b.

To begin the second iteration, compute $\mathbf{y} = (7, 1, 4, 2, 0)$ and

$$\bar{c}_{34} = 1 - (4 - 2) = -1,$$
$$\bar{c}_{42} = 4 - (2 - 1) = 3.$$

The entering arc is $(3, 4)$. In the cycle $1, 3, 4, 1$ only arc $(1, 4)$ is in the reverse direction, so $\lambda = x_{14} = 15$. Sending a flow of 15 around the cycle results in the flows and new spanning tree shown in Figure 11.3c.

Start the third iteration by computing $\mathbf{y} = (6, 0, 3, 2, 0)$ and

$$\bar{c}_{14} = 5 - (6 - 2) = 1,$$
$$\bar{c}_{42} = 4 - (2 - 0) = 2.$$

Since the reduced costs are nonnegative, the flow in Figure 11.3c is optimal.

11.2.4 Transportation Problem

The transportation problem described in Section 2.6 is an important special case. In fact, any network flow problem can be transformed into a larger transportation problem (see (Ahuja et al., 1993)). When one applies the network flow simplex algorithm to a transportation problems, some simplifications are possible. First, there is a very easy method to find an initial bfs. There is also a simpler way to keep track of the calculations at each iteration. The basic idea is that, because the arcs are from supply nodes to demand nodes, we can use a matrix with rows for supply nodes and columns for demand nodes to represent the flow $\mathbf{x}$ and the reduced costs.

It is convenient to number the supply nodes $1, \ldots, m$ and the demand nodes $1, \ldots, n$. The costs and flows on arcs can be recorded in matrices with elements c_{ij} and x_{ij} for supply node i and demand node j. The correct terminology for, say, the arc from supply node 2 to demand node 1 would be (S_2, D_1) if we use the unique node names

$$\mathcal{N} = \{S_1, \ldots, S_m, D_1, \ldots, D_n\}.$$

Figure 11.5 A transportation problem with three supply nodes and two demand nodes.

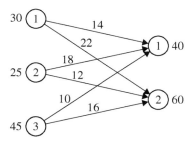

We will follow convention and refer to this arc as $(2, 1)$, but should note that this notation only applies when we have already partitioned the nodes into two subsets and is not standard graph notation. Such graphs are called bipartite. Each supply node has a supply $s_i > 0$; each demand node has a demand $d_j > 0$. With these definitions, the transportation problem is

$$\min \sum_{i=1}^{m} \sum_{j=1}^{n} c_{ij} x_{ij}$$

$$\text{s.t.} \sum_{j=1}^{n} x_{ij} \leq s_i, \ i = 1, \dots, m \qquad (11.16)$$

$$\sum_{i=1}^{m} x_{ij} \geq d_j, \ j = 1, \dots, n$$

$$x_{ij} \geq 0.$$

The assumption that the network is balanced implies that $\sum_{i=1}^{m} s_i = \sum_{j=1}^{n} d_j$. For example, Figure 11.5 shows a transportation problem with three supply nodes and two demand nodes where total supply and total demand are balanced at 100. As discussed in Section 2.6, problems where total supply exceeds total demand can be converted to a balanced problem by adding a dummy demand node.

Here is a greedy constructive algorithm for finding an initial bfs, called the *minimum cost method*. At each iteration, choose a remaining arc with minimal cost c_{ij}. Set $x_{ij} = \min\{s_i, d_j\}$, which is the maximum flow that can be assigned and be part of a feasible solution. Reduce s_i and d_j by x_{ij}, so that they represent the remaining unused supply and unmet demand. At least one will be zero. If $s_i = 0$, delete all remaining arcs starting at i; otherwise, delete all remaining arcs ending at j. Either way, we have deleted all arcs at one node. Repeat until there are no arcs remaining.

Example 11.5. Consider the transportation problem in Figure 11.5. We will find an initial bfs using the minimum cost method.

1. Choose the least expensive arc $(3, 1)$ and set $x_{31} = \min\{s_3, d_1\} = 40$. Update $s_3 = 45 - 40 = 5$ and $d_1 = 40 - 40 = 0$. Delete arcs $(1, 1)$, $(2, 1)$, and $(3, 1)$ because they end at demand node 1.
2. The remaining arcs are $(1, 2)$, $(2, 2)$, and $(3, 2)$. Choose $(2, 2)$ and set $x_{22} = \min\{s_2, d_2\} = 25$. Update $s_2 = 25 - 25 = 0$ and $d_2 = 60 - 25 = 35$. Delete arc $(2, 2)$, the only remaining arc starting at supply node 2.
3. Choose $(3, 2)$ and set $x_{32} = \min\{s_3, d_2\} = 5$. Update $s_3 = 5 - 5 = 0$ and $d_2 = 35 - 5 = 30$. Delete arc $(3, 2)$.
4. Choose $(1, 2)$ and set $x_{12} = \min\{s_1, d_2\} = 30$. Delete this arc. There are no arcs remaining.

The bfs found is shown in Figure 11.6.

At each step, the minimum cost method deletes the remaining arcs incident to one node, except for the last step, which does this for two nodes: the supply and demand nodes for the selected arc. Thus, it selects $m + n - 1$ arcs, which is the correct number of basic arcs (one less than the number of nodes). However, some of these basic arcs could have flows of zero, indicating a degenerate bfs.

Now we describe how to adapt Algorithm 11.2 for transportation problems. Let u_i be the dual variable associated with supply node i and v_j with demand node j. Then the dual of (11.16) is

$$\max \sum_{i=1}^{m} s_i u_i + \sum_{j=1}^{n} d_i v_i$$

$$\text{s.t. } u_i + v_j \le c_{ij}, \ i = 1, \dots, m, \ j = 1, \dots, n$$

$$u_i, v_j \text{ u.r.s.}$$

Given a basis, or spanning tree, $\mathcal{B}$ for the primal problem, we compute the corresponding dual solution as follows. Set one of them to a fixed value, say, $v_n = 0$. The dual constraints corresponding to basic variables are active,

$$u_i + v_j = c_{ij}, \ (i, j) \in \mathcal{B},$$

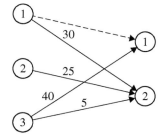

Figure 11.6 Initial tree solution for Figure 11.5.

and can be used to find the other dual variables. Given the dual solution, the reduced costs are

$$\bar{c}_{ij} = c_{ij} - (u_i + v_j).$$

Because we only need the values of the basic variables and reduced costs of the nonbasic variables, we can combine the two in one $m \times n$ matrix or table.

The flow can be updated in Step 3 as before. Cycles alternate between supply and demand nodes, so they must contain an even number of arcs, and at least four. Exactly half the arcs will be in the reverse direction from the entering arc.

Example 11.6. Consider the transportation problem in Example 11.5 and the initial bfs shown in Figure 11.6. The corresponding dual solution is $v_2 = 0$,

$$u_1 = c_{12} - v_2 = 22 - 0 = 22,$$
$$u_2 = c_{22} - v_2 = 12 - 0 = 12,$$
$$u_3 = c_{32} - v_2 = 16 - 0 = 16,$$
$$v_1 = c_{31} - u_3 = 10 - 16 = -6.$$

The nonbasic reduced costs are

$$\bar{c}_{11} = c_{11} - (u_1 + v_1) = 14 - (22 + (-6)) = -2,$$
$$\bar{c}_{21} = c_{21} - (u_2 + v_1) = 18 - (12 + (-6)) = 12.$$

These are recorded in the first table, where reduced costs are in parentheses to distinguish them from the flows. The entering arc is $(1, 1)$, shown by the dotted line in Figure 11.6. The cycle $S_1 \to D_1 \to S_3 \to D_2 \to S_1$ contains arcs $(3, 1)$ and $(1, 2)$ in reverse direction, so $\lambda = \min\{x_{31}, x_{12}\} = \min\{40, 30\} = 30$ and the leaving arc is $(1, 2)$. Sending a flow of 30 around the cycle, the new spanning tree and flow are shown in Figure 11.7.

To begin the second iteration, we record the new flow in the second table and compute the new u_i, v_j, and $\bar{c}_{ij}$. Since all reduced costs are nonnegative, this solution is optimal, with an optimal cost of 1380.

Figure 11.7 Final tree solution for Figure 11.5.

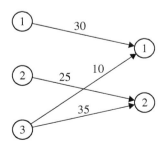

$u_i \backslash v_j$	−6	0	s_i
22	(−2)	30	30
12	(12)	25	25
16	40	5	45
d_j	40	60	

$u_i \backslash v_j$	−6	0	s_i
20	30	(2)	30
12	(12)	25	25
16	10	35	45
d_j	40	60	

Of course, it is possible to perform the calculations without drawing the graph. In particular, there are methods to find the cycle in the graph when the entering arc is added. The tables earlier, with arc costs also recorded in each cell, are called *transportation simplex tableaus* and are convenient for hand calculations.

There is another sense in which network flow problems are easy: if the external supplies are integers and an optimal solution exists, then the optimal solution found will have integer flows. Thus, even if an application requires integer solutions, e.g. the number of trucks, we can still solve a linear program, not an integer program. Section 12.3 explores this property.

11.3 Primal-Dual Interior Point Method

The simplex method visits extreme points of the feasible region, moving along edges to adjacent extreme points. As discussed in Section 9.4, the path to an optimal vertex may contain a large number of edges and the algorithm typically does not travel the path with the fewest edges. Another approach is to move through the interior of the feasible region in directions that are more aligned with the gradient, potentially moving toward optimality in fewer iterations. Interior point methods stay in the interior, avoiding the difficulty of dealing with inequality constraints.

The interior point method that we present starts with a feasible solution to the primal problem and to the dual problem that do not satisfy complimentary slackness. Both solutions are in the interior of their feasible regions; no inequality constraints are active. Each iteration stays in the interior, but seeks to move closer to complimentary slackness. As the violation of complementary slackness decreases, the distance to the inequality constraints is also allowed to decrease, so that these solutions can approach the optimal solution, where some of the inequality constraints are active.

The algorithm is applied to a problem in canonical form,

$$
\begin{aligned}
\max \quad & \mathbf{c}^T\mathbf{x} \\
\text{s.t.} \quad & \mathbf{A}\mathbf{x} = \mathbf{b} \\
& \mathbf{x} \geq \mathbf{0}.
\end{aligned}
\tag{11.17}
$$

Its dual is

$$
\begin{aligned}
\min \quad & \mathbf{b}^T\mathbf{y} \\
\text{s.t.} \quad & \mathbf{A}^T\mathbf{y} - \mathbf{w} = \mathbf{c} \\
& \mathbf{w} \geq \mathbf{0},
\end{aligned}
\tag{11.18}
$$

where $\mathbf{w}$ contains the dual surplus variables. The complementary slackness condition is $x_j w_j = 0, j = 1, \ldots, n$. A necessary and sufficient condition for optimality is primal and dual feasibility and complimentary slackness:

$$
\begin{aligned}
\mathbf{A}\mathbf{x} &= \mathbf{b} \\
\mathbf{x} &\geq \mathbf{0} \\
\mathbf{A}^T\mathbf{y} - \mathbf{w} &= \mathbf{c} \\
x_j w_j &= 0, \quad j = 1, \ldots, n \\
\mathbf{w} &\geq \mathbf{0}.
\end{aligned}
$$

We will consider solutions that are strictly feasible, meaning that they are feasible and $\mathbf{x}, \mathbf{w} > \mathbf{0}$ but they do not satisfy complementary slackness. Instead, they have a duality gap

$$
\mathbf{y}^T\mathbf{b} - \mathbf{c}^T\mathbf{x} = \mathbf{y}^T\mathbf{b} - (\mathbf{y}^T\mathbf{A} - \mathbf{w}^T)\mathbf{x} = \mathbf{w}^T\mathbf{x}.
$$

Although we call them interior points, these solutions are not actually in the interior of the feasible regions because the regions do not have full dimension. The correct statement is that $\mathbf{x}$ is in the relative interior of the feasible region, which means that any point close to $\mathbf{x}$ in the solution set of $\mathbf{A}\mathbf{x} = \mathbf{b}$ is in the feasible region.

The key idea is to choose a target $x_j w_j = \mu > 0$ for how close we would like to be to complimentary slackness, move close to that target, and then reduce the target. Note that using the same target μ for each variable is arbitrary. To express the targets in matrix form, let $\mathbf{X} = \text{diag}(x_j)$ and $\mathbf{W} = \text{diag}(w_j)$. Given a $\mu > 0$, we are seeking a solution to

$$
\begin{aligned}
\mathbf{A}\mathbf{x} &= \mathbf{b} \\
\mathbf{A}^T\mathbf{y} - \mathbf{w} &= \mathbf{c} \\
\mathbf{X}\mathbf{W}\mathbf{1} &= \mu\mathbf{1}
\end{aligned}
\tag{11.19}
$$

with $\mathbf{x}, \mathbf{w} \geq \mathbf{0}$. The complimentary slackness constraint is nonlinear, so we use an approximate method to solve (11.19). We will perform one iteration of Newton's method to move from the current solution $(\mathbf{x}, \mathbf{y}, \mathbf{w})$ closer to a solution of (11.19).

11.3.1 Newton's Method

For a single equation $f(x) = 0$ in one variable, Newton's method uses the linear approximation of f at the current solution x_k to find the next solution x_{k+1}. The formula for x_{k+1} is found by solving

$$f(x) \approx f(x_k) + f'(x_k)(x - x_k) = 0$$

for x. For a system of n equations $\mathbf{F}(\mathbf{x}) = (f_1(\mathbf{x}), \dots, f_n(\mathbf{x})) = \mathbf{0}$ in n variables, the Newton step $\mathbf{d} = \mathbf{x}^{k+1} - \mathbf{x}^k$ satisfies

$$\mathbf{F}(\mathbf{x}^k + \mathbf{d}) \approx \mathbf{F}(\mathbf{x}^k) + \nabla\mathbf{F}(\mathbf{x}^k)\mathbf{d} = \mathbf{0},$$

where $\nabla\mathbf{F}(\mathbf{x})$ is the $n \times n$ Jacobian matrix of partial derivatives $\frac{\partial f_i}{\partial x_j}$.

For our problem, we will find the Newton step $\mathbf{d} = (\mathbf{d}^x, \mathbf{d}^y, \mathbf{d}^w)$ at the current solution $(\mathbf{x}, \mathbf{y}, \mathbf{w})$. Substituting $(\mathbf{x}, \mathbf{y}, \mathbf{w}) + (\mathbf{d}^x, \mathbf{d}^y, \mathbf{d}^w)$ in (11.19), the only nonlinear functions of $\mathbf{d}$ are the quadratic terms in $\mathbf{XW}$. Newton's method replaces these with a linear approximation

$$(x_j + d_j^x)(w_j + d_j^w) \approx x_j w_j + x_j d_j^w + w_j d_j^x = \mu,$$

or in matrix form

$$\mathbf{X}\mathbf{d}^w + \mathbf{W}\mathbf{d}^x = \mu\mathbf{1} - \mathbf{XW1}.$$

Thus, $\mathbf{d}$ is a solution of the following linear system:

$$\mathbf{A}\mathbf{d}^x = \mathbf{0},$$
$$\mathbf{A}^T\mathbf{d}^y - \mathbf{d}^w = \mathbf{0}, \tag{11.20}$$
$$\mathbf{X}\mathbf{d}^w + \mathbf{W}\mathbf{d}^x = \mathbf{v}(\mu),$$

where

$$\mathbf{v}(\mu) = \mu\mathbf{1} - \mathbf{XW1}. \tag{11.21}$$

Writing (11.20) as a $(2n + m) \times (2n + m)$ linear system,

$$\begin{bmatrix} \mathbf{A} & \mathbf{0} & \mathbf{0} \\ \mathbf{0} & \mathbf{A}^T & -\mathbf{I} \\ \mathbf{W} & \mathbf{0} & \mathbf{X} \end{bmatrix} \begin{bmatrix} \mathbf{d}^x \\ \mathbf{d}^y \\ \mathbf{d}^w \end{bmatrix} = \begin{bmatrix} \mathbf{0} \\ \mathbf{0} \\ \mathbf{v}(\mu) \end{bmatrix}. \tag{11.22}$$

We could solve this system numerically, but it is more efficient to first use substitution, obtaining

$$\mathbf{d}^y = (\mathbf{AXW}^{-1}\mathbf{A}^T)^{-1}\mathbf{AW}^{-1}\mathbf{v}(\mu),$$
$$\mathbf{d}^w = \mathbf{A}^T\mathbf{d}^y, \qquad (11.23)$$
$$\mathbf{d}^x = \mathbf{W}^{-1}(\mathbf{v}(\mu) - \mathbf{Xd}^w).$$

This form requires inverting only an $m \times m$ matrix (because $\mathbf{W}$ is diagonal, finding $\mathbf{W}^{-1}$ is easy).

11.3.2   Step Size

Next we modify the length of the step to maintain strict feasibility. It is possible that the step (11.23) would lead to an infeasible point. We can adjust the primal and dual components separately and still satisfy the equality constraints:

$$\mathbf{x}^{\text{new}} = \mathbf{x} + \beta_P\mathbf{d}^x,$$
$$\mathbf{y}^{\text{new}} = \mathbf{y} + \beta_D\mathbf{d}^y, \qquad (11.24)$$
$$\mathbf{w}^{\text{new}} = \mathbf{w} + \beta_D\mathbf{d}^w.$$

For $0 < \alpha < 1$, the following formulas reduce the step size to guarantee each x_j and w_j not only is nonnegative but shrinks by no more than a factor of $1 - \alpha$:

$$\beta_P = \min\left\{1, \alpha\min\left\{\frac{x_j}{-d_j^x} : d_j^x < 0\right\}\right\}, \qquad (11.25)$$
$$\beta_D = \min\left\{1, \alpha\min\left\{\frac{w_j}{-d_j^w} : d_j^w < 0\right\}\right\}.$$

For example, if $x_1 = 100$, $d_1^x = -200$, and $\alpha = 0.95$, if x_1 achieves the minimum then

$$\beta_P = \min\left\{1, 0.95\left(\frac{100}{-(-200)}\right)\right\} = 0.475$$

and $x_1^{\text{new}} = x_1 + \beta_P d_1^x = 100 - 0.475(200) = 5$.

11.3.3   Choosing the Complementary Slackness Parameter

Although the step size maintains strict feasibility, μ also plays a role in keeping the new point away from the constraints $\mathbf{x} \geq \mathbf{0}$, $\mathbf{w} \geq \mathbf{0}$ by influencing $\mathbf{d}$. We want μ to approach 0 as the algorithm progresses because as $\mu \to 0$ the solution of (11.20) approaches an optimal solution of the problem. Since the step $\mathbf{d}$ approximately moves to a complementary slackness violation of $\mu = x_j w_j$ for each j, it is reasonable to make μ somewhat smaller than the average value of $x_j w_j$ at the current solution. Using

$$\mu = \gamma \frac{\mathbf{x}^T \mathbf{w}}{n}, \tag{11.26}$$

where $0 < \gamma < 1$, makes μ a fraction γ of this average value. Since $\mathbf{x}^T \mathbf{w}$ is the duality gap, we can interpret (11.26) as seeking to multiply the duality gap by γ. This method performs very well in practice. We will use this idea to initialize μ but will simply multiply μ by γ at each iteration. Although both of these choices of μ work well in practice, a modified formula for μ is used to prove that the algorithm converges to an optimal solution.

The algorithm is applied to a problem in canonical form whose **A** matrix has full row rank.

Algorithm 11.3 (Primal-Dual Path Following Algorithm)

Step 0. Initialization. Start with a strictly feasible solution **x** to the primal problem, a strictly feasible solution $(\mathbf{y}, \mathbf{w})$ to the dual problem, a step size parameter $0 < \alpha < 1$, a duality gap parameter γ, and an optimality tolerance $\varepsilon > 0$. Set the initial μ, e.g. using (11.26).
Step 1. Optimality test. If $\mathbf{x}^T \mathbf{w} < \varepsilon$, stop. The current solution has a value within ε of optimal.
Step 2. Compute Newton step. Compute the step **d** using (11.21) and (11.23).
Step 3. Compute step sizes β_P and β_D using (11.25).
Step 4. Update solution. Update the solution $(\mathbf{x}, \mathbf{y}, \mathbf{w})$ using (11.24) and update $\mu^{new} = \gamma \mu$. Go to Step 1.

Example 11.7. We will apply the algorithm to the problem

$$
\begin{array}{llll}
\max & 4x_1 + 3x_2 \\
\text{s.t.} & 1.5x_1 + x_2 + s_1 & = 16 \\
& x_1 + x_2 & + s_2 & = 12 \\
& x_2 & + s_3 = 10 \\
& x_j, s_i \geq 0,
\end{array}
$$

which we solved in Section 9.1. We will write $\mathbf{x} = (x_1, x_2, s_1, s_2, s_3)$. The dual problem, after adding surplus variables, is

$$
\begin{aligned}
\min \quad & 16y_1 + 12y_2 + 10y_3 \\
\text{s.t.} \quad & 1.5y_1 + y_2 \quad\quad - w_1 \quad\quad\quad\quad\quad\quad = 4 \\
& y_1 + y_2 + y_3 \quad\quad - w_2 \quad\quad\quad\quad = 3 \\
& y_1 \quad\quad\quad\quad\quad\quad\quad - w_3 \quad\quad\quad = 0 \\
& y_2 \quad\quad\quad\quad\quad\quad\quad\quad - w_4 \quad\quad = 0 \\
& y_3 \quad\quad\quad\quad\quad\quad\quad\quad\quad - w_5 = 0 \\
& w_i \geq 0.
\end{aligned}
$$

The last three constraints are equivalent to nonnegativity of y. However, because the algorithm assumes the primal is in canonical form and y is unrestricted in sign, they must be kept as functional constraints. We use the initial solutions $\mathbf{x} = (3, 3)$ and $\mathbf{y} = (2, 2, 1)$, $\alpha = 0.9$, and $\gamma = 0.25$. These solutions are strictly feasible, with slacks $\mathbf{s} = (8.5, 6, 7)$ and $\mathbf{w} = (1, 2, 2, 2, 1)$ and duality gap $\mathbf{x}^T \mathbf{w} = 45$, or an average of $\mathbf{x}^T \mathbf{w}/n = 9$. We chose an initial $\mu = 5$, somewhat smaller than this average. Next, we compute $v_j(\mu) = \mu - x_j w_j$ and set up the matrices

$$
\mathbf{v}(\mu) = \begin{bmatrix} 2 \\ -1 \\ -12 \\ -7 \\ -2 \end{bmatrix}, \quad
\mathbf{X} = \begin{bmatrix} 3 & 0 & 0 & 0 & 0 \\ 0 & 3 & 0 & 0 & 0 \\ 0 & 0 & 8.5 & 0 & 0 \\ 0 & 0 & 0 & 6 & 0 \\ 0 & 0 & 0 & 0 & 7 \end{bmatrix}, \quad
\mathbf{W} = \begin{bmatrix} 1 & 0 & 0 & 0 & 0 \\ 0 & 2 & 0 & 0 & 0 \\ 0 & 0 & 2 & 0 & 0 \\ 0 & 0 & 0 & 2 & 0 \\ 0 & 0 & 0 & 0 & 1 \end{bmatrix}.
$$

Solving (11.22),

$$
\mathbf{d}^x = \begin{bmatrix} 3.149 \\ 0.268 \\ -4.992 \\ -3.418 \\ -0.268 \end{bmatrix}, \quad
\mathbf{d}^y = \begin{bmatrix} -0.237 \\ -0.027 \\ -0.247 \end{bmatrix}, \quad
\mathbf{d}^w = \begin{bmatrix} -0.383 \\ -0.512 \\ -0.237 \\ -0.027 \\ -0.247 \end{bmatrix}.
$$

The step sizes are

$$
\beta_P = \min\left\{1, 0.9 \min\left\{\frac{8.5}{4.992}, \frac{6}{3.418}, \frac{7}{0.268}\right\}\right\} = \min\{1, 1.53\} = 1,
$$

$$
\beta_D = \min\left\{1, 0.9 \min\left\{\frac{1}{0.383}, \frac{2}{0.512}, \frac{2}{0.237}, \frac{2}{0.027}, \frac{1}{0.247}\right\}\right\}
$$

$$
= \min\{1, 2.35\} = 1.
$$

Then we update the solution to

$$
\mathbf{x}^{\text{new}} = \begin{bmatrix} 3 \\ 3 \\ 8.5 \\ 6 \\ 7 \end{bmatrix} + 1 \begin{bmatrix} 3.149 \\ 0.268 \\ -4.992 \\ -3.418 \\ -0.268 \end{bmatrix} = \begin{bmatrix} 6.149 \\ 3.268 \\ 3.508 \\ 2.582 \\ 6.732 \end{bmatrix},
$$

Table 11.2 Iterations of the path following algorithm.

Iteration	μ	Gap $x^T w$	Objective	(x_1, x_2)
1	5	25	34.40	(6.15, 3.27)
2	1.25	6.80	40.79	(7.82, 3.17)
3	0.313	1.77	43.31	(8.72, 2.82)
4	0.105	0.53	43.74	(8.34, 3.46)
5	0.206	0.103	43.959	(8.019, 3.961)
6	0.00488	0.024	43.990	(8.005, 3.991)

$$\mathbf{y}^{\text{new}} = \begin{bmatrix} 2 \\ 2 \\ 1 \end{bmatrix} + 1 \begin{bmatrix} -0.237 \\ -0.027 \\ -0.247 \end{bmatrix} = \begin{bmatrix} 1.763 \\ 1.973 \\ 0.753 \end{bmatrix},$$

$$\mathbf{w}^{\text{new}} = \begin{bmatrix} 1 \\ 2 \\ 2 \\ 2 \\ 1 \end{bmatrix} + 1 \begin{bmatrix} -0.383 \\ -0.512 \\ -0.237 \\ -0.027 \\ -0.247 \end{bmatrix} = \begin{bmatrix} 0.617 \\ 1.488 \\ 1.763 \\ 1.973 \\ 0.753 \end{bmatrix}.$$

After the first iteration, $x^T\mathbf{w}/n = 5$, which happens to exactly match μ. For the next iteration, $\mu = 0.25(5) = 1.25$. The results of the next few iterations are shown in Table 11.2. The solution converges quickly to (8, 4), which is optimal. Note that the objective function value always differs from the optimal value of 44 by less than the optimality gap. As shown in Figure 11.8, each iteration moves closer to the boundary of the feasible region but not directly toward the optimal point.

To use this algorithm we must find strictly feasible solutions for the primal and the dual. In our examples it is easy to construct strictly feasible solutions. However, as we saw in Section 9.3, finding a feasible solution can be as hard as solving a linear program. Fortunately, there are ways to use the same algorithm to find strictly feasible solutions, so that performance guarantees for the algorithm also apply to finding the strictly feasible solutions. The key ideas behind several different methods are as follows:

- Construct a linear program with *artificial primal and dual variables* measuring the infeasibility in the constraints and a large penalty in the primal objective function. Solving this problem drives the artificial variables to zero, resulting in an optimal solution to both problems. Rather than one artificial variable for each constraint, only one variable for each type of constraint (five variables total) are needed.

Figure 11.8 Progress of algorithm for Example 11.7.

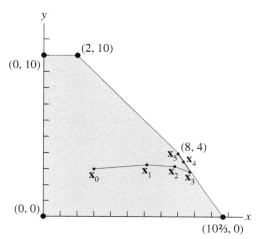

- *Self-dual methods*, developed by Ye et al. (1994): Construct a linear program with an artificial variable measuring the infeasibility in all the primal constraints and minimize just this variable. The problem is constructed so that it is its own dual (self-dual) and so that the algorithm either finds an optimal solution to both problems or determines that the original problem is infeasible.
- *Infeasible primal-dual path following methods*, developed by Kojima et al. (1993): Apply the primal-dual algorithm but don't require the initial solutions to be feasible. The only change to the algorithm is that in the derivation of the Newton step $\mathbf{d}$ the equality constraints are not assumed to hold.

The last two methods have the advantage of applying to any problem, without assuming that a feasible or optimal solution exists.

Analysis of this algorithm is very different than the analysis of the simplex method. For one thing, the algorithm does not reach an optimal bfs. Instead, it converges to an optimal solution. That is, if $\mathbf{x}^k$ is the solution after k iterations, the sequence $\mathbf{x}^k \to \mathbf{x}^*$ for some optimal solution $\mathbf{x}^*$. Performance guarantees describe the rate at which the objective function value approaches optimal. More precisely, for a slightly modified version of this algorithm, a bound can be found on the number of iterations needed to reduce the duality gap by a certain ratio R. Since the suboptimality $\mathbf{c}^T\mathbf{x}^* - \mathbf{c}^T\mathbf{x}^k$ is less than or equal to the duality gap, the bound also applies to suboptimality. The bound is $O(\sqrt{n}\ln(1/R))$; see Den Hertog (2012). Average behavior has been observed to be faster, depending less on n.

An interesting feature of this algorithm is that it is not guaranteed to improve the objective function at each iteration–it could decrease. However, even if that occurs, the performance guarantee tells us that after some number of iterations the objective function improves.

11.3.4 Barrier Function Interpretation

The primal-dual path following algorithm can be interpreted as what is called a *barrier function* method. In these methods, a penalty is included in the objective function that grows to infinity as each inequality constraint is approached. Optimizing the barrier function leads to a solution where no inequality constraint is active, so these constraints can be ignored. As the algorithm progresses, the weight assigned to the penalty is decreased so that the solution using the barrier function approaches the optimal solution to the linear program, where some of the inequality constraints are active. The algorithm never exactly solves the barrier function problem. Instead, one Newton step is taken for a given barrier function, then the weight changes. However, considering the solution to the barrier function problem helps in understanding the algorithm.

For $\mu > 0$, define the *barrier function*

$$B_\mu(\mathbf{x}) = \mathbf{c}^T\mathbf{x} + \mu \sum_{j=1}^n \ln x_j.$$

Notice that $B_\mu(\mathbf{x})$ approaches $-\infty$ as $x_j \to 0^+$. For the primal problem (11.17), the associated *barrier problem* is the nonlinear program

$$\max \ B_\mu(\mathbf{x}) \tag{11.27}$$
$$\text{s.t.} \quad \mathbf{Ax} = \mathbf{b}.$$

We have omitted the nonnegativity constraints because the domain of $B_\mu(\mathbf{x})$ imposes the implicit constraint $\mathbf{x} > \mathbf{0}$ (equivalently, we can set $B_\mu(\mathbf{x}) = -\infty$ if $x_j \le 0$ for some j). Assume that (11.17) has an optimal solution and has strictly feasible solutions $\mathbf{x} > \mathbf{0}$. Then it is not hard to show that the barrier problem has an optimal solution, denoted by $\mathbf{x}(\mu)$, for each $\mu > 0$. Further, since $B_\mu(\mathbf{x})$ is strictly concave, it cannot have multiple optimal solutions.

Now we apply the method of Lagrange multipliers, discussed in Section 8.4, to the barrier problem. The Lagrangian function is

$$L(\mathbf{x}, \mathbf{y}) = \mathbf{c}^T\mathbf{x} + \mu \sum_{j=1}^n \ln x_j + \mathbf{y}^T(\mathbf{b} - \mathbf{Ax}),$$

where y_i is the Lagrange multiplier or dual variable associated with constraint i. For a fixed $\mathbf{y}$, we maximize the Lagrangian by setting the partial derivative with respect to each x_j to zero:

$$\frac{\partial L}{\partial x_j} = c_j + \frac{\mu}{x_j} - \mathbf{y}^T\mathbf{a}_j = 0, \ j = 1, \dots, n. \tag{11.28}$$

These equations, combined with the primal constraints $\mathbf{Ax} = \mathbf{b}, \mathbf{x} \ge \mathbf{0}$, are satisfied by $\mathbf{x}(\mu)$; in fact, they are necessary and sufficient conditions for optimality of the barrier problem. But if we set $x_j = \mu/w_j$, (11.28) is also the dual constraints! Thus, the optimality conditions for the barrier problem can be written

$$\mathbf{Ax} = \mathbf{b}$$
$$\mathbf{y}^T \mathbf{a}_j - w_j = c_j, \ j = 1, \dots, n$$
$$x_j w_j = \mu, \ j = 1, \dots, n$$

with $\mathbf{x}, \mathbf{w} \geq \mathbf{0}$. These are identical to (11.19), the problem that the primal-dual path following algorithm seeks to solve. Thus, we can interpret the complimentary slackness target μ as the weight in the barrier function and the iterations of the algorithm as approximate solutions of the barrier problem.

The optimal solutions of the barrier problem, $\{\mathbf{x}(\mu) : \mu > 0\}$, are called the *central path*. It can be shown that $\lim_{\mu \to 0} \mathbf{x}(\mu) = \mathbf{x}^*$, where $\mathbf{x}^*$ is an optimal solution to the primal problem. As the algorithm finds approximate solutions of the barrier problem for smaller and smaller μ, it is approximately following the central path. A barrier problem can also be defined for the dual using the same weight μ. Its optimality condition is the same as for the primal barrier problem, so the dual solutions $(\mathbf{y}(\mu), \mathbf{w}(\mu))$, where $w_j(\mu) = \mu/x_j(\mu)$, associated with $\mathbf{x}(\mu)$ also form a central path for the dual. That is why the algorithm is called primal-dual path following.

Example 11.8. We will find the central path for the problem

$$
\begin{aligned}
\max \quad & x_1 + x_2 \\
\text{s.t.} \quad & x_1 && \leq 1 \\
& x_1 + x_2 && \leq 3 \\
& x_j && \geq 0.
\end{aligned}
$$

First we add slack variables s_i, but then use $s_1 = 1 - x_1$ and $s_2 = 3 - x_1 - x_2$ to eliminate them from the barrier function:

$$B_\mu(\mathbf{x}) = x_1 + x_2 + \mu \ln x_1 + \mu \ln x_2 + \mu \ln(1 - x_1) + \mu \ln(3 - x_1 - x_2).$$

Setting the partial derivatives to zero, the optimality conditions for the barrier problem are

$$1 + \frac{\mu}{x_1} - \frac{\mu}{1 - x_1} - \frac{\mu}{3 - x_1 - x_2} = 0,$$

$$1 + \frac{\mu}{x_2} - \frac{\mu}{3 - x_1 - x_2} = 0.$$

Numerical solutions over a range of μ are listed in Table 11.3. For large μ, the solution is near the "center" of the feasible region, far from all boundaries. As μ decreases, the solution approaches the boundary $x_1 + x_2 = 3$. All points on this boundary between $(1, 2)$ and $(0, 3)$ are optimal.

Table 11.3 Points on central path.

μ	$\mathbf{x}(\mu)$
10 000	(0.41, 1.30)
1	(0.44, 1.91)
0.5	(0.44, 2.15)
0.1	(0.45, 2.45)
0.01	(0.451, 2.539)
0.001	(0.451, 2.548)

The central path in this example approaches the midpoint of the optimal line segment. In general, it approaches the "center" of the optimal set because of the penalty for approaching boundaries. If there is a unique optimal extreme point, it approaches that point. Note that the algorithm only finds a bfs if there is a unique optimal bfs.

Problems

1 Solve the linear program in Example 11.3 using the dual simplex method, starting with the solution given there.

2 For the following linear program,
(a) Convert to canonical form.
(b) Solve using the dual simplex method, starting with the solution $\mathbf{x} = (0, 0)$.

$$\begin{aligned}
\min \quad & 4x_1 + 3x_2 \\
\text{s.t.} \quad & x_1 + 2x_2 \geq 8 \\
& 3x_1 + x_2 \geq 12 \\
& x_1, x_2 \geq 0.
\end{aligned}$$

3 Add the constraint $x_1 + 4x_2 \leq 12$ to the linear program in Exercise 11.2, and resolve using the dual simplex method starting at the optimal solution $\mathbf{x} = (3.2, 2.4)$ to Exercise 11.2.

4 Consider the canonical form linear program in Example 11.1 and the equality form of its dual given just before Example 11.3.
(a) How many basic variables are there in the primal? How many nonbasic?
(b) How many basic variables are there in the dual? How many nonbasic?

(c) Of the dual surplus variables w_i, how many are basic? How many are nonbasic?

5   For the equality form of the dual in (11.6), show that the reduced costs of **y** are zero.
Hint: Start with $\bar{\mathbf{c}}^T = \hat{\mathbf{c}}^T - \mathbf{c_B}^T \hat{\mathbf{B}}^{-1} \hat{\mathbf{A}}$.

6   Solve the network flow problem defined by the network in Figure 11.1 and the external supplies **b** = (40, −50, −10, 45, −25). Use the initial spanning tree in Figure 11.1 and the network simplex method.

7   Consider the network flow problem with arcs, costs, and supplies shown as follows.

Arc	Cost	Node	b
(1, 2)	2	1	10
(2, 5)	1	2	−30
(3, 4)	4	3	10
(3, 6)	1	4	−5
(4, 1)	1	5	−10
(4, 5)	2	6	25
(5, 3)	6		
(6, 1)	4		
(6, 2)	7		

(a) Draw the network, labeling nodes with supplies and arcs with costs.
(b) Solve using the network simplex method and initial spanning tree (1, 2), (2, 5), (3, 4), (4, 1), (6, 2).

8   Exercise 2.23 considered grain shipments by the U.N. World Food Program.
(a) Find a feasible solution to the transportation problem in Exercise 2.23(c) using the minimum cost method.
(b) For the solution in (a) earlier, send a flow around the only cycle in the network to find the optimal solution. Using this optimal solution to the transportation problem, state the optimal shipping and trucking plan.

9   For the transportation problem shown
(a) Find an initial feasible solution using the minimum cost method.
(b) Solve using the network simplex method.
Source: (Rader, 2010, Exercise 11.6).

Supply	Arc costs				
	Demand point				
point	1	2	3	4	Supply
1	8	6	12	9	35
2	9	12	13	7	45
3	14	9	16	5	40
Demand	40	20	30	30	–

10 For the transportation problem shown:
(a) Find an initial feasible solution using the minimum cost method.
(b) Solve using the network simplex method.

Supply	Arc costs				
	Demand point				
point	1	2	3	4	Supply
1	45	50	50	100	75
2	35	40	70	80	125
3	100	70	40	85	100
Demand	80	65	70	85	–

11 Convert Exercise 11.6 to a transportation problem and draw the transportation network.

12 Convert the network flow problem of Figure 11.4, with costs shown in Figure 11.1, to a transportation problem. Draw the transportation network and show that it is infeasible.

13 Consider the linear program

$$\max \quad x_1 + 2x_2$$
$$\text{s.t.} \quad x_1 + x_2 \leq 2$$
$$-x_1 + x_2 \leq 1$$
$$x_1, x_2 \geq 0.$$

We will apply the primal-dual path following algorithm with $\gamma = 0.25$ and $\alpha = 0.9$.

(a) Convert to canonical form, find the dual, and add surplus variables.

(b) Using the initial primal solution $\mathbf{x} = (1, 0.5)$ and the dual solution $\mathbf{y} = (2, 0.5)$, find the values of the primal slack and dual surplus variables. Find the duality gap.

(c) Using an initial $\mu = 0.25$, write out the matrices $\mathbf{A}$, $\mathbf{X}$, and $\mathbf{W}$ used in (11.20) and find $v(\mu)$.

(d) Complete the first iteration, finding $\mathbf{d}$, β_P, β_D, and the new solutions.

(e) Perform iterations 2 through 5. Make a table showing $\mathbf{x}$, $\mathbf{x}^T\mathbf{w}$, $\mathbf{c}^T\mathbf{x}$, and μ at each iteration.

14 Consider the linear program

$$\max\ 7x_1 + 5x_2$$
$$\text{s.t.}\quad 6x_1 + 4x_2 \leq 8$$
$$4x_1 + 2x_2 \leq 14$$
$$8x_1 + 2x_2 \leq 10$$
$$x_1, x_2 \geq 0.$$

(a) Perform two iterations of the primal-dual path following algorithm. Use the initial solutions $\mathbf{x} = (0.5, 0.5)$ and $\mathbf{y} = (1, 1, 0.5)$, initial $\mu = 10$, $\gamma = 0.25$, and $\alpha = 0.9$.

(b) How do $\mathbf{x}^T\mathbf{w}$ and $\mathbf{c}^T\mathbf{x}$ change at each iteration?

15 Perform six iterations of the primal-dual path following algorithm for the linear program

$$\max\ 6x_1 + 4x_2 + 3x_3 + 2x_4$$
$$\text{s.t.}\quad 4x_1 + 3x_2 + 2x_3 + x_4 = 48$$
$$6x_1 + 3x_2 + x_3 + x_4 = 60$$
$$x_j \geq 0.$$

Use the initial solutions $\mathbf{x} = (8, 2, 4, 2)$ and $\mathbf{y} = (2, 2)$, initial $\mu = 50$, $\gamma = 0.2$, and $\alpha = 0.9$. Make a table showing $\mathbf{x}$, $\mathbf{x}^T\mathbf{w}$, and μ at each iteration.

16 Consider the linear program

$$\max\ x_1 + 2x_2$$
$$\text{s.t.}\quad 0 \leq x_j \leq 1.$$

(a) Show that the central path is

$$x_1(\mu) = \frac{1 - 2\mu + \sqrt{1 + 4\mu^2}}{2},$$

$$x_2(\mu) = \frac{1 - \mu + \sqrt{1 + \mu^2}}{2}.$$

     (b) Show that $\lim\limits_{\mu \to 0} x(\mu)$ is the optimal solution.

     (c) Find $\lim\limits_{\mu \to \infty} x(\mu)$.

17   Consider the linear program

$$\max\ x_1 + x_2$$
$$\text{s.t.}\ \ x_1 + x_2 + x_3 = 1$$
$$x_j \geq 0.$$

     (a) Find the central path.

     (b) Find $\lim\limits_{\mu \to 0} x(\mu)$. Is it an optimal solution? A basic solution?

12

Integer Programming Theory

Now we turn our attention from solving linear programs with continuous variables to solving integer programs, where the variables are restricted to only have integer values. In some ways, solving an integer program is very different. Instead of moving in an improving feasible direction, one must identify and compare discrete points. Linear programs have simple optimality conditions, but integer programs generally do not, making them much harder to solve than linear programs. Still, there are relationships with linear programs that are used in integer programming algorithms. This chapter describes the relationship and the importance of how the integer program is formulated.

Section 12.1 introduces some terminology and examines the relationship between integer and linear programs. Using this relationship, Section 12.2 defines what is meant by a formulation of an integer program and discusses strategies for obtaining good formulations. A class of "easy" integer programs is identified in Section 12.3, which can be solved as a linear program because both have the same optimal solution.

Throughout this chapter we consider the following problems.

A **mixed integer program** is a linear program with the additional constraint that some of the variables must be an integer. If J is the set of indices of the integer variables and the linear program is in standard form, then the problem can be written as

$$\max \quad \mathbf{c}^T \mathbf{x}$$
$$\text{s.t.} \quad \mathbf{A}\mathbf{x} \leq \mathbf{b}$$
$$\mathbf{x} \geq \mathbf{0}$$
$$x_j \text{ integer}, \ j \in J.$$

If all the variables must have integer values, it is an **integer program**. If all the variables must take on the values 0 or 1, it is a **binary integer program**.

Linear and Convex Optimization: A Mathematical Approach, First Edition. Michael H. Veatch.
© 2021 John Wiley & Sons, Inc. Published 2021 by John Wiley & Sons, Inc.
Companion website: www.wiley.com/go/veatch/convexandlinearoptimization

We assume that the data $\mathbf{A}, \mathbf{b}, \mathbf{c}$ for the (mixed) integer program has integer values. This is not a major restriction, since fractional data can be converted to integers by multiplying by an appropriate integer. Also, any discrete optimization problem can be converted to one where the variables are restricted to integers. For example, if the possible values of x are $0, \frac{1}{2}, 1, \frac{3}{2}, \dots$ then $y = 2x$ must be an integer. More important is the assumption that the objective function and constraints are linear. To emphasize this, some authors call these problems *mixed integer linear programs*. Unlike the situation for linear programs, there are methods to convert a nonlinear integer program to a linear integer program. However, they often require adding more integer variables.

12.1 Linear Programming Relaxations

The relationship between an integer program and the associated linear program without the integer constraints is shown in the following example.

Example 12.1. Consider the integer program

$$\begin{array}{rl}
\max & 5x + 2y \\
\text{s.t.} & 3x + 11y \leq 44 \\
& 2x - 11y \leq 2 \\
& -9x + 14y \leq 21 \\
& x, y \geq 0, \text{ integer.}
\end{array}$$

The feasible region consists of the circled integer points in Figure 12.1. The optimal solution is $(x, y) = (7, 2)$ with value $z_{\text{IP}} = 39$. Without the integer constraints, the problem is a linear program and the feasible region is a polyhedron, namely, the region bounded by all five constraints in Figure 12.1. The optimal solution of the linear program, found by solving the first two constraint equations, is $(x, y) = (46/5, 82/55) \approx (9.2, 1.49)$ with value $z_{\text{LP}} = 2694/55 \approx 48.98$.

We see from this example that the optimal solution of the integer program cannot be found by rounding the optimal solution of the linear program. Indeed, in this example rounding up or down does not even give feasible solutions. For a different objective function, the corner point $(1, 0)$, which is integer, could have been the optimal solution to both the integer program and the linear program. In either situation, the optimal value of the linear program is as good or better than that of the integer program because its feasible region includes the feasible region of the linear program. These observations are summarized in the following definition and theorem.

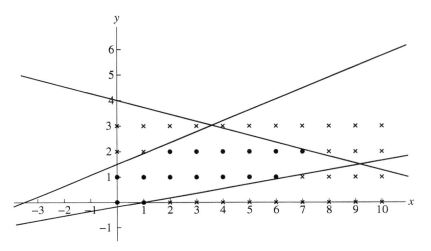

Figure 12.1 Feasible region for Example 12.1.

Given the optimization problem (P) $\max_{\mathbf{x}\in S} f(\mathbf{x})$, the problem (R) $\max_{\mathbf{x}\in S'} f(\mathbf{x})$ is a **relaxation** of (P) if $S \subset S'$. The **linear programming relaxation** of a mixed integer program is the problem obtained by removing the integrality constraints. For binary integer programs, the bounds $0 \le x_j \le 1$ are added for each variable.

Since the relaxation has a larger feasible region, it has at least as good an optimal value as the original problem. We will use this fact repeatedly.

Lemma 12.1: *Let z_{IP} be the optimal value of an integer program and z_{LP} be the optimal value of its linear programming relaxation. For maximization problems, $z_{LP} \ge z_{IP}$. For minimization problems, $z_{LP} \le z_{IP}$.*

12.2 Strong Formulations

While there are many equivalent formulations of a given linear program, we usually do not need to be concerned about which one we use to model a situation because there are efficient algorithms to solve linear programs. Sparsity (having fewer variables in each constraint) is important for solving very large problems, but is not affected much by the choice of formulation. When formulating integer programs, the situation is very different. There are several considerations that can have a major impact on how hard it is to solve:

- The number of integer variables.
- Bounds on the integer variables that limit the number of possible values of the variable.
- The geometry of the constraints; in particular, the relationship between the feasible integer points and the feasible region of the linear programming relaxation.

For example, a problem with 10 integer variables, each with bounds $2 \leq x_j \leq 5$, has at most $4^{10} \approx 10^6$ possible solutions. If the bounds are $0 \leq x_j \leq 9$, it has at most 10^{10} possible solutions.

To explore the third consideration geometrically, let S be the feasible region of the integer program, which is the set of integer points contained in the feasible region P of the linear programming relaxation. For a given S, there are many possible sets of constraints with different polyhedral regions P. Each is called a formulation.

A polyhedron P is a **formulation** of a set of integral points S, both in $\mathbb{R}^n$, if the set of integral points in P is S.

We see in Figure 12.1 that

$$S = \{(0,0), (1,0), (1,0), (1,1), (1,2), (1,3), (1,4), (1,5), (1,6),$$
$$(2,2), (2,3), (2,4), (2,5), (2,6), (2,7)\}.$$

One formulation of this set is to use the five constraints in Example 12.1, which define the five-sided region graphed there. This formulation has the disadvantage that it extends well beyond S. As a result, the linear programming relaxation z_{LP} is not a very close bound for the integer program z_{IP}. We shall see in Chapter 13 that the quality of the linear programming relaxation bound is important to the performance of algorithms for integer programs.

Now suppose we add the constraint $y \leq 2$, cutting off the top part of the pentagon in Figure 12.1. All points in S satisfy this constraint and are in the new polygon, so it is also a formulation of S. The new polygon is smaller, but the new linear program has the same optimal solution as before. By adding more constraints, the smallest possible polygon containing S can be constructed. It is called an ideal formulation and has the following characterization.

Let S be a finite set of feasible integer solutions. Given two polyhedral formulations P_1 and P_2 for S, P_1 is a **stronger formulation** than P_2 if $P_1 \subset P_2$. A formulation P is **ideal** if each extreme point of P is in S.

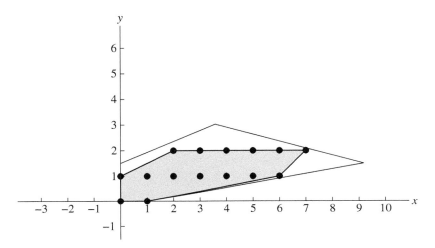

Figure 12.2 Feasible region and convex hull for Example 12.1.

Example 12.2. For the integer program of Example 12.1, the polyhedron defined by

$$\begin{aligned}
x - 5y &\leq 1 \\
x - y &\leq 5 \\
y &\leq 2 \\
-x + 2y &\leq 2 \\
x, y &\geq 0
\end{aligned}$$

is ideal. Figure 12.2 shows that this region has extreme points $(0,0), (1,0), (0,1),$ $(2,2), (6,1), (7,2)$ which are all feasible for the integer program.

It is clear from Figure 12.2 that the polyhedron with these extreme points is the smallest polyhedron containing all of the feasible integer solutions. Section 6.1 described the smallest polyhedron containing a finite set of points as their convex hull. We repeat the definition here for convenience.

For the set of points $\mathbf{v}_1, \ldots, \mathbf{v}_k \in \mathbb{R}^n$ and scalars α_i satisfying $\sum_{i=1}^{k} \alpha_i = 1$ and $\alpha_i \geq 0$, $i = 1, \ldots, k$, $\sum_{i=1}^{k} \alpha_i \mathbf{v}_i$ is a **convex combination** of $\mathbf{v}_1, \ldots, \mathbf{v}_k$.
The **convex hull** $\text{conv}\{\mathbf{v}_1, \ldots, \mathbf{v}_k\}$ is the set of all convex combinations of $\mathbf{v}_1, \ldots, \mathbf{v}_k$.

As discussed there, a bounded polyhedron is the convex hull of its extreme points and this is the smallest convex set containing the extreme points. Adding the feasible integer solutions that are not extreme points does not change the convex hull;

they are convex combinations of the extreme points. Thus, the ideal formulation is the convex hull of the feasible integer solutions and is the smallest polyhedron containing them.

The ideal formulation reveals an important connection between integer programs and linear programs. Because the ideal formulation is bounded, the linear programming relaxation using this formulation has an extreme point optimal solution. Since its extreme points are integral, these optimal solutions are also optimal for the integer program. Thus, if we can find the convex hull of the feasible integer solutions, *we can solve an integer program by solving this linear program*. Of course, it is hard to find the convex hull and it may require a large number of constraints (if that were not the case, integer programs would not be much harder to solve than linear programs). A more practical approach is to create stronger formulations that more closely approximate this convex hull.

The following theorem summarizes these observations.

Theorem 12.1: *Consider the integer program* $\max_{\mathbf{x} \in S} \mathbf{c}^T \mathbf{x}$, *where S is finite and nonempty. Let* $P = conv(S)$ *and consider the linear program* $\max_{\mathbf{x} \in P} \mathbf{c}^T \mathbf{x}$. *Then*

(i) *P is an ideal formulation of S (extreme points of P are in S).*

(ii) *P is stronger than any other formulation of S (i.e., it is the smallest polyhedron containing S).*

(iii) *The integer and linear programs have the same optimal value* $z_{IP} = z_{LP}$ *and any optimal solution of the integer program is optimal for the linear program.*

Proof:

(i) Let $\mathbf{x}$ be an extreme point of P. To be in P it must be a convex combination $\mathbf{x} = \sum_{i=1}^{k} \alpha_i \mathbf{v}_i$, where $S = \{\mathbf{v}_1, \dots, \mathbf{v}_k\} \subset P$. To be an extreme point, only one of $\{\alpha_i\}$ may be nonzero, i.e. $\mathbf{x} = \mathbf{v}_i$ for some i.

(ii) Exercise 6.10 shows that conv(S) is the smallest convex set containing S. Thus, it is the smallest polyhedron containing S.

(iii) Since S is finite and nonempty, P is finite and nonempty. By the Fundamental Theorem of Linear Programming (Theorem 7.3), there is an extreme point optimal solution $\mathbf{x}_{LP}$ to the linear program. By (i), $\mathbf{x}_{LP} \in S$, so the optimal values satisfy $z_{LP} \le z_{IP}$. But since the linear program is a relaxation, $z_{LP} \ge z_{IP}$ and we conclude $z_{LP} = z_{IP}$. □

12.2.1 Aggregate Constraints

Because of the importance of having a good linear programming bound when solving an integer program, a stronger formulation may be preferable even if it has many more constraints. One strategy for obtaining a stronger formulation is to avoid aggregate logical constraints, as the following example illustrates.

Example 12.3 (Rapid deployment team). Disaster response organizations keep some of their employees ready to deploy on short notice to a disaster site. When a disaster occurs, they must choose which of the available employees to deploy. There is a set of skills, unique to the type of disaster, that the team should possess. Each employee has some of these skills and may have them to varying degree. One approach to choosing a team is to require each skill to be possessed by the team and to minimize the cost of sending the team plus penalties when the person assigned a task is not fully qualified. This approach ignores workload, since it does not consider how many people have a skill or how many tasks may fall on one person. Suppose there are n available employees, m skills, a cost c_j of sending employee j, and a penalty d_{ij} of assigning skill i to employee j. If the employee is fully trained in the skill then $d_{ij} = 0$ and if they do not possess the skill then $d_{ij} = \infty$.

The decision variables are y_j, which equals 1 if employee j is selected and 0 otherwise, and x_{ij}, which equals 1 if skill i is assigned to employee j and 0 otherwise. One formulation of the problem is

$$\min \sum_{j=1}^{n} c_j y_j + \sum_{i=1}^{m} \sum_{j=1}^{n} d_{ij} x_{ij}$$

$$\text{s.t.} \sum_{j=1}^{n} x_{ij} = 1, \ i = 1, \ldots, m$$

$$x_{ij} \leq y_j, \ i = 1, \ldots, m, \ j = 1, \ldots, n$$

$$x_{ij}, y_j \text{ binary.}$$

The constraint $x_{ij} \leq y_j$ requires that if employee j is not selected ($y_j = 0$), then no skills may be assigned to them ($x_{ij} = 0$).

Notice that the skill constraints are equivalent to the aggregate constraints

$$\sum_{i=1}^{m} x_{ij} \leq m y_j, \ j = 1, \ldots, n.$$

For example, if employee 1 has skills 1 and 2, then the aggregate constraint

$$x_{11} + x_{21} \leq 2 y_1 \tag{12.1}$$

is equivalent to the individual constraints

$$x_{11} \leq y_1 \text{ and } x_{21} \leq y_1. \tag{12.2}$$

Both require $x_{11} = x_{21} = 0$ if $y_1 = 0$. However, in the linear programming relaxations with $0 \le x_{ij}, y_j \le 1$, (12.1) is a relaxation of (12.2) because it adds to the feasible region. Thus, the individual constraints are preferable to the fewer aggregate constraints. This form of problem appears in many contexts, such as the facility location problem, where the decisions are which facilities to open and which clients to assign to each facility.

12.2.2 Bounding Integer Variables

Example 12.2 suggests that we can approximate the convex hull of the feasible integer solutions by adding constraints that cut off part of the feasible region for the current formulation and pass through one or more integer solutions. A general method for finding such constraints is discussed in Section 13.2. Here we present an easy method to add bounds. It is one of several preprocessing techniques that solvers commonly use before applying an iterative algorithm.

Example 12.4. Suppose an integer program has constraints

$$
\begin{aligned}
6x_1 - 3x_2 + 9x_3 &\le 12 \\
9x_1 + 4x_2 + 3x_3 &\le 18 \\
x_1 + x_2 + x_3 &\le 7 \\
x_1 + x_2 + x_3 &\ge 5 \\
x_j &\ge 0.
\end{aligned}
$$

Solving the second constraint for x_1,

$$
x_1 \le \frac{18 - 4x_2 - 3x_3}{9} \le 2
$$

for feasible x_2 and x_3. Similarly, $x_2 \le \frac{18}{4}$, so for integer values $x_2 \le 4$, and $x_3 \le 6$. With these bounds, examining the first constraint,

$$
x_1 \le \frac{12 + 3x_2 - 9x_3}{6} \le \frac{12 + 3(4)}{6} = 4,
$$

which does not improve on the bound earlier. The other bounds from the first constraint are $x_2 \ge -4$, which is not useful, and

$$
x_3 \le \frac{12 - 6x_1 + 3x_2}{9} \le \frac{12 + 3(4)}{9} = \frac{24}{9},
$$

giving the bound $x_3 \le 2$. Finally, with these bounds, the fourth constraint gives the bound $x_2 \ge 5 - 2 - 2 = 1$. To summarize, the bounds

$$
\begin{aligned}
x_1 &\le 2 \\
1 \le x_2 &\le 4 \\
x_3 &\le 2
\end{aligned}
$$

can be added.

12.3 Unimodular Matrices

The last section considered the situation where the linear programming relaxation did not have all integer extreme points and constraints were added to make the formulation stronger, or closer to an ideal formulation. This section describes a class of linear programs where the coefficient matrix has a simple structure that guarantees integer extreme points as long as the right-hand side is integral. A necessary and sufficient condition on the $\mathbf{A}$ matrix will be given that is called unimodularity. Notably, the condition applies to minimum cost network flow problems. For example, the assignment problem is a network flow problem in which the supplies and demands are integers and the solution must also be integers. Because it satisfies the unimodularity condition, it can be solved as a linear program rather than an integer program. The (basic) optimal solution will be integral.

The property that all extreme points have integer values is easiest to understand for a canonical form linear program

$$\max \quad \mathbf{c}^T\mathbf{x}$$
$$\text{s.t.} \quad \mathbf{Ax} = \mathbf{b}$$
$$\mathbf{x} \geq \mathbf{0}.$$

Extreme points are basic feasible solutions $\mathbf{x}_B = \mathbf{B}^{-1}\mathbf{b}$ for some basis with basis matrix $\mathbf{B}$. We assume $\mathbf{A}$ and $\mathbf{b}$ are integral. By Cramer's rule,

$$x_{B(k)} = \frac{\det(\mathbf{B}_k^*)}{\det(\mathbf{B})},$$

where $\mathbf{B}_k^*$ is the matrix formed by replacing the kth column of $\mathbf{B}$ by $\mathbf{b}$. Now, because $\mathbf{B}_k^*$ is integral, its determinant is an integer. Thus, to guarantee that $\mathbf{x}$ is integral, we would like $\det(\mathbf{B})$ to equal 1 or -1 for each basis. In terms of $\mathbf{A}$, we would like all nonsingular square matrices formed from m columns of $\mathbf{A}$ to have determinant either 1 or -1.

12.3.1 Network Flow Problems

Next we show that the minimum cost network flow problem

$$\min \quad \sum_{(i,j)\in\mathcal{A}} c_{ij}x_{ij}$$
$$\text{s.t.} \quad \sum_{j:(i,j)\in\mathcal{A}} x_{ij} - \sum_{k:(k,i)\in\mathcal{A}} x_{ki} = b_i, \ i \in \mathcal{V} \qquad (12.3)$$
$$x_{ij} \geq 0.$$

with arc set $\mathcal{A}$ and node set $\mathcal{V}$ from Section 2.6 has this property.

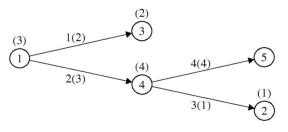

Figure 12.3 Renumbering a spanning tree. Node and arc numbers in parentheses are renumbered.

Lemma 12.2: *For the minimum cost network flow problem (12.3) on a connected graph with m nodes, after eliminating one constraint, the remaining $m - 1$ constraints are linearly independent and every basis matrix* **B** *has determinant either* 1 *or* -1.

Proof: Recall that for the network flow problem **A** is the node-arc incidence matrix, with rows representing nodes and columns representing arcs. We know from Lemma 11.1 that bases are tree solutions, so the columns of **B** correspond to the $m - 1$ arcs of a spanning tree. Assume that the constraint for node m was eliminated, so that the rows of **B** are for nodes $1, \ldots, m - 1$. We will show that we can reorder the rows and columns of **B** to obtain a lower triangular matrix **L**. This reordering may multiply the determinant by -1 but otherwise does not change it, so $\det(\mathbf{B}) = \pm \det(\mathbf{L})$. Further, the determinant of a lower triangular matrix is the product of the diagonal entries. Since **B** only contains the entries $1, 0$, and -1, and it is nonsingular, its determinant must be 1 or -1.

Renumber nodes 1 to $m - 1$ so that, in the spanning tree of this basis, the numbers increase along any path ending at node m. See Figure 12.3 for an example. Then, using the new node numbers, assign arc (i, j) in the tree the number $\min\{i, j\}$. These are unique numbers because there is only one path from each node to node m. Form the matrix **L** by reordering the rows and then the columns of **B** according to the new numbering. The column of **L** for arc (i, j) has nonzero entries only in rows i and j (there is no row m, so it may have only one nonzero). Its new column number is $\min\{i, j\}$, so it has one entry on the diagonal and (possibly) one below the diagonal making **L** lower triangular. □

To illustrate the reordering to a lower triangular matrix, the basis matrix for the spanning tree in Figure 12.3 is

$$
\mathbf{B} = \begin{bmatrix} 1 & 1 & 0 & 0 \\ 0 & 0 & -1 & 0 \\ -1 & 0 & 0 & 0 \\ 0 & -1 & 1 & 1 \end{bmatrix}.
$$

The constraint row for node 5 has been omitted, giving four linearly independent constraints. The new node ordering is 2, 3, 1, 4, 5. Reordering the rows of **B** gives

$$\begin{bmatrix} 0 & 0 & -1 & 0 \\ -1 & 0 & 0 & 0 \\ 1 & 1 & 0 & 0 \\ 0 & -1 & 1 & 1 \end{bmatrix}.$$

The arcs in this basis were $(1, 3), (1, 4), (4, 2), (4, 5)$. With the new node numbers, these arcs are $(3, 2), (3, 4), (4, 1), (4, 5)$. To reorder them, these arcs receive the unique numbers 2, 3, 1, 4. Their new order is $(4, 1), (3, 2), (3, 4), (4, 5)$. Reordering the columns of the matrix earlier gives

$$\mathbf{L} = \begin{bmatrix} -1 & 0 & 0 & 0 \\ 0 & -1 & 0 & 0 \\ 0 & 1 & 1 & 0 \\ 1 & 0 & -1 & 1 \end{bmatrix},$$

which is lower triangular.

It follows from this lemma that the minimum cost network flow problem with integer supplies and demands **b** has integer optimal solutions. Interestingly, this result can also be established by arguing that the network simplex algorithm can be initialized at an integer solution and that its iterations preserve the integer solution property. As explained in Bertsimas and Tsitsiklis (1997, Theorem 7.5), $\mathbf{B}^{-1}$ is an integer matrix and, if the costs **c** are integers, then all of the computations in the network simplex algorithm can be done using integer, as opposed to floating point, arithmetic.

12.3.2 Unimodularity

Now consider a linear program in standard form

$$\begin{aligned} \max \quad & \mathbf{c}^T \mathbf{x} \\ \text{s.t.} \quad & \mathbf{Ax} \le \mathbf{b} \\ & \mathbf{x} \ge \mathbf{0}. \end{aligned}$$

If we convert to canonical form, a basis matrix **B** consists of linearly independent columns of $[\mathbf{A}|\mathbf{I}]$, including k columns from **A** corresponding to variables x_j and $m - k$ columns of **I** corresponding to slack variables s_i. Let $\mathbf{A}_1$ be the $k \times k$ submatrix of **A** containing the basic columns and the rows corresponding to basic variables x_j and let $\mathbf{A}_2$ be the submatrix of **A** containing the basic columns and the rows not in $\mathbf{A}_1$. Then, with the basis ordered appropriately,

$$\mathbf{B} = \begin{bmatrix} \mathbf{A}_1 & \mathbf{0} \\ \mathbf{A}_2 & \mathbf{I} \end{bmatrix}.$$

Since $\det(\mathbf{B}) = \det(\mathbf{A}_1)$, the requirement for integer solutions is that $\det(\mathbf{A}_1)$ equal 1 or -1. Noting that $\mathbf{A}_1$ could be any square invertible submatrix of $\mathbf{A}$, we have the following property.

A matrix $\mathbf{A}$ is **totally unimodular** if every square submatrix of $\mathbf{A}$ has determinant 1, -1, or 0.

The determinant is 0 if the submatrix cannot be the x-variable part of a basis, i.e. cannot be the matrix $\mathbf{A}_1$ described earlier. For any basis, the determinant is 1 or -1, and the solution is integral. This result is due to Hoffman and Kruskal (1956), who also show that it is a sufficient condition.

Theorem 12.2: *Let $\mathbf{A}$ be a matrix with all integer entries. The feasible region $\{\mathbf{x} : \mathbf{Ax} \leq \mathbf{b}, \mathbf{x} \geq \mathbf{0}\}$ has integer extreme points for every integer vector $\mathbf{b}$ if and only if $\mathbf{A}$ is totally unimodular.*

The matrix $\mathbf{A}$ for the minimum cost network flow problem is totally unimodular; most other linear programs used in applications are not. For $\mathbf{A}$ to be totally unimodular, all entries must be 1, -1, or 0. Typically, each column must contain at most two nonzero entries. There are methods to check whether a matrix with these properties is totally unimodular. See Schrijver (1986) for more discussion.

Problems

1 Consider a minimization integer program with optimal value z_{IP} and its linear programming relaxation with optimal value z_{LP}.
   (a) If the relaxation has a unique, noninteger optimal solution, how are z_{IP} and z_{LP} related? Give a reason.
   (b) Give an example with two variables where $z_{\text{LP}} = z_{\text{IP}}$ but there is not an integer extreme point optimal solution to the linear program.

2 Find an ideal formulation of the solutions of
$$
\begin{aligned}
x - y &\geq -1 \\
2x + 4y &\geq 3 \\
5x + 2y &\leq 14 \\
x, y \geq 0, \text{ integer.}
\end{aligned}
$$

3 Find an ideal formulation of the solutions of

$$4x + 3y \leq 12$$
$$x + 2y \leq 6$$
$$x, y \geq 0, \text{ integer.}$$

4 Find an ideal formulation of the solutions of

$$6x + 3y \geq 9$$
$$8x - 5y \leq 16$$
$$2y \leq 5$$
$$x, y \geq 0, \text{ integer.}$$

5 If we extend the definition of ideal formulations to infinite sets, not all of them have ideal formulations using a finite number of linear inequalities. Find an ideal formulation of the solutions of

$$-2x + 2y \leq 1$$
$$2x - 4y \leq 3$$
$$x, y \geq 0, \text{ integer.}$$

6 Show that the aggregate constraints

$$x_1 + x_2 \leq 2y$$
$$0 \leq x_1, x_2, y \leq 1$$

are a weaker formulation than the individual constraints

$$x_1 \leq y$$
$$x_2 \leq y$$
$$0 \leq x_1, x_2, y \leq 1$$

for the same integer feasible region obtained by adding the constraint that x_1, x_2, y are binary.

7 Which of the following is a stronger formulation, P_1, P_2, or neither?

$$P_1: \quad x - y \leq 1$$
$$2x + y \geq 3$$
$$y \leq 2,$$
$$P_2: \quad x - y \leq 1$$
$$3x + 2y \geq 5$$
$$y \leq 2.$$

8 For Example 12.1, find an objective function for which the linear program and integer program have the same optimal solution.

9 Use the method of Example 12.4 to find bounds on each variable given the constraints

$$
\begin{aligned}
2x_1 + 5x_2 + 6x_3 &\le 20 \\
4x_1 - x_2 - 2x_3 &\le 8 \\
3x_1 + x_2 + x_3 &\ge 12 \\
x_j \ge 0, \text{ integer.}
\end{aligned}
$$

10 Consider a feasible network flow problem on a connected graph with node-arc incidence matrix $\mathbf{A}$. If the direction of all arcs is reversed, a new problem is created. Call its coefficient matrix $\overline{\mathbf{A}}$.
(a) How is $\overline{\mathbf{A}}$ related to $\mathbf{A}$?
(b) Use the fact that $\mathbf{A}$ is totally unimodular to show that $\overline{\mathbf{A}}$ is totally unimodular.
(c) The new problem might not be feasible. How can the supplies and demands $\mathbf{b}$ be changed to make sure it is feasible? With this change, assuming $\mathbf{b}$ is an integer, does the new problem have an integer optimal solution?

11 Show that the following matrix is unimodular using the definition.

$$
\begin{bmatrix}
1 & 0 & -1 \\
0 & -1 & 1 \\
-1 & 1 & 0
\end{bmatrix}.
$$

13

Integer Programming Algorithms

This chapter discusses algorithms for solving integer programs. In contrast to linear programming, success in solving integer programs or other discrete optimization problems often requires methods that are tailored to an individual model or application. No efficient generic algorithm is available. Thus, while linear programs are usually solved by simply running a solver, an integer programming model might require reformulating the problem as discussed in Chapter 12, selecting a good algorithm, or even customizing an algorithm for the model. Accordingly, two key ideas are presented that can be combined in different ways when seeking to find a good algorithm for a particular model. Examples of tailored approaches can be found in Bertsimas and Tsitsiklis (1997, Chapter 12).

Section 13.1 begins by describing a fundamental strategy, called branch and bound, to avoid having to enumerate all the possible values of the integer variables. This "divide and conquer" approach relies on using relaxations to quickly compute bounds on the optimal objective function value. Its success depends largely on the quality of the bounds. We use linear programming relaxations for bounds, but also describe a Lagrangian relaxation. Section 13.2 presents a method of successively improving the quality of the formulation by adding constraints called cutting planes. With these added constraints, the linear programming relaxation provides better bounds.

13.1   Branch and Bound Methods

As long as the variables have upper and lower bounds, an integer program has a finite number of possible solutions that can be enumerated. A brute-force method of solving it is to check every possible solution to see if it is feasible, evaluating the objective function and maintaining the best value. This approach, called *complete enumeration*, is too slow for even modest size problems. The number of possible solutions to check is exponential in the number of variables. Increased computing power will only allow slightly larger examples to be solved, so this is a permanent limitation. Branch and bound methods improve on complete enumeration, finding an optimal solution after only examining certain solutions. Solutions are repeatedly partitioned and sets of solutions are eliminated using objective function bounds.

To describe the algorithm conceptually, suppose we are solving (P):

$$\max \quad \mathbf{c}^T\mathbf{x}$$
$$\text{s.t.} \quad \mathbf{x} \in S,$$

where S contains the integer feasible solutions to linear constraints, so that (P) is an integer program. When we partition S into S_1 and S_2, it creates the subproblems (P1) and (P2):

$$\max \quad \mathbf{c}^T\mathbf{x}$$
$$\text{s.t.} \quad \mathbf{x} \in S_i,$$

$i = 1, 2$. The partitioning is repeated as many times as needed. For example, S_1 might be partitioned into S_3 and S_4, generating subproblems (P3) and (P4). The algorithm needs to be able to compute upper bounds on the optimal value of each subproblem. We will solve a linear programming relaxation to obtain a bound z_{LP} on subproblem i. A current lower bound L on (P) is also kept; it can be obtained from any integer feasible solution that is found. The best feasible solution found so far is called the *incumbent*. At each iteration, the steps in a generic branch and bound algorithm are as follows:

Step 1. **Subproblem Selection.** Choose one of the active subproblems.
Step 2. **Relaxation.** Solve the linear programming relaxation of the current subproblem, either showing it to be infeasible or obtaining z_{LP}.
Step 3. **Infeasibility.** If the relaxation is infeasible, so is the subproblem–delete it.
Step 4. **Bound.** If $z_{LP} \leq L$, the subproblem cannot improve on the current lower bound–delete it.
Step 5. **Subproblem Solution.** If $z_{LP} > L$ and the optimal solution to the linear programming relaxation is integral, then a better feasible solution has been found. Update the lower bound and incumbent solution and delete the subproblem.
Step 6. **Branch.** If $z_{LP} > L$ but the linear programming optimal solution is not integral, then further partition its feasible set. Replace the subproblem with these new subproblems on the active list.

A complete algorithm must also specify (i) how to choose an active subproblem and (ii) how to further partition a subproblem. We address (ii) first.

13.1.1   Branching and Trees

We will use the following method to partition a subproblem. If the optimal solution $\mathbf{x}^*$ to the linear programming relaxation of the subproblem is not integer, we choose an index j for which x_j^* is not integer as the *branching variable* and create two subproblems. For example, if $x_2^* = 3.7$, we add the constraints $x_2 \leq 3$ or $x_2 \geq 4$. Because $x_2^* = 3.7$ is infeasible for the relaxations of the subproblems, their optimal solutions will differ from $\mathbf{x}^*$. This method of branching not only partitions the feasible integer solutions, but also eliminates part of the feasible region of the linear programming relaxation. If more than one variable has a fractional value in $\mathbf{x}^*$, we will choose the one closest to an integer value. For example, if $\mathbf{x}^* = (0, 3.7, 1.2, 1)$ then x_3 is chosen as the branching variable, the constraint $x_3 \leq 1$ is added to one subproblem and $x_3 \geq 2$ to the other. One could also justify branching on the variable that is farthest from an integer value, since it forces a larger change in $\mathbf{x}^*$ and could lead to a larger decrease in the linear program bound z_{LP}. In practice, solvers do more computation to choose a branching variable.

It is helpful to think of (P) and the subproblems as nodes in a decision tree. The tree starts with (P), called the *root node*. When we branch from a node, two nodes are added below it, each called a *child* and the original node, called their *parent*, is no longer active. Of the two child nodes, the *nearest child* is the one whose constraint is closest to the value of the branching variable in the parent. In the example previously, where the parent has solution $\mathbf{x}^* = (0, 3.7, 1.2, 1)$ and x_3 is the branching variable, the nearest child is the subproblem with $x_3 \leq 1$ added.

At each iteration, the *active nodes* are some of the "leaves" of the tree. A *leaf node* is connected only to one node above it, with no nodes below it. Inactive leaf nodes have a special meaning: they do not need to be explored further because they are infeasible, do not contain solutions better than the incumbent, or have been solved. Some authors call these nodes *fathomed*, alluding to having gone as far down the tree as needed, or *pruned*, because the potential subproblems that could be formed as children of this node are not needed.

13.1.2   Choosing a Subproblem

The order in which subproblems are examined will affect the speed of the algorithm. Two considerations when choosing a subproblem are how far down the tree it is and how good a bound it has. Since the relaxations of active subproblems have not been solved yet, the best bound on a subproblem is z_{LP} for its parent. Notice that as one moves up the tree, the linear programs associated with the nodes are relaxations of the ones at the nodes below, so the bound for the parent (and its parent, etc.) are also upper bounds for the nodes below. Two common rules for choosing a subproblem are

1. *Best-first search.* Choose the subproblem with the best bound z_{LP} of its parent among all active subproblems. For a maximization problem, this means choosing the largest parent z_{LP}. If both child nodes are active, choose the nearest child.

2. *Depth-first search.* This method moves down the tree from parent to child as far as possible, then moves back up the tree until a node is reached that has active leaf nodes below it. More specifically, if the current subproblem was branched from, then choose its nearest child from the two nodes that were just created. If it was made inactive without branching, move back up the tree to a node that has an active leaf node below it (its child, or child of a child, etc.), then down to this leaf node.

Both methods choose the original problem (P) in the first iteration, when it is the only active node. Depth-first prioritizes nodes further down the tree that have the advantage that they are closer to being solved, in that they have smaller feasible regions. However, they have the disadvantage of exploring a smaller part of the feasible region, so that more nodes, or iterations, might be needed to solve the problem. Best-first also has the disadvantage of needing to store a larger number of subproblems and their bounds. One approach that has worked well is to use depth-first search until a feasible solution and bound L is found, then use best-first search.

For example, in the tree in Figure 13.1, the leaves are 2, 5, 7, 8, 9, 10. Of these, the active nodes are 2, 5, 8, 10. Node 4 is the child of node 1 and the parent of 7 and 8. If the last node deleted was 9, then depth-first search from this point would choose node 10 next. After 10 (and any children created and their children, etc.) is chosen and deleted, node 5 is chosen, then 8, and finally 2.

Now we can describe the complete algorithm for solving a maximization integer program. We assume the problem is bounded.

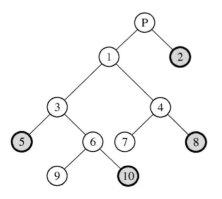

Figure 13.1 A branch and bound tree.

Algorithm 13.1 (Branch and Bound)

Step 0. Initialization. Make (P) the only active subproblem. Set the lower bound $L = -\infty$.

Step 1. Stopping and subproblem selection. If there are no active subproblems, stop. If there is an incumbent solution, it is optimal. Otherwise, the problem is infeasible. If there are active subproblems, choose one.

Step 2. Linear programming relaxation. Solve the linear programming relaxation of the current subproblem, either showing it to be infeasible or finding the solution $\mathbf{x}^*$ with value z_{LP}.

Step 3. Infeasibility. If the relaxation is infeasible, delete the current subproblem. Go to Step 1.

Step 4. Bound. If $z_{LP} \leq L$, the subproblem cannot improve on the lower bound. Delete it and go to Step 1.

Step 5. Subproblem solution. If $z_{LP} > L$ and $\mathbf{x}^*$ is integer, then a better feasible solution has been found. Update the lower bound L to z_{LP} and make $\mathbf{x}^*$ the incumbent. Delete the current subproblem and go to Step 1.

Step 6. Branch. If $z_{LP} > L$ and $\mathbf{x}^*$ is not integer, then choose a fractional variable x_j^* that is closest to integer. Replace the subproblem with two new subproblems on the active list, one with the constraint $x_j \leq \lfloor x_j^* \rfloor$ added and one with the constraint $x_j \geq \lceil x_j^* \rceil$ added. Go to Step 1.

If the objective function coefficient vector $\mathbf{c}$ is integer, then we can improve the bound z_{LP} slightly by rounding it down. If $z_{\mathrm{LP}} = 48.7$, for example, then the best possible integer solution in its subproblem has $z = 48$. Rounding down should be used in the exercises.

Example 13.1. Consider the integer program

$$\begin{aligned}
\max \quad & 5x + 2y \\
\text{s.t.} \quad & 3x + 11y \leq 44 \\
& 2x - 11y \leq 2 \\
& -9x + 14y \leq 21 \\
& x, y \geq 0, \text{ integer,}
\end{aligned}$$

which was solved in Example 12.1. We will apply the branch and bound algorithm with depth-first search. Solving the linear program for the root node (the original problem),

(P) $z_{\mathrm{LP}} = 48.98, x = 9.2, y = 1.49$

The solution is not integer and, since $L = -\infty$, it cannot be bounded by an incumbent. We branch on x because it is closest to an integer, creating subproblems (P1)

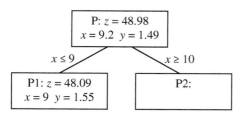

Figure 13.2 Branch and bound tree for Example 13.1 after the first branching.

with $x \leq 9$ and (P2) with $x \geq 10$, which are the active subproblems. The value $x = 9.2$ is closer to the constraint $x \leq 9$, so (P1) is the nearest child. Thus, we choose (P1) and solve its linear program:

(P1) $(x \leq 9) : z_{LP} = 48.09, x = 9, y = 1.55$
(P2) $(x \geq 10)$

At this point we have the tree shown in Figure 13.2. Again the solution is not integer and we branch on y, creating (P3) and (P4). The active subproblems are (P2), (P3), and (P4). Depth-first chooses (P4) as the closest child of (P1) and solves its linear program:

(P3) $(x \leq 9, y \leq 1)$
(P4) $(x \leq 9, y \geq 2) : z_{LP} = 40.67, x = 7.33, y = 2$

The solution is not integer and we branch on x, creating (P5) and (P6). The linear program is solved for (P5), the closest child of (P4):

(P5) $(x \leq 9, y \geq 2, x \leq 7) : z_{LP} = 39.18, x = 7, y = 2.09$
(P6) $(x \leq 9, y \geq 2, x \geq 8)$

The solution is still not integer and we branch on y, creating (P7) and (P8) and choosing (P7) as the closest child of (P5):

(P7) $(x \leq 9, y \geq 2, x \leq 7, y \leq 2) \equiv (x \leq 7, y = 2) : z_{LP} = 39, x = 7, y = 2$
(P8) $(x \leq 9, y \geq 2, x \geq 8, y \geq 3)$

Its linear program has an integer solution, so (P7) is deleted and $x = 7$, $y = 2$ becomes the incumbent with $L = z_{LP} = 39$. The active subproblems are the leaf nodes (P2), (P3), (P6), and (P8). Using depth-first, the nearest is (P8), which is infeasible, so it is deleted. Next, we select (P6). Its linear program is also infeasible, so it is deleted. Moving up to (P3),

(P3) $(x \leq 9, y \leq 1) : z_{LP} = 34.5$

This subproblem is deleted because it is bounded by the incumbent, $z_{LP} < L = 39$. Finally, the linear program for (P2) is infeasible. There are no more active subproblems, so the incumbent $(x, y) = (7, 2)$ with value $z_{IP} = 39$ is optimal. Figure 13.3 shows the complete branch and bound tree.

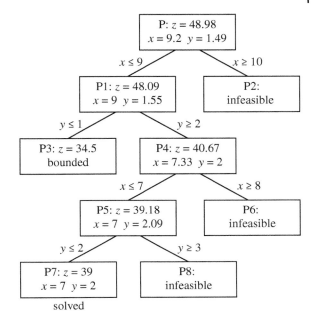

Figure 13.3 Branch and bound tree for Example 13.1.

Notice that the subproblems are not evaluated in the same order that they are created, so they are not solved in numerical order. For this example, they are evaluated in the order 1, 4, 5, 7, 8, 6, 3, 2. In a more simplistic version of the algorithm, subproblems are evaluated when they are created: when branching in Step 6, the linear programming relaxation is solved and Steps 3–5 performed for both subproblems created. Then Step 1 would select a subproblem to branch from whose linear program has already been solved.

Although it is useful for illustration, the algorithm is not very effective on this example. Eight subproblems were evaluated, while there are only 15 feasible integer solutions. Next we consider a knapsack problem, introduced in Example 5.2, with five binary variables. When a constraint is added for a binary variable, it fixes the value at 0 or 1. Thus, rather than writing the constraint as, e.g. $x_5 \geq 1$, we write $x_5 = 1$. Also, each level of the tree fixes one binary variable. The depth of the largest possible tree is the number of variables – five in our example, with at most 2^5 leaves. The goal of the branch and bound method is to avoid exploring the whole tree. An advantage of using the branch and bound algorithm on a knapsack problem is that the linear programming relaxation is very easy to solve; its closed-form solution is given in (5.2).

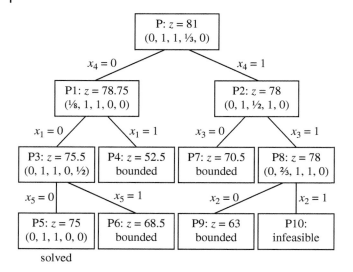

Figure 13.4 Branch and bound tree for Example 13.2.

Example 13.2. Consider the knapsack problem

$$\max 30x_1 + 45x_2 + 30x_3 + 18x_4 + 10x_5$$
$$\text{s.t.} \quad 8x_1 + 6x_2 + 4x_3 + 3x_4 + 2x_5 \le 11$$
$$x_j \ge 0, \text{binary.}$$

Applying the branch and bound algorithm with depth-first search produces the tree shown in Figure 13.4. The subproblems are examined in the order 1, 3, 5, 6, 4, 2, 7, 8, 10, 9. The first incumbent solution found, $x = (0, 1, 1, 0, 0)$ in (P5), turns out to be optimal.

The branch and bound tree in Figure 13.4 has a depth of three (three fixed variables), compared with the possible depth of five. The tree has six leaves, compared with the 2^5 leaves of a complete enumeration tree for five binary variables. The optimal solution happened to be found at the third subproblem examined, which explains why many subproblems were bounded.

If best-first search is used to select subproblems in this example, they are examined in the order 1, 2, 3, 4, 7, 8, 9, 10, 3, 5, 6. Because (P4), (P7), and (P9) are evaluated before the bound in (P5) is obtained, they would not be deleted. Branching at these three nodes creates six more subproblems not shown in Figure 13.4. These six subproblems would then be evaluated and each found to be bounded or infeasible. Thus, best-first search takes longer on this problem. However, it also suggests an improvement to the algorithm: when a new incumbent solution is found with value L, delete all active nodes whose parent bound z_{LP} is no better, i.e. $\lfloor z_{LP} \rfloor \leq L$.

Next we mention two variations of the branch and bound algorithm.

13.1.3   Mixed Integer Programs

The branch and bound algorithm can readily be applied to mixed integer programs. Let J be the set of indices j for which x_j is an integer variable. We interpret the requirement that $\mathbf{x}^*$ is integer in Steps 5 and 6 as x_j is integer for $j \in J$. When branching, we branch on a variable $x_j, j \in J$, that has a fractional value x_j^*.

13.1.4   Integer Lagrangian Relaxations

As discussed in Section 12.2, the linear programming relaxation may or may not provide a good bound, depending on how the integer program is formulated. The *Lagrangian relaxation* discussed in Section 8.4 is a popular alternative that gives better bounds for some problems. Instead of relaxing the integer constraints, a Lagrangian relaxation "dualizes" some of the linear constraints by replacing them with penalty terms in the objective function, called the Lagrangian. For an integer program in the form

$$z_{IP} = \max \quad \mathbf{c}^T \mathbf{x}$$
$$\text{s.t.} \quad \mathbf{Ax} \leq \mathbf{b}$$
$$\mathbf{Dx} \leq \mathbf{d}$$
$$\mathbf{x} \geq \mathbf{0}, \text{integer,}$$

we dualize the constraints $\mathbf{Ax} \leq \mathbf{b}$, giving the Lagrangian relaxation

$$\max \quad \mathbf{c}^T \mathbf{x} + \lambda^T (\mathbf{b} - \mathbf{Ax})$$
$$\text{s.t.} \quad \mathbf{Dx} \leq \mathbf{d} \tag{13.1}$$
$$\mathbf{x} \geq \mathbf{0}, \text{integer.}$$

Let $g(\lambda)$ be its optimal objective function value. Here λ_i is the Lagrange multiplier associated with the ith constraint in $\mathbf{Ax} \leq \mathbf{b}$. We would like (13.1) to be an upper bound for the integer program. To accomplish this, we restrict $\lambda \geq \mathbf{0}$. Letting $\mathbf{x}^*$ be an optimal solution to the original integer program,

$$g(\lambda) \geq \mathbf{c}^T \mathbf{x}^* + \lambda^T (\mathbf{b} - \mathbf{Ax}^*) \geq \mathbf{c}^T \mathbf{x}^* = z_{IP}.$$

The first inequality holds because $\mathbf{x}^*$ is feasible for the Lagrangian relaxation, the second because it is feasible for the integer program.

To use a Lagrangian relaxation in the branch and bound algorithm, we must

- Relax enough constraints $\mathbf{Ax} \le \mathbf{b}$ so that the remaining constraints in (13.1) are tractable. The Lagrangian relaxation is still an integer program, so the remaining constraints generally need to have a special structure to be tractable.
- Choose λ so that the bound $g(\lambda)$ is strong, that is, fairly close to z_{IP}. A search is usually required to find useful values of the Lagrange multipliers.

There is a theoretical justification for using the Lagrangian relaxation of an integer program, based on integer programming duality. As in linear programming, Lagrange multipliers can be interpreted as dual variables. The optimal value of the Lagrangian relaxation, $g(\lambda)$, is the dual function and minimizing it to give the tightest bound, $z_D = \min_{\lambda \ge 0} g(\lambda)$, is the dual problem. The fact that $z_{IP} \le g(\lambda)$ means that weak duality holds for integer programs. Strong duality does not hold. In general, there is a duality gap $z_D - z_{IP} > 0$ and no choice of λ in the Lagrangian dual gives a perfect bound. However, the optimal dual value can be characterized as the optimal value of a linear program with a large number of constraints. As in Section 12.2, the key idea is an ideal formulation of the constraints. In this case, the ideal formulation is for the constraints $\mathbf{Dx} \le \mathbf{d}, \mathbf{x} \ge \mathbf{0}$ that are not relaxed. See Schrijver (1986) or Bertsimas and Weismantel (2005). The Lagrangian relaxation method was introduced in Held and Karp (1970, 1971) for the travelling salesperson problem.

Lagrangian duality will also be described for convex programs in Chapter 14.

13.2 Cutting Plane Methods

Given a noninteger optimal solution $\mathbf{x}^*$ to the linear programming relaxation, the branch and bound method adds a constraint that is satisfied by only some integer feasible solutions, partitioning them. Another approach is to add a constraint that is satisfied by all integer feasible solutions, but not by $\mathbf{x}^*$. Adding such a constraint cuts off part of the feasible region of the linear program, including $\mathbf{x}^*$. In the terminology of Section 12.2, adding this constraint gives a stronger formulation.

A **valid inequality** for a set of integer solutions S is one that is satisfied by all of the solutions in S.

Consider a polyhedron P that is a formulation of S, that is, $S \subset P$ and every integral point in P is in S. For the integer program $\max_{\mathbf{x} \in S} \mathbf{c}^T \mathbf{x}$, let $\mathbf{x}^*$ be an optimal solution of the linear programming relaxation $\max_{\mathbf{x} \in P} \mathbf{c}^T \mathbf{x}$. A *cutting plane* is a valid linear inequality for S that is not satisfied by $\mathbf{x}^*$. In Figure 13.5, P is the shaded

Figure 13.5 A cutting plane for **x***.

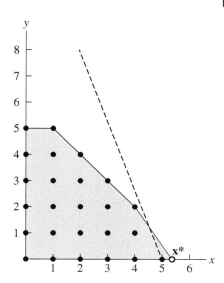

region and S contains the integer points in P. The dashed line represents the valid inequality $11x + 4y \leq 55$, which is a cutting plane for $\mathbf{x}^* = (5\frac{1}{3}, 0)$. This region is from (1.5), where $\mathbf{x}^*$ is an optimal solution of the linear programming relaxation.

A cutting plane can be thought of as a separating hyperplane: since it also is satisfied by the convex hull of S, it separates this polyhedron from $\mathbf{x}^*$. Cutting plane algorithms successively add cutting planes to the integer program formulation and solve the linear programming relaxation until its solution is integer.

In this section we describe a general cutting plane algorithm, then present in detail one method of finding a cutting plane, called a Gomory cut. Although Gomory cuts are of some use, other cutting planes are often used. Next we describe a general method of constructing valid inequalities. To close the section, a stronger type of cut is presented for binary variables.

A general cutting plane algorithm has the following steps:

Step 1. *Linear programming relaxation.* Solve the linear programming relaxation of the current formulation. If it is infeasible, stop. The problem is infeasible. Otherwise, call the optimal solution $\mathbf{x}^*$.

Step 2. *Optimality check.* If $\mathbf{x}^*$ is integer, stop. It is an optimal solution.

Step 3. *Cutting plane.* Find one or more valid inequalities $\mathbf{A}_{\text{new}}\mathbf{x} \leq b_{\text{new}}$ for the integer program that are not satisfied by $\mathbf{x}^*$. Add them to the current formulation. Go to Step 1.

Although a linear program must be solved at each iteration, it can be solved fairly easily by starting with the solution of the previous linear program. When the constraint(s) are added, $\mathbf{x}^*$ becomes infeasible but can be used as an initial solution in the dual simplex method. The number of constraints grows as the algorithm progresses, but the speed of the dual simplex method is generally not too sensitive to the number of constraints.

13.2.1 Gomory Cutting Plane Method

The first algorithm for integer programming shown to take a finite number of steps was Gomory's cutting plane algorithm (Gomory, 2010). However, because they can converge very slowly to optimal, cutting plane methods were not used for many years. More recently, these and other cutting planes have proven useful in algorithms that combine them with other approaches; see Cornuéjols (2007). Gomory cuts make use of the information computed in the simplex method. For this reason, we consider the canonical form problem

$$\begin{aligned} \max \quad & \mathbf{c}^T \mathbf{x} \\ \text{s.t.} \quad & \mathbf{Ax} = \mathbf{b} \\ & \mathbf{x} \geq \mathbf{0}, \text{integer}. \end{aligned} \tag{13.2}$$

Let $\mathbf{x}^*$ be an optimal basic feasible solution to the linear programming relaxation with basis matrix $\mathbf{B}$. Partitioning into basic and nonbasic variables (see Section 7.3), the equality constraint can be written

$$\mathbf{Bx}_B + \mathbf{Nx}_N = \mathbf{b},$$

or

$$\mathbf{x}_B + \mathbf{B}^{-1}\mathbf{Nx}_N = \mathbf{B}^{-1}\mathbf{b}.$$

Now, if $\mathbf{x}^*$ is not integer, let $x^*_{B(i)}$ be fractional, where $x^*_{B(i)}$ is the ith component of $\mathbf{x}^*_B$. Let $\overline{\mathbf{A}} = \mathbf{B}^{-1}\mathbf{A}$ and $\overline{\mathbf{b}} = \mathbf{B}^{-1}\mathbf{b}$. Then the ith equation is

$$x_{B(i)} + \sum_{j \in \mathcal{N}} \overline{a}_{ij} x_j = \overline{b}_i \tag{13.3}$$

and $\overline{b}_i$ is fractional. Assuming the problem is feasible, (13.3) is satisfied by some integral $\mathbf{x}$, so some of the coefficients $\overline{a}_{ij}$ must also be fractional. Since feasible $\mathbf{x}$ satisfy $\mathbf{x} \geq \mathbf{0}$, we can round down $\overline{a}_{ij}$ and obtain the valid inequalities

$$x_{B(i)} + \sum_{j \in \mathcal{N}} \lfloor \overline{a}_{ij} \rfloor x_j \leq \overline{b}_i$$

and, rounding down $\overline{b}_i$,

$$x_{B(i)} + \sum_{j \in \mathcal{N}} \lfloor \overline{a}_{ij} \rfloor x_j \leq \lfloor \overline{b}_i \rfloor. \tag{13.4}$$

This inequality is violated by $\mathbf{x}^*$ because $\overline{b}_i$ is fractional, so that $\lfloor \overline{b}_i \rfloor < \overline{b}_i$, and $\mathbf{x}_N = \mathbf{0}$. Thus, it is a cutting plane. Another implication is that an integer solution satisfying (13.3) cannot have $\mathbf{x}_N = \mathbf{0}$; some of the nonbasic variables must be nonzero. That implies that

$$\sum_{j \in \mathcal{N}} x_j \geq 1$$

is also a valid inequality violated by $\mathbf{x}^*$.

We describe the algorithm for a canonical form problem (13.2) with n variables. For an integer program with integer data, we can convert to canonical form by adding integer slack variables.

Algorithm 13.2 (Gomory's Cutting Plane Algorithm)

Step 1. Linear programming relaxation. Solve the linear programming relaxation of the current formulation. If it is infeasible, stop. The problem is infeasible. Otherwise, find an optimal basic feasible solution $\mathbf{x}^*$ with basis matrix $\mathbf{B}$.

Step 2. Optimality check. If $\mathbf{x}^*$ is integer, stop. It is an optimal solution.

Step 3. Cutting plane. Choose a basic variable $x_{B(i)}$ that is fractional and farthest from integer. Compute row i of $\overline{\mathbf{A}} = \mathbf{B}^{-1}\mathbf{A}$ and let $\overline{b}_i = x_{B(i)}^*$. At iteration k, add the constraint

$$x_{B(i)} + \sum_{j \in \mathcal{N}} \lfloor \overline{a}_{ij} \rfloor x_j + x_{n+k} = \lfloor \overline{b}_i \rfloor$$

and the integer slack variable x_{n+k} to the current formulation. Go to Step 1.

The slack variable is named so that it can be appended to the current vector $\mathbf{x}$. For example, if the initial variables are (x_1, x_2, x_3), then x_4 is added in the first iteration, x_5 in the second, etc. Choosing a variable farthest from integer is rather arbitrary. One could also add more than one cut at an iteration. The coefficients $\overline{a}_{ij}$ needed in the algorithm are already available if the simplex method is used to solve the linear programs.

Example 13.3. Consider the integer program

$$\begin{aligned}
\min \quad & 9x_1 + 12x_2 \\
\text{s.t.} \quad & 3x_1 + x_2 \geq 4 \\
& x_1 + 2x_2 \geq 6 \\
& x_j \geq 0, \text{ integer.}
\end{aligned}$$

To apply Gomory's cutting plane algorithm, first convert to the canonical form

$$\min 9x_1 + 12x_2$$
$$\text{s.t. } 3x_1 + x_2 - x_3 = 4$$
$$x_1 + 2x_2 \qquad -x_4 = 6$$
$$x_j \geq 0, \text{ integer.}$$

Solving the linear programming relaxation gives $\mathbf{x}^* = (0.4, 2.8, 0, 0)$ with basis (x_1, x_2). We choose x_1 because it is farthest from integer. To form its cut, we need row 1 of

$$\mathbf{B}^{-1}\mathbf{N} = \begin{bmatrix} -0.4 & 0.2 \\ 0.2 & -0.6 \end{bmatrix}.$$

The cut is

$$x_1 + \lfloor -0.4 \rfloor x_3 + \lfloor 0.2 \rfloor x_4 + x_5 = \lfloor 0.4 \rfloor$$

or

$$x_1 - x_3 + x_5 = 0.$$

For iteration 2, we add this constraint to the linear program, giving $\mathbf{x}^* = (\frac{2}{3}, \frac{8}{3}, \frac{2}{3}, 0, 0)$ with basis (x_1, x_2, x_3). We choose x_2, which is tied in distance from an integer. Using row 2 of

$$\mathbf{B}^{-1}\mathbf{N} = \begin{bmatrix} \dfrac{1}{3} & -\dfrac{2}{3} \\ -\dfrac{2}{3} & \dfrac{1}{3} \\ \dfrac{1}{3} & -\dfrac{5}{3} \end{bmatrix},$$

the cut is

$$x_2 + \left\lfloor -\dfrac{2}{3} \right\rfloor x_4 + \left\lfloor \dfrac{1}{3} \right\rfloor x_5 + x_6 = \left\lfloor \dfrac{8}{3} \right\rfloor$$

or

$$x_2 - x_4 + x_6 = 2.$$

For iteration 3, we add this constraint to the linear program, giving $\mathbf{x}^* = (2, 2, 4, 0, 2, 0)$. This solution is integer, so $(x_1, x_2) = (2, 2)$ is optimal.

13.2.2 Generating Valid Inequalities by Rounding

Gomory cuts may only cut off a very small portion of the feasible region of the linear program. Much better cuts are available for some types of problems, and somewhat better cuts for general integer programs. To develop an appreciation for these other methods, we present a general approach to finding valid inequalities

Figure 13.6 Feasible region for Example 13.4. The dashed line is a C–G inequality that defines an edge of conv(S).

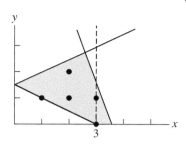

called *rounding*. The idea is to create a linear combination of constraints that has fractional coefficients and then round them down.

Example 13.4. Consider the set S of integer points satisfying the constraints

$$
\begin{aligned}
-2x + 3y &\le -6 \\
-x + 2y &\le 3 \\
5x + 2y &\le 18 \\
x, y &\ge 0, \text{integer}
\end{aligned}
$$

and the corresponding linear relaxation P without the integer constraints, shown in Figure 13.6. Multiplying the second constraint by $\frac{1}{4}$, multiplying the third constraint by $\frac{1}{4}$, and adding,

$$
x + y \le \frac{21}{4}.
$$

This inequality holds for all points in P. If we consider only integer solutions, we can round down and obtain a valid inequality for S,

$$
x + y \le 5,
$$

which is not valid for P; it cuts off the top of the region in Figure 13.6. Multiplying the third constraint by 0.2,

$$
x + 0.4y \le 3.6.
$$

For $(x, y) \in P, y \ge 0$ so we can round the left-hand side down, and $x \le 3.6$ is valid for P. Rounding down the right-hand side, $x \le 3$ is valid for S.

Notice that it would not be correct to round down the right-hand side before the left. In particular, the inequality

$$
x + 0.4y \le 3
$$

is violated by the point $(2, 3)$ that satisfies the third constraint. The left-hand side is not integer, so it is not valid to round down the right-hand side. To generalize, consider the set S of integer points satisfying the constraints

$$\mathbf{Ax} \leq \mathbf{b}$$

$$\mathbf{x} \geq \mathbf{0}, \text{integer}$$

and the corresponding linear relaxation P without the integer constraints. For any $\mathbf{u} \geq \mathbf{0}$,

$$\mathbf{u}^T \mathbf{Ax} \leq \mathbf{u}^T \mathbf{b}$$

holds for all $\mathbf{x} \in P$. Writing $\lfloor \mathbf{x} \rfloor$ to denote rounding down each component of $\mathbf{x}$ to the nearest integer,

$$\lfloor \mathbf{u}^T \mathbf{A} \rfloor \mathbf{x} \leq \mathbf{u}^T \mathbf{Ax} \leq \mathbf{u}^T \mathbf{b} \tag{13.5}$$

for $\mathbf{x} \in P$ (because $\mathbf{x} \geq \mathbf{0}$). Then, for $\mathbf{x} \in S$, the left-hand side is integer, so we can round down the right-hand side. Thus, for any $\mathbf{u} \geq \mathbf{0}$,

$$\lfloor \mathbf{u}^T \mathbf{A} \rfloor \mathbf{x} \leq \lfloor \mathbf{u}^T \mathbf{b} \rfloor \tag{13.6}$$

is a valid inequality for S. Equation (13.6) is known as a *Chvátal–Gomory inequality*, or *C–G inequality*, developed in Gomory (2010) and Chvátal (1973). It is a very general approach, since any $\mathbf{u} \geq \mathbf{0}$ can be used. Gomory cuts are a special case of C–G inequalities where the constraint multipliers $\mathbf{u}$ are a row of $\mathbf{B}^{-1}$, where $\mathbf{B}$ is a basis matrix.

The goal when adding valid inequalities is to move closer to the ideal formulation, where the linear relaxation is conv(S). In Example 13.4, the first C–G inequality does not intersect conv(S), but the second, $x \leq 3$ shown in Figure 13.6, defines an edge of the polygon conv(S). It is one of the inequalities needed to define conv(S). Similarly, for a polyhedron of dimension three some valid inequalities have boundary planes that intersect the polyhedron at a face of the polyhedron, which is a polygon of dimension two, and these inequalities define the polyhedron. In contrast, if the boundary plane of a valid inequality intersects the polyhedron at an edge (dimension of one), an extreme point (dimension of zero), or not at all, it is not needed to define the polyhedron. The nomenclature for an n-dimensional polyhedron is that the intersection of the polyhedron and the hyperplane of a valid inequality is called the *face* defined by the inequality, regardless of its dimension. A *facet* is a face of dimension $n - 1$ and a *facet defining inequality* is one whose face is a facet. Using this terminology, the facets of a three-dimensional polyhedron are its two-dimensional faces; extreme points and edges are also called faces.

Not all such inequalities can be obtained as C–G inequalities: in general, the solution set to all possible C–G inequalities is larger than conv(S). This set is called the first Chvátal closure and is a polyhedron. It is shown in Chvátal (1973) that, as long as S is bounded, then by repeating the process (i.e. form C–G inequalities from the original inequalities plus the C–G inequalities already generated) one can obtain all of the facet defining hyperplanes and conv(S); see Schrijver (1986) for more discussion.

The C–G inequality can be strengthened somewhat using *mixed integer rounding*, introduced in Nemhauser and Wolsey (1988). Notice that we first weakened the inequality by rounding the left-hand side in (13.5) then strengthened it by rounding the right-hand side. Instead of rounding the left down to integers, we can just decrease it enough to allow the right-hand side to be rounded. For example, if

$$x + 0.8y \leq 3.6$$

is a valid inequality for some $S \subset \mathbb{Z}_+^n$, then

$$x + 0.5y \leq 3 \tag{13.7}$$

is also valid. To see that (13.7) is not more restrictive, observe that for $y = 0$ the integer solutions to both inequalities have $x \leq 3$, for $y = 1$ they have $x \leq 2$, and for $y \geq 2$ the left-hand side of (13.7) is at least $(0.8 - 0.5)2 = 0.6$ less than that of the original inequality. See Marchand and Wolsey (2001) for details.

13.2.3   Valid Inequalities for Binary Variables

Binary variables are easier to handle than general integer variables in several ways. Next we show how to generate stronger cuts for problems with binary variables and nonnegative constraint coefficients. The method generates cuts based on individual constraints, so it can be understood in the context of a knapsack problem with the constraint

$$a_1 x_1 + a_2 x_2 + \cdots + a_n x_n \leq b.$$

We assume $a_j > 0$ and x_j binary for all j. Recall that the linear relaxation replaces the binary constraints with $0 \leq x_j \leq 1$. A set $C \subset \{1, \ldots, n\}$ of items that does not fit in the knapsack,

$$\sum_{j \in C} a_j > b,$$

is called a *cover*. Since they don't all fit, we must choose less than $|C|$ of them, so

$$\sum_{j \in C} x_j \leq |C| - 1 \tag{13.8}$$

is a valid inequality, known as a cover inequality. They can be strengthened as follows. The same limit on the number of items applies if we also consider other items that are at least as large as the largest item in C. We add these items to each cover, defining the *extended cover*

$$E(C) = C \cup \{k : a_k \geq a_j \text{ for all } j \in C\}.$$

The valid inequality is

$$\sum_{j \in E(C)} x_j \leq |C| - 1. \tag{13.9}$$

Incidentally, these are also valid for the linear relaxation. Covers that are too large give weaker constraints; we only need to consider *minimal covers*, namely, one for which removing any item gives a set that is not a cover (the items fit in the knapsack).

Example 13.5. The binary integer program in Example 13.2 had the constraint

$$3x_1 + 5x_2 + 4x_3 + x_4 + 4x_5 + x_6 + 3x_7 \leq 10.$$

A cover for this problem is a set of indices $j \in \{1, \ldots, 7\}$ whose coefficients sum to more than 10. The minimal covers include $\{2, 3, 5\}$, $\{3, 5, 7\}$, and $\{1, 3, 6, 7\}$ with extended covers $\{2, 3, 5\}$, $\{2, 3, 5, 7\}$, and $\{1, 2, 3, 5, 6, 7\}$, respectively. Their valid inequalities are

$$x_2 + x_3 + x_5 \leq 2$$
$$x_2 + x_3 + x_5 + x_7 \leq 2$$
$$x_1 + x_2 + x_3 + x_5 + x_6 + x_7 \leq 3.$$

While cover inequalities are useful, we should note that they do not give the ideal formulation, i.e. they do not include all of the facet defining hyperplanes of the convex hull. The number of minimal covers grows exponentially in the number of items, so rather than generating all inequalities of the form (13.9), they are used in a cutting plane algorithm. Given a noninteger solution $\mathbf{x}^*$ for some linear programming relaxation, one finds a minimal cover inequality that is violated at $\mathbf{x}^*$. This cutting plane is added to the formulation and the process repeated. The following example illustrates the process of adding cover inequalities.

Example 13.6. Consider the knapsack problem

$$
\begin{aligned}
\max \quad & 12x_1 + 16x_2 + 10x_3 + 2x_4 + 4x_5 + x_6 + 2x_7 \\
\text{s.t.} \quad & 3x_1 + 5x_2 + 4x_3 + x_4 + 4x_5 + x_6 + 3x_7 \leq 10 \\
& x_j \geq 0, \text{integer.}
\end{aligned}
$$

The items have been sorted so that the ratios c_j/a_j are in decreasing order. Thus, the linear programming relaxation chooses the items in order and has optimal solution $\mathbf{x}^* = (1, 1, 0.5, 0, 0, 0, 0)$ with optimal value 33. This solution uses the items in the cover $C = \{1, 2, 3\}$ and violates the constraint for the cover. There are no other items with $a_j \geq a_2 = 5$, so the extended cover is just $E(C) = C$. We add the constraint

$$x_1 + x_2 + x_3 \leq 2$$

and resolve the relaxation, giving $\mathbf{x}^* = (1, 1, 0, 1, 0.25, 0, 0)$. This solution uses the items $\{1, 2, 4, 5\}$, but this is not a minimum cover. Eliminating item 4, the minimum cover is $\{1, 2, 5\}$. Its extended cover is just $\{1, 2, 5\}$ with constraint

$$x_1 + x_2 + x_5 \leq 2.$$

Adding this constraint, the solution to the relaxation is $\mathbf{x}^* = (1, 1, 0, 1, 0, 1, 0)$ with optimal value 31. Since it is integral, it is optimal for the original problem.

For a binary integer program with more than one constraint and nonnegative coefficients, the method can be applied to each constraint separately. Cutting plane methods are good at improving formulations, so that the upper bound from the relaxation is tighter, but often cannot find an optimal solution in a practical amount of time. Branch and cut algorithms combine them with branch and bound methods, obtaining much better performance than either method used alone.

Problems

1 Solve the integer program

$$\begin{aligned}
\max \quad & 2x_1 + 5x_2 \\
\text{s.t.} \quad & x_1 + 3x_2 \leq 10 \\
& 9x_1 + 6x_2 \leq 46 \\
& x_j \geq 0, \text{ integer}
\end{aligned}$$

using branch and bound with depth-first search. Show the order in which you solved the subproblems.

2 Repeat Exercise 13.1 using best-first search.

3 Solve the integer program

$$\begin{aligned}
\max \quad & 7x + 3y \\
\text{s.t.} \quad & 2x + 5y \leq 28 \\
& 8x + 3y \leq 48 \\
& x, y \geq 0, \text{ integer}
\end{aligned}$$

using branch and bound with depth-first search. Show the order in which you solved the subproblems.

4 Repeat Exercise 13.3 using the best-first search.

5 Solve the integer program

$$
\begin{aligned}
\min\ & 3x + 6y \\
\text{s.t.}\ & 7x + 2y \geq 28 \\
& x + 6y \geq 12 \\
& x, y \geq 0,\ \text{integer}
\end{aligned}
$$

using branch and bound with depth-first search. Show the order in which you solved the subproblems.

6 Solve the knapsack problem in Example 13.2 using complete enumeration.

7 Solve the knapsack problem

$$
\begin{aligned}
\max\ & 15x_1 + 25x_2 + 9x_3 + 21x_4 \\
\text{s.t.}\ & 3x_1 + 5x_2 + 2x_3 + 4x_4 \leq 6 \\
& x_j\ \text{binary}
\end{aligned}
$$

using branch and bound with depth-first search. Show the order in which you solved the subproblems.

8 Solve the knapsack problem in Exercise 13.7 using complete enumeration.

9 Which of the following are valid inequalities for the integer program in Exercise 13.1?
(a) $9x_1 + 6x_2 \leq 45$
(b) $x_2 \leq 4$
(c) $2x_1 + 4x_2 \leq 12$.

10 For the linear programming relaxation of Exercise 13.3,
(a) Find the C–G inequality for the multipliers $u = (\frac{1}{10}, \frac{1}{6})$
(b) Find the C–G inequality for the multipliers $u = (\frac{1}{5}, 0)$
(c) Find a feasible point that violates (b).

11 It is easier to find a Gomory cut when the tableau simplex method (Section 9.4) is used. The linear program

$$\max \ 2x_1 - x_2 + x_3$$
$$\text{s.t.} \quad x_1 - x_2 + x_3 \leq 4$$
$$2x_1 + x_2 \qquad \leq 10$$
$$x_1 - x_2 - x_3 \leq 7$$
$$x_j \geq 0$$

after adding slack variables x_4, x_5, x_6 has optimal tableau

x_1	x_2	x_3	x_4	x_5	x_6	rhs	
0	0	3	4/3	1/3	0	26/3	
1	0	1	1/3	1/3	0	14/3	x_1
0	1	-2	-2/3	1/3	0	2/3	x_2
0	0	-2	-1	0	1	3	x_6

The constraint rows are in the form (13.3); rounding gives a Gomory cut. Find the Gomory cut for each fractional variable.

12 Use Gomory's cutting plane algorithm to solve the integer program, adding a cut for each fractional variable at each iteration.

$$\max \ 9x_1 + 12x_2$$
$$\text{s.t.} \quad 3x_1 + \ x_2 \leq 4$$
$$x_1 + 2x_2 \leq 6$$
$$x_j \geq 0, \ \text{integer}.$$

13 Use Gomory's cutting plane algorithm to solve the integer program

$$\max \ -x_1 + 2x_2$$
$$\text{s.t.} \quad -4x_1 + 6x_2 \leq 10$$
$$x_1 + \ x_2 \leq 4$$
$$x_j \geq 0, \ \text{integer}.$$

To break ties when choosing a variable for which to add a cut, choose the variable with largest index.

14 For a binary integer program with constraint

$$x_1 + 2x_2 + 4x_3 + 5x_4 + 10x_5 \leq 11.$$

(a) Find all minimum covers and their cover inequalities.
(b) Find the extended cover for each minimal cover found in (a).

15 For a binary integer program with constraint

$$4x_1 + 7x_2 + 3x_3 + 2x_4 + 9x_5 \leq 15.$$

(a) Find all minimum covers and their cover inequalities.
(b) Find the extended cover for each minimal cover found in (a).

16 Consider a binary integer program with constraint

$$10x_1 + 5x_2 + 4x_3 + 7x_4 + 3x_5 + 2x_6 \leq 18.$$

Find a cover inequality that is violated by each of the following feasible solutions to its linear programming relaxation

(a) $\mathbf{x} = (0.6, 1, 0, 1, 0, 0)$
(b) $\mathbf{x} = (0, 0.8, 1, 1, 1, 0)$.

14

Convex Programming: Optimality Conditions

In this chapter we extend the linear programming optimality conditions to non-linear programs, especially convex programs. The approach is an extension of the Lagrange multiplier method, which applies to equality constraints. In Section 8.4 we showed how to extend the method to linear inequality constraints. Now we consider nonlinear constraints and objectives. Section 14.1 presents optimality conditions, called the Karush–Kuhn-Tucker (KKT) conditions, for convex and for general nonlinear problems. They generalize Theorems 7.5 and 7.6 from linear programming. However, instead of the approach of Chapter 7 based on improving feasible directions, the approach uses the Lagrangian and duality.

Next, Section 14.2 provides a brief introduction to duality theory for convex programs. The concepts of dual variables, sensitivity, duality gap, and complementary slackness all have counterparts in the nonlinear setting. One major difference is that strong duality – which asserts that the primal and dual optimal objective function values are the same – does not always hold. However, strong duality does hold for convex programs. As in the linear case, duality and optimality conditions are useful in the design of algorithms.

14.1   KKT Optimality Conditions

This section presents the KKT optimality conditions for nonlinear programs. The method of Lagrange multipliers is reviewed and its connection with linear programming duality is traced. They can be viewed as an extension of the method of Lagrange multipliers to inequality constraints. They are related to the optimality condition of Section 8.3 for linear programs, namely, primal and dual feasibility and complementary slackness. The KKT conditions are important largely because of their role in algorithm design. In Section 11.3, we saw that an interior point algorithm for linear programming attempts to approximately solve the optimality

Linear and Convex Optimization: A Mathematical Approach, First Edition. Michael H. Veatch.
© 2021 John Wiley & Sons, Inc. Published 2021 by John Wiley & Sons, Inc.
Companion website: www.wiley.com/go/veatch/convexandlinearoptimization

conditions, including complementary slackness. For convex programming, many algorithms attempt to solve the KKT conditions.

In this chapter we consider the nonlinear optimization problems

$$\max \quad f(\mathbf{x}) \tag{14.1}$$
$$\text{s.t.} \quad h_i(\mathbf{x}) = b_i, \ i = 1, \dots, m$$

with equality constraints and

$$\max \quad f(\mathbf{x}) \tag{14.2}$$
$$\text{s.t.} \quad g_i(\mathbf{x}) \leq b_i, \ i = 1, \dots, m$$

with inequality constraints. We assume that f, g_i, and h_i are differentiable. It will be clear how to combine equality and inequality constraints; we separate them for notational simplicity. We have chosen notation and conventions (maximization, "$\leq$" constraints) close to our linear program standard form. We will explain how the results can be translated to minimization problems.

14.1.1 Equality Constraints

We begin by reviewing the method of Lagrange multipliers, which was also mentioned in Section 8.4. For (14.1), define the Lagrangian

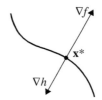

Figure 14.1
Optimality
condition with
one constraint
$g(\mathbf{x}) = b.$

$$L(\mathbf{x}, \lambda) = f(\mathbf{x}) + \sum_{i=1}^{m} \lambda_i(b_i - h_i(\mathbf{x})),$$

where λ_i is called the *Lagrange multiplier* for constraint i. To understand the role of the Lagrangian, consider a problem with one constraint h and two variables. The constraint gradient $\nabla h(\mathbf{x})$ is perpendicular to the curve $h(\mathbf{x})$ at any $\mathbf{x}$. At a local maximum of f, ∇f must also be perpendicular, so ∇f and ∇h must be parallel as shown in Figure 14.1. Here the vectors ∇f and ∇h have been translated to $\mathbf{x}^*$.

Similarly, for a problem with two constraints and three variables, if it is feasible the constraints intersect in a smooth curve. The constraint gradients ∇h_1 and ∇h_2 are perpendicular to this curve at any point on the curve. At a local maximum $\mathbf{x}^*$, let $\mathbf{y}$ be a vector in the direction of the tangent line to the curve. Then $\nabla h_1(\mathbf{x}^*)$, $\nabla h_2(\mathbf{x}^*)$, and $\nabla f(\mathbf{x}^*)$ are all perpendicular to $\mathbf{y}$, so they lie in a plane with normal vector $\mathbf{y}$. Assuming $\nabla h_1(\mathbf{x}^*)$ and $\nabla h_2(\mathbf{x}^*)$ are linearly independent, $\nabla f(\mathbf{x}^*)$ must be a linear combination of them, which we write as

$$\nabla f(\mathbf{x}^*) = \lambda_1 \nabla h_1(\mathbf{x}^*) + \lambda_2 \nabla h_2(\mathbf{x}^*).$$

Combining this equation with the gradient of the Lagrangian, we get

$$\nabla_x L(\mathbf{x}^*, \lambda) = \nabla f(\mathbf{x}^*) - \lambda_1 \nabla h_1(\mathbf{x}^*) - \lambda_2 \nabla h_2(\mathbf{x}^*) = \mathbf{0}.$$

Thus, the first-order condition for a local maximum is that the partial derivatives of the Lagrangian are zero. The following theorem extends this observation to any number of constraints, under a condition on the constraints known as a constraint qualification.

Theorem 14.1: *(Lagrange multipliers):* *Let* $\mathbf{x}^*$ *be a local extremum of (14.1) for which the constraint gradients* $\nabla h_i(\mathbf{x}^*)$ *are linearly independent. Then, for some* $\lambda_1, \ldots, \lambda_m$,

$$\nabla f(\mathbf{x}^*) - \sum_{i=1}^{m} \lambda_i \nabla h_i(\mathbf{x}^*) = \mathbf{0}. \tag{14.3}$$

Proof: For a proof see (Luenberger and Ye, 2016, Section 11.3). □

Often, if the gradients are linearly dependent, a linearly independent subset can be found and we can omit the other constraints without changing the local extremum.

Example 14.1. For the convex program

$$\begin{aligned} \max \ & 50 - x_1^2 - 2x_2^2 - x_3^2 \\ \text{s.t.} \quad & x_1 + x_2 + x_3 = 9 \\ & x_1 + x_2 \quad\ = 6 \end{aligned}$$

the Lagrangian is

$$L(\mathbf{x}, \lambda) = 50 - x_1^2 - 2x_2^2 - x_3^2 + \lambda_1(9 - x_1 - x_2 - x_3) + \lambda_2(6 - x_1 - x_2).$$

Setting the partial derivatives to zero as in (14.3),

$$\begin{aligned} -2x_1 - \lambda_1 - \lambda_2 &= 0, \\ -4x_2 - \lambda_1 - \lambda_2 &= 0, \\ -2x_3 - \lambda_1 &= 0. \end{aligned}$$

Solving these and the two constraints, $\mathbf{x} = (4, 2, 3)$ and $\lambda = (-6, -2)$. The problem has an optimal solution, so by Theorem 14.1, this is the optimal solution with value $z^* = 17$.

14.1.2 Inequality Constraints

Now consider (14.2) with inequality constraints. The Lagrangian is

$$L(\mathbf{x}, \lambda) = f(\mathbf{x}) + \sum_{i=1}^{m} \lambda_i(b_i - g_i(\mathbf{x})). \tag{14.4}$$

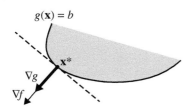

$g(\mathbf{x}) = b$

∇g

∇f

$\mathbf{x}^*$

Figure 14.2 Optimality condition with one constraint $g(\mathbf{x}) \leq b$.

As noted in Section 8.4 for linear problems, $L(\mathbf{x}, \lambda)$ will not have a maximum unless we restrict $\lambda \geq \mathbf{0}$. To understand the role of the Lagrangian geometrically, consider a problem with one active constraint g at the local maximum $\mathbf{x}^*$ and two variables. Now $\nabla h(\mathbf{x}^*)$ and $\nabla f(\mathbf{x}^*)$ must be parallel and must point in the same direction. As shown in Figure 14.2, the improving direction $\nabla f(\mathbf{x}^*)$ points outside of the feasible region and is perpendicular to the tangent line to the constraint at $\mathbf{x}^*$.

The improving directions at $\mathbf{x}^*$ lie on the same side of the tangent line as $\nabla f(\mathbf{x}^*)$, in a (open or closed) half plane, while the feasible directions lie on the opposite side in an open half plane. Thus, there is no improving feasible direction at $\mathbf{x}^*$. If $\nabla f(\mathbf{x}^*)$ did not point in the same direction as $\nabla g(\mathbf{x}^*)$ there would be an improving feasible direction and $\mathbf{x}^*$ could not be a local maximum.

Now consider a problem with two active constraints at the local maximum $\mathbf{x}^*$ and two variables. In Figure 14.3, $\nabla f(\mathbf{x}^*)$ must lie between $\nabla g_1(\mathbf{x}^*)$ and $\nabla g_2(\mathbf{x}^*)$; otherwise there would exist a feasible improving direction. Any such vector can be written

$$\nabla f(\mathbf{x}^*) = \lambda_1 \nabla g_1(\mathbf{x}^*) + \lambda_2 \nabla g_2(\mathbf{x}^*)$$

for some $\lambda_1, \lambda_2 \geq 0$. As long as $\nabla g_1(\mathbf{x}^*)$ and $\nabla g_2(\mathbf{x}^*)$ are linearly independent, this optimality condition allows $\nabla f(\mathbf{x}^*)$ to lie in a set of full dimension, namely, the cone generated by $\nabla g_1(\mathbf{x}^*)$ and $\nabla g_2(\mathbf{x}^*)$. Improving directions at $\mathbf{x}^*$ must be on the $\nabla f(\mathbf{x}^*)$ side of the line through $\mathbf{x}^*$ perpendicular to $\nabla f(\mathbf{x}^*)$. Again, this does not include any feasible directions. The same concept was illustrated for linear programs in Farkas' lemma and Figure 8.2. There the constraint rows $\mathbf{A}_i$ play the role of ∇g_i and $\nabla f = \mathbf{c}$.

Note that the constraints in both figures are active at the local maximum $\mathbf{x}^*$. A constraint that is not active should not influence whether $\mathbf{x}^*$ is a local maximum.

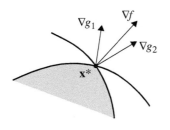

∇g_1

∇f

∇g_2

$\mathbf{x}^*$

Figure 14.3 Optimality condition with two inequality constraints.

We can remove inactive constraints from the optimality condition by setting their $\lambda_i = 0$. In general, when there are p linearly independent active constraints, their gradients form a cone of dimension p in which ∇f must lie. When $p = 1$, the optimality condition requires ∇f to be parallel to ∇g (perpendicular to the constraint). Also, if $p = 0$ optimality requires $\nabla f = \mathbf{0}$. Thus, this optimality condition for constrained optimization reduces to the first-order condition for unconstrained optimization at interior points where no constraints are active. These optimality conditions are called KKT conditions. They were first published by Kuhn and Tucker (1951), but are also named after Karush, whose unpublished 1939 master's thesis is closely related; see Kuhn (1976).

Theorem 14.2: *(**KKT conditions**):* *Let* $\mathbf{x}^*$ *be a local maximum of (14.2) for which the constraint gradients* $\nabla g_i(\mathbf{x}^*)$ *are linearly independent. Then, for some* $\lambda_1, \ldots, \lambda_m \geq 0$,

$$\nabla f(\mathbf{x}^*) - \sum_{i=1}^{m} \lambda_i \nabla g_i(\mathbf{x}^*) = \mathbf{0} \tag{14.5}$$

$$\lambda_i(b_i - g_i(\mathbf{x}^*)) = 0, \quad i = 1, \ldots, m. \tag{14.6}$$

We recognize the second condition, which eliminates inactive constraints from the first condition, as complementary slackness: either constraint i is active or $\lambda_i = 0$. Indeed, the Lagrange multiplier λ_i will be interpreted as the dual variable associated with constraint i. Duality is discussed in the next section. Weaker constraint qualifications can be used instead of linear independence of the gradients; see, e.g. Bertsekas (1999) and Boyd and Vandenberghe (2004). For linear programs, no constraint qualification is needed.

14.1.3 Convexity

Now we consider convex, or in some cases concave, functions f and g_i. With these assumptions, we can state necessary and sufficient conditions for global optima. Convexity of a multivariable function is defined in Section 6.2. For a function f of one variable, if $f''(x) \geq 0$ (≤ 0) for all $x \in S$, then f is convex (concave) on S. In the multivariable setting, if the matrix of second partial derivatives, or Hessian matrix, $\mathbf{H} = [\frac{\partial^2 f}{\partial x_i \, \partial x_j}]$ is positive semidefinite for all $\mathbf{x} \in S$ then f is convex on S. If $\mathbf{H}$ is negative semidefinite for all $\mathbf{x} \in S$ then f is concave on S. Section A.8 describes how to check if a matrix is positive semidefinite.

The KKT conditions are necessary conditions for a local maxima, analogous to the first-order conditions $\nabla f(\mathbf{x}) = \mathbf{0}$ for unconstrained optimization. Points that meet these conditions are called critical points and those that meet the KKT

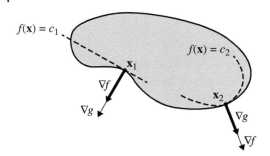

Figure 14.4 Points x_1 and x_2 satisfy the KKT conditions but are not local maxima.

conditions are called KKT points; they are potential local maxima. For functions of one variable, adding the assumption that $f''(x^*) < 0$ makes the condition $f'(x^*) = 0$ necessary and sufficient for a local maximum. Figure 14.4 illustrates why Theorem 14.2 is not a sufficient condition. There is a single constraint g. Both x_1 and x_2 satisfy the KKT condition: for each there is a $\lambda > 0$ such that $\nabla f = \lambda \nabla g$. Because g is not convex, the feasible region is not a convex set. There are feasible points in any neighborhood of x_1 on the "wrong" side of the tangent line to g at x_1, namely, a point x_0 for which $x_0^T \nabla g > x_1^T \nabla g$. Suppose f is linear near x_1. Then

$$f(x_0) - f(x_1) = (x_0 - x_1)^T \nabla f(x_1) = \lambda (x_0 - x_1)^T \nabla g(x_1) > 0,$$

contradicting x_1 being a local maximum. At x_2, because f is not concave there, the level set where $f(x) \geq f(x_2)$ is not a convex set. There are feasible points in any neighborhood of x_2 on the "wrong" side of the level curve $f(x) = f(x_2)$, namely, a point x_0 for which $f(x_0) > f(x_2)$, so x_2 is not a local minimum.

Figure 14.4 suggests that we need f to be concave and all active constraints g_i to be convex in a neighborhood of x^*. Since we don't know x^* and which constraints are active in advance, it is more practical to make these assumptions for all feasible x and all constraints.

Convex Program

> max $f(x)$
>
> s.t. $g_i(x) \leq b_i, \ i = 1, \ldots, m$

is a **convex program** if f is concave and all g_i are convex.

For a minimization problem to be a convex program, f must be convex; for constraints $g_i(x) \geq b_i$, g_i must be concave.

As we saw in Lemma 6.1 and Theorem 6.2, these assumptions also guarantee that a local maximum is a global maximum.

Theorem 14.3: *(Necessary and sufficient KKT conditions):* *Consider a feasible solution* $\mathbf{x}^*$ *to (14.2) at which the constraint gradients are linearly independent.*

Maximization: Suppose f is concave and g_i, $i = 1, \ldots, m$, are convex. Then $\mathbf{x}^*$ is a global maximum if and only if (14.5) and (14.6) hold for some $\lambda \geq \mathbf{0}$.

Minimization: Suppose f and g_i, $i = 1, \ldots, m$, are convex. Then $\mathbf{x}^*$ is a global minimum if and only if (14.5) and (14.6) hold for some $\lambda \geq \mathbf{0}$.

To summarize, the KKT conditions for the maximization problem (14.2) are that $(\mathbf{x}, \lambda)$ satisfy

$$\nabla f(\mathbf{x}) - \sum_{i=1}^{m} \lambda_i \nabla g_i(\mathbf{x}) = \mathbf{0}$$

$$\lambda_i \geq 0 \qquad\qquad (14.7)$$

$$g_i(\mathbf{x}) \leq b_i$$

$$\lambda_i(b_i - g_i(\mathbf{x})) = 0, \quad i = 1, \ldots, m.$$

The first constraint involves the gradient of the Lagrangian and has one equation for each variable x_j, the third is feasibility, and the fourth is complementary slackness. If some of the constraints are equality constraints, we omit $\lambda_i \geq 0$ for these constraints; complementary slackness for these constraints is redundant.

14.1.4 KKT Conditions with Nonnegativity Constraints

To write the KKT conditions for a minimization problem, one could convert it to a maximization. For convenience, we describe the conditions for the problem

$$\min \quad f(\mathbf{x})$$
$$\text{s.t.} \quad g_i(\mathbf{x}) \geq b_i, \quad i = 1, \ldots, m.$$

The "$\geq$" constraints are used for consistency with the dual standard form linear program. The KKT conditions for this minimization problem are the same as (14.7) except that feasibility changes to $g_i(\mathbf{x}) \geq b_i$.

Now suppose we add nonnegativity constraints to (14.2). We could use the KKT conditions earlier by treating nonnegativity as additional functional constraints g_i. However, the conditions can be simplified. For the problem

$$\max f(\mathbf{x})$$
$$\text{s.t.} \ g_i(\mathbf{x}) \leq b_i, \ i = 1, \ldots, m$$
$$\mathbf{x} \geq \mathbf{0}$$

the KKT conditions simplify to

$$\frac{\partial f(\mathbf{x})}{\partial x_j} - \sum_{i=1}^{m} \lambda_i \frac{\partial g_i(\mathbf{x})}{\partial x_j} \leq 0, \ j = 1, \ldots, n,$$

$$\left[\frac{\partial f(\mathbf{x})}{\partial x_j} - \sum_{i=1}^{m} \lambda_i \frac{\partial g_i(\mathbf{x})}{\partial x_j} \right] x_j = 0, \ j = 1, \ldots, n,$$

$$\lambda_i \geq 0, \ i = 1, \ldots, m,$$

$$g_i(\mathbf{x}) \leq b_i, \ i = 1, \ldots, m,$$

$$\lambda_i(b_i - g_i(\mathbf{x})) = 0, \ i = 1, \ldots, m,$$

$$\mathbf{x} \geq \mathbf{0}.$$

The first equation writes out each component of the constraint involving gradients. The second equation is a complementary slackness condition for the nonnegativity constraints. Similarly, for the problem

$$\min \quad f(\mathbf{x})$$
$$\text{s.t.} \quad g_i(\mathbf{x}) \geq b_i, \ i = 1, \ldots, m$$
$$\mathbf{x} \geq \mathbf{0}$$

the KKT conditions simplify to

$$\frac{\partial f(\mathbf{x})}{\partial x_j} - \sum_{i=1}^{m} \lambda_i \frac{\partial g_i(\mathbf{x})}{\partial x_j} \geq 0, \ j = 1, \ldots, n,$$

$$\left[\frac{\partial f(\mathbf{x})}{\partial x_j} - \sum_{i=1}^{m} \lambda_i \frac{\partial g_i(\mathbf{x})}{\partial x_j} \right] x_j = 0, \ j = 1, \ldots, n,$$

$$\lambda_i \geq 0, \ i = 1, \ldots, m,$$

$$g_i(\mathbf{x}) \geq b_i, \ i = 1, \ldots, m,$$

$$\lambda_i(b_i - g_i(\mathbf{x})) = 0, \ i = 1, \ldots, m,$$

$$\mathbf{x} \geq \mathbf{0}.$$

14.1.5 Examples

Next we present several examples.

Example 14.2. Single-variable optimization. The KKT conditions give the correct procedure for maximizing a function of one variable on an interval. For the problem

$$\max \quad f(x)$$
$$\text{s.t.} \quad a \leq x \leq b,$$

rewriting the constraints as $-x \leq -a$ and $x \leq b$, the KKT conditions are

$$f'(x) + \lambda_1 - \lambda_2 = 0,$$
$$\lambda_1(-a + x) = 0,$$
$$\lambda_2(b - x) = 0.$$

The first equation is for the variable x; the last two are complementary slackness for the two constraints. The inequalities $\lambda_i \geq 0$ must also hold. A useful strategy is to consider cases where each $\lambda_i = 0$ or $\lambda_i > 0$.

Case 1. $\lambda_1 = \lambda_2 = 0$. The equations reduce to $f'(x) = 0$, which is a valid optimality condition for any feasible x.

Case 2. $\lambda_1 = 0$, $\lambda_2 > 0$. The equations reduce to $x = b$ and $f'(b) = \lambda_2 > 0$.

Case 3. $\lambda_1 > 0$, $\lambda_2 = 0$. The equations reduce to $x = a$ and $f'(a) = -\lambda_1 < 0$.

Case 4. $\lambda_1 > 0$, $\lambda_2 > 0$. The last two equations are $x = a$ and $x = b$, a contradiction, so this case cannot occur.

Cases 1–3 are the necessary conditions for interior and end point local maxima.

Example 14.3. For the problem

$$\begin{array}{ll} \max & x_1 + 4x_2 \\ \text{s.t.} & x_1^2 + x_2^2 \leq 100 \\ & x_1 + x_2 \leq 13, \end{array}$$

the KKT conditions are

$$1 - 2x_1\lambda_1 - \lambda_2 = 0$$
$$4 - 2x_2\lambda_1 - \lambda_2 = 0$$
$$\lambda_1(100 - x_1^2 - x_2^2) = 0$$
$$\lambda_2(13 - x_1 - x_2) = 0$$

and $\lambda_i \geq 0$. The first equation is for x_1, the second is for x_2, and the last two are complementary slackness for the two constraints. If we try $\lambda_1 = 0$, $\lambda_2 > 0$ the first two equations give a contradiction. Next try $\lambda_1 > 0$, $\lambda_2 = 0$. From the first two equations,

$$x_1 = \frac{1}{2\lambda_1} \quad \text{and} \quad x_2 = \frac{2}{\lambda_1}.$$

Substituting into

$$100 - x_1^2 - x_2^2 = 0,$$

$\lambda_1 = \sqrt{17}/20$, $x_1 = 10/\sqrt{17}$, and $x_2 = 40/\sqrt{17}$. We know this solution satisfies the first constraint, so we check the second constraint, $x_1 + x_2 = 50/\sqrt{17} \approx 12.13 < 13$. Since the objective is linear and the constraints are convex, this point is optimal.

Example 14.4. Projects with diminishing returns. Consider the problem of investing in a set of n projects to maximize the total return on a fixed investment. If we scale the amounts invested so that the budget is 1, the problem is

$$\max \sum_{j=1}^{n} f_j(x_j)$$

$$\text{s.t.} \sum_{j=1}^{n} x_j \leq 1$$

$$x_j \geq 0.$$

Here $f_j(x_j)$ is the return on project j when x_j is invested. We assume that f_j is increasing and concave, which means that the projects have diminishing return on investment. Using the KKT conditions for problems with nonnegativity constraints and denoting the one multiplier λ, we have

$$f_j'(x_j) - \lambda \leq 0, \ j = 1, \dots, n,$$

$$[f_j'(x_j) - \lambda]x_j = 0, \ j = 1, \dots, n,$$

$$\lambda \left(1 - \sum_{j=1}^{n} x_j \right) = 0,$$

$\lambda \geq 0$, and feasibility. The second equation requires all projects with nonzero investment to have the same marginal rate of return on investment λ. If it is optimal to not invest in project j, then from the first equation $f_j'(x_j) \leq \lambda$.

For problems with a linking constraint like Example 14.4, the multiplier λ^* is needed to compute the optimal solution $\mathbf{x}^*$. If we have not solved the overall problem, we could still pick a required return λ_0 and solve $f_j'(x_j) = \lambda_0$ for each project. However, because $\lambda_0 \neq \lambda^*$, the total investment would not be correct. This approximate solution is an example of a single-pass constructive algorithm.

14.2 Lagrangian Duality

This section introduces a duality theory for nonlinear programs. The approach is the same as Section 8.4, where a Lagrangian dual is constructed for linear programs. We will consider the inequality constrained maximization problem (14.2) with optimal solution $\mathbf{x}^*$ and the Lagrangian (14.4). Consider the unconstrained problem

$$\max_{\mathbf{x}} L(\mathbf{x}, \lambda) = \max_{\mathbf{x}} f(\mathbf{x}) + \sum_{i=1}^{m} \lambda_i(b_i - g_i(\mathbf{x})). \tag{14.8}$$

To create an upper bound on (14.2), we require $\lambda \geq 0$. For any $\lambda \geq 0$, define the *dual function*

$$g_0(\lambda) = \max_{\mathbf{x}} L(\mathbf{x}, \lambda) \geq L(\mathbf{x}^*, \lambda) = f(\mathbf{x}^*) + \sum_{i=1}^{m} \lambda_i(b_i - g_i(\mathbf{x}^*)) \geq f(\mathbf{x}^*). \qquad (14.9)$$

We have relaxed, or *dualized*, the constraints. Even when (14.2) is not a convex program, $g_0(\lambda)$ is convex in λ.

Lemma 14.1: *The dual function $g_0(\lambda)$ is convex.*

Proof: For each $\mathbf{x}$, $L(\mathbf{x}, \lambda)$ is linear in λ. The point-wise maximum of linear functions is convex. □

Examples of the maximum of linear functions are given in Section 3.2.
    The Lagrangian dual finds the tightest upper bound.

The **Lagrangian dual** of (14.2) is

$$w^* = \min_{\lambda \geq 0} \max_{\mathbf{x}} L(\mathbf{x}, \lambda) = \min_{\lambda \geq 0} g_0(\lambda). \qquad (14.10)$$

Note that feasibility for the dual is just $\lambda \geq 0$. It follows from (14.9) that weak duality holds.

Theorem 14.4: *(**Weak duality**):* *If $\mathbf{x}$ is feasible for the primal problem (14.2) and λ is feasible for the dual problem, then $f(\mathbf{x}) \leq g_0(\lambda)$.*

Strong duality does not hold in general; however, it does for convex programs.

Theorem 14.5: *(**Strong duality**):* *If f is concave, g_i, $i = 1, \ldots, m$ is convex, and the constraint gradients $\nabla g_i(\mathbf{x}^*)$ are linearly independent at some optimal solution $\mathbf{x}^*$ for the primal problem (14.2), then there exists an optimal solution to the dual problem (14.10) and $z^* = w^*$.*

Comparing these with the duality theorems for linear programming, the results are the same for the convex program case except that a constraint qualification is needed. The nonconvex case is different in that there may be a duality gap $w^* - z^* > 0$.

When strong duality holds, the inequalities in (14.9) become equalities at optimal solutions $\mathbf{x}^*$ and λ^*, so

$$\sum_{i=1}^{m} \lambda_i^*[b_i - g_i(\mathbf{x}^*)] = 0. \tag{14.11}$$

Since $\lambda_i^* \geq 0$ and $g_i(\mathbf{x}^*) \leq b_i$, (14.11) implies complimentary slackness:

$$\lambda_i^*[b_i - g_i(\mathbf{x}^*)] = 0, \quad i = 1, \dots, m.$$

Either constraint i is active or its dual variable $\lambda_i^* = 0$. Under the assumptions of the previous section (f and g_i are differentiable and a constraint qualification holds), the other KKT condition (14.6) holds. Thus, the optimal dual variables λ_i^* are also the Lagrange multipliers that satisfy the KKT conditions.

14.2.1 Geometry of Primal and Dual Functions

To understand why nonconvexity leads to a duality gap, consider the optimal value of the primal when parameterized by the right-hand side

$$z^*(\mathbf{u}) = \max \quad f(\mathbf{x})$$
$$\text{s.t.} \quad g_i(\mathbf{x}) \leq u_i, \quad i = 1, \dots, m.$$

When $\mathbf{u} = \mathbf{b}$, the original problem (14.2) is recovered and $z^*(\mathbf{b}) = z^*$. If the problem is infeasible for some $\mathbf{u}$, then set $z^*(\mathbf{u}) = -\infty$. The optimal value $z^*(\mathbf{u})$ is called the *primal function*. To illustrate a duality gap, consider a problem with one constraint function g. For simplicity, assume that the original primal problem had $b = 0$, corresponding to the optimal value $z^*(0)$. The graph of $z^*(u)$ has constraint values u on the horizontal axis and optimal objective values z on the vertical. It is useful to consider its hypograph $G = \{(u, z) : z \leq z^*(u)\}$, i.e. all points on or below the function, as shown in Figure 14.5. Each $\mathbf{x}$ corresponds to a point $(g(\mathbf{x}), f(\mathbf{x})) \in G$. With this correspondence, we can write the Lagrangian as $L(\mathbf{x}, \lambda) = f(\mathbf{x}) - \lambda g(\mathbf{x}) = z - \lambda u$.

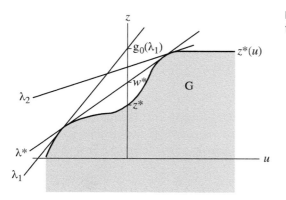

Figure 14.5 Primal and dual functions with a duality gap.

An optimal point $\mathbf{x}'$ for a given right-hand side u' lies on the primal function; however, if there is slack in the constraint, $g(\mathbf{x}') < u'$, then $z^*(\cdot)$ is horizontal to the right of the point $(g(\mathbf{x}'), f(\mathbf{x}'))$. If $\mathbf{x}$ is not optimal for any right-hand side u, then $(g(\mathbf{x}), f(\mathbf{x}))$ lies below the graph of $z^*(\cdot)$. If the problem is infeasible for $u < u'$, then $\mathcal{G}$ lies to the right of $u = u'$. Now we can express the dual function as

$$g_0(\lambda) = \max_{\mathbf{x}} L(\mathbf{x}, \lambda) = \max_{(u,z) \in \mathcal{G}} z - \lambda u.$$

Although $\mathcal{G}$ contains points (u, z) that do not correspond to any $\mathbf{x}$, these points lie below the primal function and do not affect the maximum, which occurs on the primal function at $z = z^*(u)$. Since $g_0(\lambda)$ is the maximum,

$$z - \lambda u \le g_0(\lambda)$$

for $(u, z) \in \mathcal{G}$.

For a fixed λ, we can visualize the line $z - \lambda u = g_0(\lambda)$ by sweeping a line of slope λ upwards until it touches the last point in $\mathcal{G}$. Rearranging as $z = g_0(\lambda) + \lambda u$, its vertical intercept is $g_0(\lambda)$. Figure 14.5 shows several such lines for different λ. In this example, $z^*(\cdot)$ is differentiable and these are tangent lines. The smallest vertical intercept is the optimal value of the dual,

$$w^* = \min_{\lambda \ge 0} g_0(\lambda).$$

From the definition of $\mathcal{G}$, the optimal value of the primal, $z^* = z^*(0)$, is the largest value on the vertical axis that is in $\mathcal{G}$. Since $z^* < w^*$, there is a duality gap.

In this example, $z^*(u)$ is not concave, making a duality gap possible. However, in Figure 14.6, $z^*(u)$ is concave, so the smallest vertical intercept of the Lagrangian lines is $z^*(0) = z^* = w^*$ and there cannot be a duality gap. We noted in Chapter 10 that, for linear programs, a principle of diminishing returns applies: $z^*(\mathbf{u})$ is concave and nondecreasing. The same result can be established for convex problems, so strong duality applies to them.

Figure 14.6 Primal function with no duality gap.

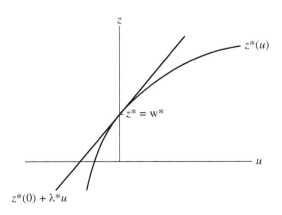

We can interpret optimal values of dual variables in the same way as for linear programs. The slope of the line in Figure 14.6 is the optimal dual variable λ^*. Assuming $z^*(u)$ is differentiable at 0, this is the tangent line at $u = 0$, making

$$\frac{\partial z^*}{\partial u} = \lambda^*.$$

Thus, as with linear programs, the optimal value of a dual variable is the shadow price, or sensitivity of the optimal value to the right-hand side of the corresponding primal constraint. However, because $z^*(u)$ generally does not have linear pieces, it is only the instantaneous rate of change, not an average rate of change over an interval.

All of these concepts can be generalized to problems with multiple constraints. We will also generalize by allowing any right-hand side $\mathbf{b}$ in the primal, instead of assuming $\mathbf{b} = \mathbf{0}$. The primal solution $\mathbf{x}$ is mapped into a point $(g_i(\mathbf{x}), \ldots, g_m(\mathbf{x}), f(\mathbf{x})) \in \mathcal{G}$. The dual function is

$$g_0(\lambda) = \max_{\mathbf{x}} L(\mathbf{x}, \lambda) = \max_{(\mathbf{u},z) \in \mathcal{G}} z + \lambda^T(\mathbf{u} - \mathbf{b}),$$

with

$$z + \lambda^T(\mathbf{u} - \mathbf{b}) \le g_0(\lambda)$$

for $(\mathbf{u}, z) \in \mathcal{G}$. This makes $z + \lambda^T(\mathbf{u} - \mathbf{b}) = g_0(\lambda)$ a supporting hyperplane of $\mathcal{G}$. The point $(\mathbf{b}, g_0(\lambda))$ lies on the hyperplane and all other points in $\mathcal{G}$ lie on or below it, i.e. if $(\mathbf{u}, z_0)$ lies on the hyperplane and $(\mathbf{u}, z) \in \mathcal{G}$, then $z \le z_0$.

14.2.2 Sensitivity Analysis

For sensitivity analysis, we compare the original problem with right-hand sides $\mathbf{b}$ to the new problem with right-hand sides $\mathbf{u}$. If $z^*(\cdot)$ is differentiable at $\mathbf{b}$ then, as argued earlier for a single constraint,

$$\frac{\partial z^*}{\partial u_i} = \lambda_i^*.$$

The interpretation is the same as in linear programming: the optimal dual variable λ_i^* is the shadow price of constraint i, namely, the sensitivity of the optimal value to the right-hand side u_i:

$$z^*(\mathbf{u}) \approx z^*(\mathbf{b}) + \lambda^{*T}(\mathbf{u} - \mathbf{b}). \tag{14.12}$$

Establishing that $z^*(\mathbf{u})$ is differentiable is rather involved. As in the case of linear programs, at a point $\mathbf{u}$ where the set of active constraints changes, $z^*(\mathbf{u})$ may have directional derivatives but not be differentiable. At such a point, the sensitivity to increases and decreases in u_i are different. For convex programs, regardless of whether $z^*(\mathbf{u})$ is differentiable, the Strong Duality Theorem implies that $z^*(\mathbf{u})$ is concave. Thus, the linear approximation is an upper bound,

$$z^*(\mathbf{u}) \le z^*(\mathbf{b}) + \lambda^{*T}(\mathbf{u} - \mathbf{b}).$$

In Figure 14.6, for example, $z^*(u)$ is concave and lies below its tangent lines. These sensitivity results require the assumption in Theorem 14.5 that the constraint gradients $\nabla g_i(\mathbf{x}^*)$ are linearly independent at some optimal solution $\mathbf{x}^*$ (or some other constraint qualification).

Example 14.5. Consider the convex program

$$\max 50 - x_1^2 - 2x_2^2$$
$$\text{s.t.} \quad x_1 + x_2 \ge 6.$$

We will find the primal and dual functions. Converting the constraint to $-x_1 - x_2 \le -6$ so that (14.7) applies and replacing the right-hand side $b = -6$ with u, the KKT conditions are

$$-2x_1 - \lambda = 0$$
$$-4x_2 - \lambda = 0$$
$$\lambda(u + x_1 + x_2) = 0$$
$$-x_1 - x_2 \le u$$

and $\lambda \ge 0$. The solution is $\lambda^* = -4u/3$ and $\mathbf{x} = (-2u/3, -u/3)$ for $u \le 0$ and $\lambda^* = 0$, $\mathbf{x} = (0,0)$ for $u > 0$. Substituting into the objective, the primal function is

$$z^*(u) = \begin{cases} 50 - \frac{2}{3}u^2, & u \le 0 \\ 50, & u > 0. \end{cases}$$

To find the dual function, we maximize the Lagrangian

$$L(\mathbf{x}, \lambda) = 50 - x_1^2 - 2x_2^2 + \lambda(-6 + x_1 + x_2)$$

over $\mathbf{x}$. The first-order conditions are the first two KKT conditions, with solution $\mathbf{x} = (\lambda/2, \lambda/4)$. Then

$$g_0(\lambda) = \max_{\mathbf{x}} L(\mathbf{x}, \lambda)$$
$$= 50 - \left(\frac{\lambda}{2}\right)^2 - 2\left(\frac{\lambda}{4}\right)^2 + \lambda\left(-6 + \frac{\lambda}{2} + \frac{\lambda}{4}\right)$$
$$= 50 - 6\lambda + \frac{3}{8}\lambda^2.$$

The dual function is minimized by $\lambda^* = 8$, consistent with that found earlier, and the optimal dual value is $w^* = 26$. Also, $z^* = z^*(-6) = 26$, confirming that there is no duality gap. Figure 14.7 shows the primal function and the optimal tangent

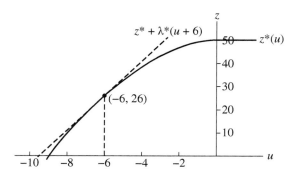

Figure 14.7 Primal function for Example 14.5 with no duality gap.

line, with slope λ^*. Optimal values occur on this graph at $u = b = -6$. Also, the sensitivity of the optimal value to the right-hand side is

$$\left.\frac{\partial z^*}{\partial u}\right|_{u=-6} = \lambda^* = 8.$$

For more difficult problems, we cannot solve the KKT conditions. Rather, as we shall see in the next chapter, they are useful for algorithms to solve convex programs.

Problems

1 Use the KKT conditions to find the optimal solution to the following convex program.

$$\begin{aligned} \max \quad & x_1 - x_2 \\ \text{s.t.} \quad & x_1^2 + 4x_2^2 \leq 16. \end{aligned}$$

2 Use the KKT conditions to find the optimal solution to the following convex program.

$$\begin{aligned} \max \quad & 54x_1 - x_1^2 + 90x_2 - 2x_2^2 - 20x_3 \\ \text{s.t.} \quad & x_1 + x_2 - x_3 \leq 0 \\ & x_j \geq 0. \end{aligned}$$

Hint: Ignore the nonnegativity constraints. The optimal solution without them is nonnegative.

3 Use the KKT conditions to find the optimal solution to the following convex program.

$$\begin{aligned} \min \quad & (x_1 - 10)^2 + (x_2 - 6)^2 \\ \text{s.t.} \quad & x_1 + x_2 \leq 14. \end{aligned}$$

4 Use the KKT conditions to find the optimal solution to the following convex program.

$$\max \quad z = x_1^{0.25} x_2^{0.75}$$
$$\text{s.t.} \quad 2x_1 + 3x_2 \leq 100$$
$$x_j \geq \quad 0.$$

Hint: Maximize $\ln z$. Ignore the nonnegative constraints.

5 Use the KKT conditions to find the optimal solution to the following convex program.

$$\min \quad -60x_1 + 0.5x_1^2 - 40x_2 + x_2^2 + 5x_3$$
$$\text{s.t.} \quad x_1 + x_3 \geq 120$$
$$2x_2 + x_3 \geq 80$$
$$x_j \geq 0.$$

Hint: Ignore the nonnegativity constraints.

6 Use the KKT conditions to find the optimal solution to the following convex program.

$$\min \quad (x_1 - 2)^2 + (x_2 - 4)^2$$
$$\text{s.t.} \quad -x_1 + x_2 = 1$$
$$x_1 + x_2 \leq 4.$$

7 Use the KKT conditions to find the optimal solution to the following convex program.

$$\min \quad e^{-x_1} + e^{-3x_2}$$
$$\text{s.t.} \quad x_1 + x_2 \leq 4.$$

8 For the optimization problem

$$\min \quad x^2 + y^2 + 2xy$$
$$\text{s.t.} \quad x^2 - y = 2.$$

(a) Show that the objective function and constraint left-hand side are convex. Why are these facts not enough to guarantee that local minima are global minima?

(b) Use the method of Lagrange multipliers to find the solutions to (14.3). Using technology will be helpful.

(c) Substitute $y = x^2 - 2$ into the objective function and solve the one-variable unconstrained problem using calculus to determine which of the solutions in (b) are optimal.

9 For the problem

$$\max \quad f(\mathbf{x})$$
$$\text{s.t.} \quad h_i(\mathbf{x}) = b_i, \ i = 1, \dots, m$$
$$g_i(\mathbf{x}) \le d_i, \ i = 1, \dots, k$$

with equality and inequality constraints, we can write the Lagrangian as

$$L(\mathbf{x}, \lambda, \nu) = f(\mathbf{x}) + \sum_{i=1}^{k} \lambda_i(b_i - h_i(\mathbf{x})) + \sum_{i=1}^{m} \nu_i(d_i - g_i(\mathbf{x})),$$

where λ_i and ν_i are the Lagrange multipliers for the h_i and g_i constraints, respectively. The KKT conditions analogous to (14.7) are

$$\nabla f(\mathbf{x}) - \sum_{i=1}^{m} \lambda_i \nabla h_i(\mathbf{x}) - \sum_{i=1}^{k} \nu_i \nabla g_i(\mathbf{x}) = \mathbf{0},$$

$$\nu_i \ge 0, \ i = 1, \dots, k,$$
$$h_i(\mathbf{x}) = b_i, \ i = 1, \dots, m,$$
$$g_i(\mathbf{x}) \le d_i, \ i = 1, \dots, k,$$
$$\nu_i(d_i - g_i(\mathbf{x})) = 0, \ i = 1, \dots, k.$$

There are no nonnegativity or complimentary slackness constraints for the multipliers λ_i associated with equality constraints. Now consider the canonical form linear program

$$\max \quad \mathbf{c}^T\mathbf{x}$$
$$\text{s.t.} \quad \mathbf{A}\mathbf{x} = \mathbf{b}$$
$$\mathbf{x} \ge \mathbf{0}.$$

(a) Find the Lagrangian dual and write it as a linear program (the dual of the original problem).

(b) Write the KKT conditions. Identify each condition as primal feasibility, dual feasibility, or complimentary slackness.

10 For the linear program

$$\max \quad \mathbf{c}^T\mathbf{x}$$
$$\text{s.t.} \quad \mathbf{A}\mathbf{x} \le \mathbf{b}.$$

(a) Find the Lagrangian dual and write it as a linear program (the dual of the original problem).

(b) Write the KKT conditions. Identify each condition as primal feasibility, dual feasibility, or complimentary slackness.

11 This exercise demonstrates that constraint qualifications are needed. For the convex program

$$\max \quad x_2$$
$$\text{s.t.} \quad (x_1 + 1)^2 + (x_2 - 1)^2 \le 1$$
$$(x_1 - 1)^2 + (x_2 - 1)^2 \le 1.$$

(a) Solve graphically for $\mathbf{x}^*$.
(b) State the KKT conditions. Show that for $\mathbf{x}^*$ found in (a) there are no values of λ that satisfy the conditions.
(c) Are the constraint gradients linearly independent at $\mathbf{x}^*$?

12 Consider the problem $\max f(x) = 3x^2 - 2x$ s.t. $x^2 \le 1$. The conditions of Theorem 14.5 (Strong duality) are not met because f is convex, not concave.
(a) Find all x, λ that satisfy the KKT conditions. Which pair is optimal?
(b) State the Lagrangian dual problem.
(c) Find the dual function and solve the Lagrangian dual problem.
(d) Using the primal solution from (a) and the dual solution from (c), check whether strong duality holds.

13 For the convex program

$$\max \quad -x^2 - 1$$
$$\text{s.t.} \quad (x - 2)(x - 4) \le 0.$$

(a) State the Lagrangian dual.
(b) Solve the primal and dual problems.
(c) Does strong duality hold?

14 For Exercise 14.1
(a) Find the dual function.
(b) State the Lagrangian dual.
(c) Show that the optimal Lagrange multiplier λ found in Exercise 14.1 is an optimal solution to the dual.
(d) Does strong duality hold?

15 Consider the problem $\max f(x)$ s.t. $x \le 1$, where

$$f(x) = \min\{4x, x^2, x + 2\}.$$

(a) Find and graph the primal function $z^*(u)$, which has constraint $x \le u$.
(b) Find the dual function, indicating the values of λ for which it is unbounded.

(c) State and solve the Lagrangian dual.

(d) Does strong duality hold?

16 A rectangular patio with a semicircular extension is to have a combined width (radius plus width of the rectangle) of 10 ft. Two widths of the rectangle, one length, and the curved boundary of the semicircle require an edge treatment, of which 40 ft is available. What dimensions maximize the combined area of the rectangle and semicircle?

15

Convex Programming: Algorithms

Solving convex programs is a broad and more advanced topic. This final chapter gives a brief introduction. Convex programs lie between linear and nonconvex programs in a certain sense. Linear programs can be efficiently solved by the simplex method. The same algorithm works well on almost every problem of practical interest. Using two other methods when appropriate – dual simplex and interior point – does even better. Specialized algorithms, such as the network simplex method, can solve even larger problems and it is clear when they can be used. Nonconvex programs are the opposite extreme. There are no general algorithms that work on all problems. Instead, there are many algorithms that might work on a given problem, but the problem might need to be reformulated, parameters of the algorithm tuned to the particular problem, and results may depend on the initial solution provided. No global optimality condition is available. For convex programs, we know from Theorem 6.2 that any local extremum is also a global optimum. Optimality conditions are available, namely, the KKT conditions of Theorem 14.3. General algorithms are available that can handle many problems. They are not as efficient as linear programming algorithms, but they are still efficient. The state of the art is described in Boyd and Vandenberghe, (2004, Section 1.3.1): interior point algorithms can reliably and quickly solve convex problems with hundreds of variables and thousands of constraints. Typically 10–100 iterations are required.

There are some practical challenges in using convex optimization:

- It can be difficult to determine if functions are convex. If a convex programming algorithm is applied to a problem that is not convex, it may find a local optimum that is not a global optimum.
- Some nonconvex problems can be transformed into convex problems; however, the transformations are more involved than the transformations for linear problems.

Linear and Convex Optimization: A Mathematical Approach, First Edition. Michael H. Veatch.
© 2021 John Wiley & Sons, Inc. Published 2021 by John Wiley & Sons, Inc.
Companion website: www.wiley.com/go/veatch/convexandlinearoptimization

- Tuning the parameters of the algorithm to a particular example may be needed to obtain good performance.
- Entering nonlinear functions in a solver requires a more involved syntax than linear functions.
- The computational effort to evaluate the nonlinear functions will be greater. Algorithms use the first and sometimes the second derivatives of these functions, which require additional computation.

Many different algorithms for general convex programs are still being studied; see Luenberger and Ye (2016). The algorithms, and particularly analysis of their rate of convergence, require more advanced mathematics. We do not discuss their rate of convergence and will present a basic version of one algorithm.

We begin by presenting some motivating examples in Section 15.1. The easier special case of separable programs is described in Section 15.2, making a connection with linear programming approximations. Unconstrained convex programs are also of interest; they are discussed in Section 15.3. Another important class of problems, quadratic programs, is introduced in Section 15.4. Section 15.5 presents the main algorithm of this chapter: a primal-dual interior point method. It is an extension of the primal-dual interior point method for linear programs from Section 11.3.

In this chapter we consider the convex optimization problem

$$\max \quad f(\mathbf{x}) \tag{15.1}$$
$$\text{s.t.} \quad g_i(\mathbf{x}) \le b_i, \quad i = 1, \dots, m,$$

where f is concave and g_i is convex. Convex programs cannot have general equality constraints because they make the feasible region nonconvex. However, linear equality constraints may be added and are easily incorporated in the algorithms. We have chosen this form (maximization, "$\le$" constraints) for consistency with the standard form of a linear program. If we start with a minimization problem, f would be convex. We assume f and g_i have continuous second derivatives.

For linear programs, the search for an optimal solution can be limited to extreme points of the feasible region. For convex programs, points in the interior of the region and points on the boundary that are not extreme points must also be considered.

Example 15.1. Consider the convex program

$$\max \quad f(\mathbf{x}) = 2x_1 x_2 + 8x_2 - x_1^2 - 2x_2^2$$
$$\text{s.t.} \qquad\qquad x_1 + 2x_2 \le 16$$
$$x_1 \le 8$$
$$x_j \ge 0.$$

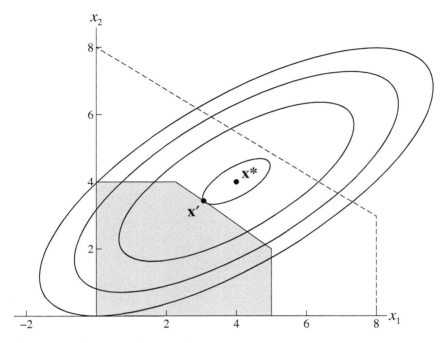

Figure 15.1 Problems with optimal solutions in the interior and on an edge.

Figure 15.1 shows this feasible region using dashed lines. The optimal solution $\mathbf{x}^* = (4, 4)$ lies in the interior. Now consider the feasible region defined by the following constraints:

$$x_1 + \frac{4}{3}x_2 \le 8,$$
$$0 \le x_1 \le 5,$$
$$0 \le x_2 \le 4.$$

The optimal solution $\mathbf{x}' = (3.31, 3.52)$ with value $f(\mathbf{x}') = 15.72$ lies on the first constraint. Figure 15.1 shows several contours of f. The KKT conditions are met at $\mathbf{x}'$; graphically, this occurs where the constraint is tangent to a contour of f.

Figure 15.1 illustrates that problems can have optimal solutions in the interior, where no constraints are active, or on a boundary where one constraint is active. An optimal solution could also have $n = 2$ active constraints; if their gradients $\nabla g_i(\mathbf{x}^*)$ are linearly independent, this solution is an extreme point. Recall from Chapter 6 that because f is concave, its level sets $\{\mathbf{x} : f(\mathbf{x}) \ge c\}$ are convex sets.

This, combined with the convexity of the feasible region, explains why a KKT point such as $\mathbf{x}'$ must be optimal. For a problem with n variables, an optimal solution may have from $k = 0$ to n linearly independent active constraints. Interior points have $k = 0$ and extreme points have $k = n$.

15.1 Convex Optimization Models

Formulating nonlinear problems often involves choosing a form of function and fitting it to data. That is beyond the scope of this book. Instead, we present models where these functions are already given.

Example 15.2. (Projects with diminishing returns). Example 14.4 gave a convex program for the problem of investing in a set of projects to maximize the total return on a fixed investment. Suppose there are three projects with returns

$$f_1(x_1) = 5(1 - e^{-2x_1}), \quad f_2(x_2) = 10(1 - e^{-0.8x_2}), \quad f_3(x_3) = 8(1 - e^{-0.5x_3}),$$

when x_j is invested in project j. The total budget is one. Maximizing total return is the convex program

$$\max \ \sum_{j=1}^{n} f_j(x_j)$$

$$\text{s.t.} \ \sum_{j=1}^{n} x_j \leq 1$$

$$x_j \geq 0.$$

The optimal solution, found from a solver, is $\mathbf{x}^* = (0.365, 0.635, 0)$ with return 9.241. At this solution the marginal returns of project 1 and 2 are equal to the Lagrange multiplier, $f_1'(0.365) = f_2'(0.635) = 4.815$. Project 3 is not used because its marginal return is smaller, $f_3'(0) = 4$.

Many applications have this structure, with one linear constraint and diminishing returns. Here is another example.

Example 15.3. (Inventory management with geometric demand). A supplier must decide how many items of each type to stock. They have agreed to provide a 95% service level, meaning that 95% of requests for an item can be met from inventory. When a request for item $j = 1, \ldots, n$ occurs, let x_j be the number of prior requests and s_j the stock level for this item. The request can be met from inventory if $x_j < s_j$; otherwise, demand exceeds inventory. The proportion of times that demand exceeds inventory is modeled as $e^{-\beta_j s_j}$. Also, the proportion of requests that are for item j is d_j and the cost of holding one unit of item j in inventory is h_j. If we ignore the integer restriction on the stock levels, the convex program to minimize holding cost is

$$\min \quad \sum_{j=1}^{m} h_j s_j$$

$$\text{s.t.} \quad \sum_{j=1}^{m} d_j e^{-\beta_j s_j} \leq 0.05$$

$$s_j \geq 0.$$

The expression $e^{-\beta_j s_j}$ applies when the number of requests follows a geometric probability distribution; it also applies for a continuous exponential distribution. Other probability distributions could be used and are more common for modeling demand. Example 15.3 could also model multi-layer security systems where s_j is the number of layers of security for asset j, an attacker chooses an asset to attack, and each level of security has the same chance of stopping the attacker. Similarly, it could model military force attrition, where s_j is the number of military units deployed to location j to defend against an enemy attack. Then the objective is to minimize the number of units deployed to achieve the desired probability (0.95 in the example) of repelling an attack. In other applications of Example 15.2, the projects could be purchasing advertising in different markets or the time spent training different employees.

Portfolio optimization models are widely used for investment management. Investors prefer higher mean rates of return and lower risk, or uncertainty, in the rate of return. These models seek an optimal trade-off between risk and mean return. In the Markowitz model, for which Harry Markowitz received the Nobel Prize in economics in 1990, risk is measured by the variance of the return.

Example 15.4. (Portfolio optimization). Let r_j be the mean rate of return on investment $j = 1, \ldots, n$. Scale the amounts invested so that the total to be invested is one. If x_j is the amount in investment j, the mean return on the portfolio is $\mathbf{r}^T \mathbf{x}$. The variance of the return has the form $\mathbf{x}^T \mathbf{Q} \mathbf{x}$, where $\mathbf{Q}$ is a symmetric positive semidefinite matrix. In multivariate probability, it is known as the covariance matrix of the vector of (random) rates of return, $\mathbf{Q}_{ii}$ is the variance of the rate of return on investment i, and $\mathbf{Q}_{ij}$ is the covariance of the rates of return on investments i and j, $i \neq j$. One way to formulate the portfolio optimization problem is to maximize mean return subject to a limit β on risk:

$$\max \quad \mathbf{r}^T\mathbf{x}$$
$$\text{s.t.} \quad \mathbf{x}^T\mathbf{Q}\mathbf{x} \leq \beta$$
$$\sum_{j=1}^{n} x_j = 1$$
$$\mathbf{x} \geq \mathbf{0}.$$

Similarly, one could minimize risk subject to a required mean return:

$$\min \quad \mathbf{x}^T\mathbf{Q}\mathbf{x}$$
$$\text{s.t.} \quad \mathbf{r}^T\mathbf{x} \geq \bar{r}$$
$$\sum_{j=1}^{n} x_j = 1$$
$$\mathbf{x} \geq \mathbf{0}.$$

The two forms give the same optimal solutions for corresponding values of the mean return and risk limits. One could also use a Lagrangian relaxation, where the risk or return constraint is moved to the objective function. In this case, a weighted average of risk and the negative of mean return is minimized.

Another way that nonlinearity arises in portfolio management is that purchasing or selling a large quantity of a stock or other investment will change its price. This is important to managers of large portfolios but not to small investors. The simplest model is that price is a linear, increasing function of the amount purchased:

$$\text{price of investment } j = p_j + b_j x_j,$$
$$\text{price impact on cost} = (p_j + b_j x_j) x_j - p_j x_j = b_j x_j^2.$$

Then the return on investment becomes

$$\mathbf{r}^T\mathbf{x} - \sum_{j=1}^{n} b_j x_j^2.$$

This quadratic function arises because price was assumed to be linear. Using supply curves, which relate price to quantity, leads to concave (maximization) objectives in many settings.

Convex optimization is widely used in statistics and geometric problems. The least squares principle in regression minimizes the sum of squares of the residuals over all data points,

$$\min_{a,b} \sum_{i=1}^{n} (y_i - a - bx_i)^2,$$

which is a convex quadratic function. Distance, or any norm, is a convex function. The classic Fermat–Weber problem finds a point $\mathbf{x} \in \mathbb{R}^n$ that minimizes the sum of the distances to the fixed points $\mathbf{c}^1, \ldots, \mathbf{c}^m$:

$$\min_{\mathbf{x}} \sum_{i=1}^{m} \|\mathbf{x} - \mathbf{c}^i\|.$$

Lastly, we give an example that can be transformed into a convex program.

Example 15.5. The *analytic center* problem finds the point $\mathbf{x} \in \mathbb{R}$ in a polyhedral set that is farthest from the m defining hyperplanes in the following sense:

$$\max_{\mathbf{x},\mathbf{s}} \prod_{i=1}^{m} s_i$$

$$\text{s.t. } \mathbf{s} = \mathbf{b} - \mathbf{A}\mathbf{x}$$

$$\mathbf{A}\mathbf{x} \le \mathbf{b}.$$

The analytic center maximizes the product of the slacks in the constraints, which is not concave. However, it is equivalent to maximize the logarithm of the objective, which is concave:

$$\max_{\mathbf{x},\mathbf{s}} \sum_{i=1}^{m} \ln s_i$$

$$\text{s.t. } \mathbf{s} = \mathbf{b} - \mathbf{A}\mathbf{x} \tag{15.2}$$

$$\mathbf{s} > \mathbf{0}.$$

The constraint $\mathbf{s} > \mathbf{0}$ may be omitted because it is implied by the domain of the objective. However, the constraint $s_i > \epsilon$ for small ϵ is useful when using a solver.

15.2 Separable Programs

A simple class of convex programs has *separable* functions.

> A function f is **separable** if it can be written in the form
> $f(\mathbf{x}) = f_1(x_1) + \cdots + f_n(x_n)$.

A separable function, then, is the sum of functions of one variable. For example, $f(x_1, x_2) = \exp(x_1) + x_2 \ln x_2$ is separable but $f(x_1, x_2) = 10x_1 + \sin x_2 + 5x_1 x_2$ is not.

One feature of separable programs is that they can be readily approximated by linear programs. Any bounded convex set can be approximated to any desired accuracy by a polyhedron and therefore by a system of linear inequalities. Furthermore, a convex objective function on a bounded domain can be approximated as the maximum of linear functions. As shown in Section 3.2, minimizing such a function is a minimax problem and can be converted to a linear program. This suggests that a linear program could be used to approximate convex programs. Even though the linear program would have more variables and constraints, it might be easier to solve. This approach is useful for separable programs.

A separable function f is convex if and only if all f_i are convex. Each of these convex functions can be approximated as the maximum of linear functions of one variable. For example, if $0 \le x_1 \le 10$, then we might use linear interpolation on 10 intervals of width 1 to approximate $f_1(x_1)$.

Example 15.6. Consider the convex program

$$\min f(\mathbf{x}) = \frac{12}{x_1 + 1} + \frac{30}{x_2 + 2}$$
$$\text{s.t.} \qquad x_1 + x_2 \le 3$$
$$x_j \ge 0.$$

The objective function is separable and convex. From the constraints, the variables have bounds $0 \le x_j \le 3$. Suppose we approximate f by linearly interpolating between the points in the following table. With this approximation, the problem can be written as the following linear program:

$$\min y_1 + y_2$$
$$\text{s.t.} \ y_1 \ge 12 - 6x_1$$
$$y_1 \ge 8 - 2x_1$$
$$y_1 \ge 6 - x_1$$
$$y_2 \ge 15 - 5x_2$$
$$y_2 \ge 12.5 - 2.5x_2$$
$$y_2 \ge 10.5 - 1.5x_2$$
$$x_1 + x_2 \le 3$$
$$x_j \ge 0.$$

The right-hand sides of the first three constraints are the piecewise linear approximation of f_1. Minimizing y_1 subject to these constraints makes y_1 equal to the approximation of f_1. The convex program has optimal solution $\mathbf{x}^* = (1.325, 1.675)$ with objective $f(\mathbf{x}^*) = 13.325$, compared with $\mathbf{x}^{\text{LP}} = (1, 2)$ and $f(\mathbf{x}^{\text{LP}}) = 13.5$ for the linear program.

x	f_1	f_2
0	12	15
1	6	10
2	4	7.5
3	3	6

Separable functions are also easier to work with because of the structure of their derivatives. For the separable function $f(\mathbf{x}) = f_1(x_1) + \cdots + f_n(x_n)$, the first partial derivatives are $\frac{\partial f(\mathbf{x})}{\partial x_1} = f_1'(x_1)$, $\frac{\partial f(\mathbf{x})}{\partial x_2} = f_2'(x_2)$, etc. Thus, less computation is required to find $\nabla f(\mathbf{x})$. The mixed second partial derivatives are zero,

$$\frac{\partial^2 f(\mathbf{x})}{\partial x_i \, \partial x_j} = 0, \ i \neq j.$$

This makes the Hessian matrix of second partial derivatives diagonal.

15.3 Unconstrained Optimization

This section considers convex optimization problems without constraints, namely, $\max f(\mathbf{x})$ where f is concave or $\min f(\mathbf{x})$ where f is convex. Although any practical optimization problem has bounds on the variables, it is often appealing to solve the problem without bounds or other constraints, then add them if they are needed. In Section 15.5 we will see an approach to solving a problem with constraints that uses a related unconstrained problem. For unconstrained convex optimization problems, the first-order condition $\nabla f(\mathbf{x}) = \mathbf{0}$ is a necessary and sufficient condition for a global optimum. Typically this is a system of n nonlinear equations that is as hard to solve as the optimization problem. Iterative search methods are needed. To determine an improving direction, we make use of the first derivatives $\nabla f(\mathbf{x})$.

15.3.1 Gradient Search

We begin with the gradient search method, also known as the method of steepest ascent/descent. Although it is not used in practice, it is simple to describe and introduces some important concepts. As discussed in Section 5.2, an improving search method can be designed by choosing an improving direction to move at each iteration. For nonlinear functions, the step size must not be too large, since the function may increase and then decrease as we move in that direction. Let $\mathbf{x}^k$ be the solution at iteration k of the algorithm. The direction $\mathbf{d} = \nabla f(\mathbf{x}^k) \neq \mathbf{0}$ is the direction with maximum initial rate of change and an obvious choice for a direction. The ideal step size λ in this direction is one that maximizes $f(\mathbf{x})$ over the line

$$\mathbf{x}^k + \lambda \mathbf{d}.$$

Writing $g(\lambda) = f(\mathbf{x}^k + \lambda \mathbf{d})$ emphasizes that we are maximizing a function of one variable. Since g is a composition of a linear and a concave function, it is concave. For some functions, the maximization can be done exactly by solving $g'(\lambda) = 0$. For more complex functions, an iterative search method is needed. This one-dimensional *line search* can be done relatively quickly; see, for example, Luenberger and Ye (2016). Putting these two ideas together is the gradient search method.

Algorithm 15.1 (Gradient Search (Maximization))

- *Initialization.* Choose an initial $\mathbf{x}^0$ and let $k = 0$.
- *Step 1.* If $\nabla f(\mathbf{x}^k) \approx \mathbf{0}$, stop. The current point is (approximately) optimal.
- *Step 2.* Set $\mathbf{d} = \nabla f(\mathbf{x}^k)$. Perform the line search $\max_{\lambda} f(\mathbf{x}^k + \lambda \mathbf{d})$ to find the step size λ^*.
- *Step 3.* Update $\mathbf{x}^{k+1} = \mathbf{x}^k + \lambda^* \mathbf{d}$ and go to Step 1.

Example 15.7. Consider the problem $\max f(\mathbf{x}) = 2x_1 x_2 + 8x_2 - x_1^2 - 2x_2^2$ from Example 15.1. The gradient is

$$\nabla f(\mathbf{x}) = \begin{bmatrix} -2x_1 + 2x_2 \\ 2x_1 - 4x_2 + 8 \end{bmatrix}.$$

From the initial point $\mathbf{x}^0 = (4, 0)$, the next two iterations are shown in the following table and in Figure 15.2. The optimal solution is $\mathbf{x}^* = (4, 4)$, which can be found by solving the first-order condition $\nabla f(\mathbf{x}) = \mathbf{0}$.

k	$\mathbf{x}^k$	$f(\mathbf{x}^k)$	$\nabla f(\mathbf{x}^k)$
0	$(4, 0)$	-16	$(-8, 16)$
1	$(2.46, 3.08)$	14.77	$(1.24, 0.6)$
2	$(4, 3.85)$	15.95	$(-0.3, 0.6)$

In this quadratic example, gradient search moves quickly toward the optimal solution. A few more iterations produce a very accurate solution. However, notice that the two steps are perpendicular. This will always be the case: the line search

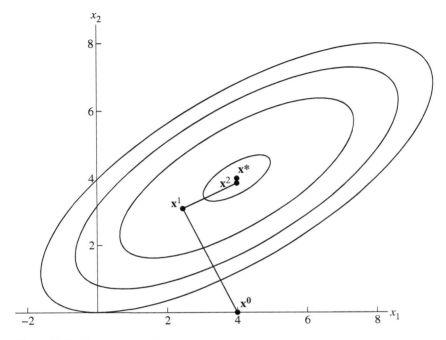

Figure 15.2   Gradient search.

finds the point where the gradient is orthogonal to the direction **d**, so the next step is orthogonal to **d**. For many functions, these perpendicular steps result in zigzagging behavior, making slow progress toward optimal.

15.3.2   Newton's Method

Several modifications of gradient search are available that can improve its performance. One approach is to use second derivative information, not just the first derivatives. The idea in Newton's method is to approximate locally the function being maximized by a quadratic function. This quadratic function is then maximized exactly. Newton's method is usually thought of as solving equations; here we are using it to maximize a function. The connection is that approximating $f(\mathbf{x})$ by a quadratic in $\max f(\mathbf{x})$ is equivalent to approximating $\nabla f(\mathbf{x})$ by a linear function in the first-order condition $\nabla f(\mathbf{x}) = \mathbf{0}$. The quadratic approximation at $\mathbf{x}^k$ is

$$f(\mathbf{x}) \approx f(\mathbf{x}^k) + \nabla f(\mathbf{x}^k)^T(\mathbf{x} - \mathbf{x}^k) + \frac{1}{2}(\mathbf{x} - \mathbf{x}^k)^T \mathbf{H}(\mathbf{x}^k)(\mathbf{x} - \mathbf{x}^k), \tag{15.3}$$

where **H** is the Hessian matrix of second partial derivatives of f. It is obtained from the Taylor series for f. Taking the derivatives of (15.3) and setting them to zero, the

Newton step satisfies

$$\nabla f(\mathbf{x}^k) + \mathbf{H}(\mathbf{x}^k)(\mathbf{x}^{k+1} - \mathbf{x}^k) = 0.$$

This is a system of n linear equations in n unknowns (the components of $\mathbf{x}^{k+1}$). Because f is concave, we know that $\mathbf{H}$ is negative semidefinite; for a minimization problem, f would be convex and $\mathbf{H}$ positive semidefinite. This is not enough to guarantee that $\mathbf{H}$ is invertible. We need the stronger condition that $\mathbf{H}$ is negative definite (for maximization) or positive definite (for minimization). Then the Newton step is

$$\mathbf{x}^{k+1} = \mathbf{x}^k - [\mathbf{H}(\mathbf{x}^k)]^{-1}\nabla f(\mathbf{x}^k). \tag{15.4}$$

There are several reasons to modify Newton's method. First, it can diverge when the initial point is far from the optimal solution. Divergence can be avoided by taking smaller steps. In particular, if we introduce a parameter α_k to scale the step size,

$$\mathbf{x}^{k+1} = \mathbf{x}^k - \alpha_k[\mathbf{H}(\mathbf{x}^k)]^{-1}\nabla f(\mathbf{x}^k), \tag{15.5}$$

a search for α_k could be done at each iteration to choose a step that guards against the possibility that the objective decreases.

A second concern is that $\mathbf{H}(\mathbf{x}^k)$ could be ill-conditioned, making it difficult to compute the Newton step without numerical errors. This issue arises if the magnitude of its largest eigenvalue is much greater than the magnitude of its smallest eigenvalue. (Since $\mathbf{H}(\mathbf{x}^k)$ is negative definite, all of its eigenvalues are negative.) It is often addressed by replacing the Hessian with $-\varepsilon_k\mathbf{I} + \mathbf{H}(\mathbf{x}^k)$ for some $\varepsilon_k > 0$. For ε_k near zero, this matrix is approximately equal to $\mathbf{H}(\mathbf{x}^k)$, representing a Newton step. However, for large ε_k, it is roughly proportional to $-\mathbf{I}$. Replacing $\mathbf{H}(\mathbf{x}^k)$ with $-\mathbf{I}$ in (15.5) gives

$$\mathbf{x}^{k+1} = \mathbf{x}^k + \alpha_k\nabla f(\mathbf{x}^k),$$

which uses the gradient as the direction of the step. Thus, this method uses a direction that is a compromise between gradient search and a pure Newton step.

15.3.3 Quasi-Newton Methods

With these modifications, Newton's method is appealing because it converges very quickly – much faster than gradient search or other methods based on first derivatives. However, the computation required at each iteration can be prohibitive because there a $n(n + 1)/2$ second derivatives to evaluate. Quasi-Newton methods approximate the inverse of the Hessian. While we will not use this approximation in the following sections, we describe the general approach. The basic idea is to use function values and first derivatives from previous iterations, not just the current iteration. Call the approximation the *deflection matrix* $\mathbf{S}_k$, since it modifies or deflects the gradient to set the search direction:

$$\mathbf{x}^{k+1} = \mathbf{x}^k - \alpha_k \mathbf{S}_k \nabla f(\mathbf{x}^k).$$

Note that $\mathbf{S}_k = \mathbf{I}$ is a gradient search.

For a function of one variable, the second derivative at a point can be approximated using two nearby derivatives, e.g.

$$f''(x^k) \approx \frac{f'(x^k) - f'(x^{k-1})}{x^k - x^{k-1}}. \tag{15.6}$$

For a function of n variables, the same idea can be used to approximate the second derivatives on the diagonal of $\mathbf{H}$ but more information is needed to estimate all of the mixed second derivatives. To incorporate more past information, one can update $\mathbf{S}_k$ at each iteration, rather than compute it just from the last two iterations. The update has the form

$$\mathbf{S}_k = \mathbf{C}_1 + \mathbf{C}_2 \mathbf{S}_{k-1}$$

for some matrices $\mathbf{C}_1$ and $\mathbf{C}_2$, i.e. it is linear in $\mathbf{S}_{k-1}$. These matrices are constructed from just the last step $\mathbf{x}^k - \mathbf{x}^{k-1}$ and the change in the gradient $\nabla f(\mathbf{x}^k) - \nabla f(\mathbf{x}^{k-1})$.

We would like $\mathbf{S}_k$ to be consistent with the last step and change in gradient in the same sense as (15.6). For a function of n variables, the analogous equation is

$$\nabla f(\mathbf{x}^k) - \nabla f(\mathbf{x}^{k-1}) \approx \mathbf{H}(\mathbf{x}^k)(\mathbf{x}^k - \mathbf{x}^{k-1}).$$

This expression is exact for quadratic functions (because the third derivatives are zero). Thus, we require the deflection matrix to satisfy the quasi-Newton condition

$$\nabla f(\mathbf{x}^k) - \nabla f(\mathbf{x}^{k-1}) = \mathbf{S}_k(\mathbf{x}^k - \mathbf{x}^{k-1}).$$

See Luenberger and Ye (2016) for how to compute the update using a method called the Broyden, Fletcher, Goldfarb, and Shanno formula.

15.4 Quadratic Programming

We saw in the last section that, for unconstrained problems, Newton's method is equivalent to optimizing a quadratic function at each iteration. This relationship can be extended to problems with equality constraints. As before, optimizing a quadratic function with linear equality constraints can be done by solving a linear system. Consider the quadratic program

$$\max \quad \frac{1}{2}\mathbf{x}^T \mathbf{Q} \mathbf{x} + \mathbf{c}^T \mathbf{x}$$
$$\text{s.t.} \quad \mathbf{A}\mathbf{x} = \mathbf{b}. \tag{15.7}$$

The Lagrangian is

$$L(\mathbf{x}, \lambda) = \frac{1}{2}\mathbf{x}^T \mathbf{Q} \mathbf{x} + \mathbf{c}^T \mathbf{x} + \lambda^T (\mathbf{b} - \mathbf{A}\mathbf{x}).$$

The first-order conditions from Section 14.1 set the gradient of $L(\mathbf{x}, \lambda)$ to zero, namely,

$$\mathbf{Q}\mathbf{x} + \mathbf{c} - \mathbf{A}^T \lambda = \mathbf{0}, \tag{15.8}$$

$$\mathbf{A}\mathbf{x} = \mathbf{b}.$$

If $\mathbf{Q}$ is invertible, these equations can be solved for $\mathbf{x}$ and λ. For the objective function to be concave, $\mathbf{Q}$ must be negative semidefinite. The additional assumption needed for $\mathbf{Q}$ to be invertible is that it is negative definite, or equivalently, that none of its eigenvalues are zero.

The same first-order conditions apply when minimizing. In this case $\mathbf{Q}$ must be positive semidefinite for the objective to be convex. If these conditions are not met, then the quadratic program is unbounded.

Example 15.8. For the quadratic program

$$\max \quad 2x_1 x_2 + 8x_2 - x_1^2 - 2x_2^2$$
$$\text{s.t.} \quad x_1 + \frac{4}{3}x_2 = 8,$$

$$\mathbf{Q} = \begin{bmatrix} -2 & 2 \\ 2 & -4 \end{bmatrix}, \quad \mathbf{c}^T = \begin{bmatrix} 0 & 8 \end{bmatrix}, \quad \mathbf{A} = \begin{bmatrix} 1 & \frac{4}{3} \end{bmatrix}, \quad b = 8.$$

The first-order conditions are

$$\begin{bmatrix} -2 & 2 \\ 2 & -4 \end{bmatrix} \begin{bmatrix} x_1 \\ x_2 \end{bmatrix} - \begin{bmatrix} 1 \\ \frac{4}{3} \end{bmatrix} \lambda = \begin{bmatrix} 0 \\ -8 \end{bmatrix},$$

$$\begin{bmatrix} 1 & \frac{4}{3} \end{bmatrix} \begin{bmatrix} x_1 \\ x_2 \end{bmatrix} = 8,$$

with solution $\mathbf{x} \approx (3.31, 3.52)$ and $\lambda \approx 0.42$. Since $\det(\mathbf{Q}) = 4 > 0$ and its leading minor is $a_{11} = -2 < 0$, it is negative definite and this is the optimal solution. Figure 15.1 shows the optimal solution, labeled $\mathbf{x}'$. The upper right boundary of the shaded region coincides with the equality constraint of this problem.

We note in passing that quadratic programs are also important in applications, such as support vector machines. The more challenging inequality form,

$$\max \quad \frac{1}{2}\mathbf{x}^T \mathbf{Q} \mathbf{x} + \mathbf{c}^T \mathbf{x}$$
$$\text{s.t.} \quad \mathbf{A}\mathbf{x} \le \mathbf{b},$$

can be solved using methods similar to the next section, but without any need to take derivatives.

15.5 Primal-dual Interior Point Method

While there are many algorithms for constrained convex programs, the one we present here performs well on many types of problems [as summarized in Boyd and Vandenberghe (2004, Section 11.7)] and has a good theoretical rate of convergence; see Luenberger and Ye (2016, Section 15.8). It also is closely related to the primal-dual interior point algorithm for linear programming from Section 11.3. As motivation for using an interior point method, Figure 15.3 shows an initial point $\mathbf{x}^0$ in the interior of the feasible region. Level curves for the objective function are shown as dashed lines. The maximum move in the direction of the gradient would move to $\mathbf{x}^1$. However, an iteration from $\mathbf{x}^1$ is complicated by the fact that an inequality constraint, say, $g_1(\mathbf{x}) \leq b_1$, is active. The gradient is not a feasible direction at $\mathbf{x}^1$; even projecting the gradient $\nabla f(\mathbf{x}^1)$ onto the tangent plane to the constraint, which has normal vector $\nabla g(\mathbf{x}^1)$, is not a feasible direction because of the curvature of the constraint. The only feasible directions are those that move back into the interior. To avoid these complexities, barrier methods impose a large cost for approaching an inequality constraint. With this term in the objective, a line search from $\mathbf{x}^0$ stops at an interior point $\mathbf{x}'$.

In addition to stopping short of the boundary, the method we present uses a form of Newton's method to adjust the search direction based on second derivative information. The combination of a barrier term and Newton's method chooses a direction that differs from the gradient, potentially moving more directly toward $\mathbf{x}^*$. At $\mathbf{x}'$, this direction would point more to the right, toward $\mathbf{x}^*$, than $\nabla f(\mathbf{x}')$ does. With each iteration, the barrier term is given less weight, allowing the iterates to approach the boundary and converge to $\mathbf{x}^*$.

The method linearizes and approximately solves the KKT conditions at each iteration. As discussed in Section 11.3, these solutions include primal and dual

Figure 15.3 Moving to the interior point $\mathbf{x}'$ or the boundary point $\mathbf{x}^1$.

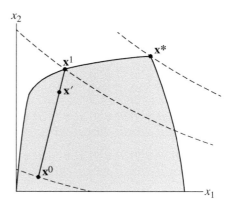

variables that are feasible but do not satisfy the complementary slackness condition for linear programming. The same is true for convex programming, using the duality theory of Section 14.2. The Lagrangian dual of the primal problem (15.1) is given by (14.8) and (14.10) as

$$w^* = \min_{\mathbf{y} \geq 0} \max_{\mathbf{x}} f(\mathbf{x}) + \sum_{i=1}^{m} y_i[b_i - g_i(\mathbf{x})]. \tag{15.9}$$

We have renamed the Lagrange multipliers $\mathbf{y}$ to emphasize that they are also dual variables. As noted there, if a constraint qualification holds and there is a primal optimal solution $\mathbf{x}^*$ with value z^*, then the optimal dual variables $\mathbf{y}^*$ are also the Lagrange multipliers that appear in the KKT conditions for (15.1). Thus, if $\mathbf{y}^*$ is optimal for the dual, then $(\mathbf{x}^*, \mathbf{y}^*)$ satisfies the KKT conditions (14.7):

$$\nabla f(\mathbf{x}) - \sum_{i=1}^{m} y_i \nabla g_i(\mathbf{x}) = \mathbf{0},$$

$$y_i \geq 0,$$

$$g_i(\mathbf{x}) \leq b_i,$$

$$y_i(b_i - g_i(\mathbf{x})) = 0, \quad i = 1, \ldots, m.$$

There is also no duality gap: by the Strong Duality Theorem (Theorem 14.5), $z^* = w^*$.

15.5.1 Barrier Problem

The *barrier problem* adds a logarithmic barrier function to the objective in (15.1):

$$\max f(\mathbf{x}) + \mu \sum_{i=1}^{m} \ln(b_i - g_i(\mathbf{x})) \tag{15.10}$$

for some $\mu > 0$. This choice of objective function has several nice properties. First, because $\ln(b_i - g_i(\mathbf{x})) \to -\infty$ as $\mathbf{x}$ approaches the ith constraint, an optimal solution to the barrier problem is strictly feasible for the original problem, so no constraints are needed. That property makes it a barrier function. Second, (15.10) is still concave. This is true because $-g_i(\mathbf{x})$ is concave and $\ln x$ is concave and nondecreasing, making their composition concave [see Boyd and Vandenberghe (2004, Section 3.2.4)]. Finally, as we will see, if we add slack variables the first-order conditions have just the right form to approximate the KKT conditions for the original problem.

It is easier to state the optimality conditions for the equality form

$$\max f(\mathbf{x}) + \mu \sum_{i=1}^{m} \ln s_i \tag{15.11}$$

$$\text{s.t. } g_i(\mathbf{x}) + s_i = b_i, \quad i = 1, \ldots, m,$$

where s_i is the slack in constraint i. The Lagrangian for this problem is

$$L(\mathbf{x}, \mathbf{s}, \mathbf{y}) = f(\mathbf{x}) + \mu \sum_{i=1}^{m} \ln s_i + \sum_{i=1}^{m} y_i [b_i - g_i(\mathbf{x}) - s_i].$$

Setting the derivatives of the Lagrangian to zero, the first-order conditions are

$$\frac{\partial L}{\partial x_j} = \frac{\partial f(\mathbf{x})}{\partial x_j} - \sum_{i=1}^{m} y_i \frac{\partial g_i(\mathbf{x})}{\partial x_j} = 0, \ j = 1, \dots, n$$

$$\frac{\partial L}{\partial s_i} = \frac{\mu}{s_i} - y_i = 0, \ i = 1, \dots, m$$

and feasibility (15.11). Multiplying the second equation by s_i, the conditions are

$$\frac{\partial f(\mathbf{x})}{\partial x_j} - \sum_{i=1}^{m} y_i \frac{\partial g_i(\mathbf{x})}{\partial x_j} = 0, \ j = 1, \dots, n,$$

$$s_i y_i = \mu, \ i = 1, \dots, m, \tag{15.12}$$

$$b_i - g_i(\mathbf{x}) - s_i = 0, \ i = 1, \dots, m.$$

To stay in the domain of (15.10), we are only interested in strictly feasible solutions with $\mathbf{s} > \mathbf{0}$; the second equation implies that $\mathbf{y} > \mathbf{0}$, so it is strictly feasible for the dual. As in (11.19) for linear programs, this equation is a modified complementary slackness condition, where μ is the amount each complementary slackness condition is violated.

15.5.2 The Newton Step

We wish to solve these nonlinear equations iteratively based on a Newton step. This method solves the linear approximation of the equations at the current solution. Denoting the current solution $(\mathbf{x}, \mathbf{s}, \mathbf{y}) = (\mathbf{x}^k, \mathbf{s}^k, \mathbf{y}^k)$, the Newton step moves to $(\mathbf{x}, \mathbf{s}, \mathbf{y}) + (\mathbf{d}^x, \mathbf{d}^s, \mathbf{d}^y)$. The linear approximation of $(s_i + d_i^s)(y_i + d_i^y)$ is $s_i y_i + y_i d_i^s + s_i d_i^y$. Also, the linear approximation of g_i is

$$g_i(\mathbf{x} + \mathbf{d}^x) \approx g_i(\mathbf{x}) + \sum_{k=1}^{n} \frac{\partial g_i(\mathbf{x})}{\partial x_k} d_k^x$$

and the linear approximation of the partial derivative of f is

$$\frac{\partial f(\mathbf{x} + \mathbf{d}^x)}{\partial x_j} \approx \frac{\partial f(\mathbf{x})}{\partial x_j} + \sum_{k=1}^{n} \frac{\partial^2 f(\mathbf{x})}{\partial x_j \, \partial x_k} d_k^x.$$

The final linear approximation needed is

$$(y_i + d_i^y) \frac{\partial g_i(\mathbf{x} + \mathbf{d}^x)}{\partial x_j} \approx y_i \left(\frac{\partial g_i(\mathbf{x})}{\partial x_j} + \sum_{k=1}^{n} \frac{\partial^2 g_i(\mathbf{x})}{\partial x_j \, \partial x_k} d_k^x \right) + \frac{\partial g_i(\mathbf{x})}{\partial x_j} d_i^y.$$

Substituting these approximations into the first-order conditions (15.12),

$$\frac{\partial f(\mathbf{x})}{\partial x_j} + \sum_{k=1}^{n} \frac{\partial^2 f(\mathbf{x})}{\partial x_j\, \partial x_k} d_k^x - \sum_{i=1}^{m} y_i \left(\frac{\partial g_i(\mathbf{x})}{\partial x_j} + \sum_{k=1}^{n} \frac{\partial^2 g_i(\mathbf{x})}{\partial x_j\, \partial x_k} d_k^x \right) - \sum_{i=1}^{m} \frac{\partial g_i(\mathbf{x})}{\partial x_j} d_i^y = 0,$$

$$s_i y_i + y_i d_i^s + s_i d_i^y = \mu,$$

$$g_i(\mathbf{x}) + \sum_{k=1}^{n} \frac{\partial g_i(\mathbf{x})}{\partial x_k} d_k^x + s_i + d_i^s = b_i.$$

Next we eliminate the variables $\mathbf{d}^s$. Rearranging the middle equation,

$$d_i^s = -\frac{s_i}{y_i} d_i^y + \frac{\mu}{y_i} - s_i.$$

Substituting this into the bottom equation and moving the constants to the right side,

$$\sum_{k=1}^{n} \left(-\frac{\partial^2 f(\mathbf{x})}{\partial x_j\, \partial x_k} + \sum_{i=1}^{m} y_i \frac{\partial^2 g_i(\mathbf{x})}{\partial x_j\, \partial x_k} \right) d_k^x + \sum_{i=1}^{m} \frac{\partial g_i(\mathbf{x})}{\partial x_j} d_i^y = \frac{\partial f(\mathbf{x})}{\partial x_j} - \sum_{i=1}^{m} y_i \frac{\partial g_i(\mathbf{x})}{\partial x_j}$$

$$\sum_{k=1}^{n} \frac{\partial g_i(\mathbf{x})}{\partial x_k} d_k^x - \frac{s_i}{y_i} d_i^y = b_i - g_i(\mathbf{x}) - \frac{\mu}{y_i}.$$

This is a system of $m + n$ linear equations in the $m + n$ unknowns $(\mathbf{d}^x, \mathbf{d}^y)$. To write these equations in matrix form, let $\mathbf{g}(\mathbf{x}) = (g_1(\mathbf{x}), \ldots, g_m(\mathbf{x}))$, $\mathbf{J}(\mathbf{x})$ be the $m \times n$ Jacobian matrix of partial derivatives $\frac{\partial g_i(\mathbf{x})}{\partial x_j}$, $\nabla^2 f(\mathbf{x})$ be the Hessian of $f(\mathbf{x})$, $\nabla^2 g_i(\mathbf{x})$ be the Hessian of $g_i(\mathbf{x})$, $\mathbf{m} = (\frac{\mu}{y_1}, \ldots, \frac{\mu}{y_m})$ and

$$\mathbf{S} = \text{diag}\left(\frac{b_1 - g_1(\mathbf{x})}{y_1}, \ldots, \frac{b_m - g_m(\mathbf{x})}{y_m} \right).$$

Also let

$$\mathbf{Q} = \nabla^2 L(\mathbf{x}, \mathbf{s}, \mathbf{y}) = -\nabla^2 f(\mathbf{x}) + \sum_{i=1}^{m} y_i \nabla^2 g_i(\mathbf{x}), \tag{15.13}$$

which is the Hessian of the Lagrangian. Then the equations can be written

$$\begin{bmatrix} \mathbf{Q} & \mathbf{J}(\mathbf{x})^T \\ \mathbf{J}(\mathbf{x}) & \mathbf{S} \end{bmatrix} \begin{bmatrix} \mathbf{d}^x \\ \mathbf{d}^y \end{bmatrix} = \begin{bmatrix} \nabla f(\mathbf{x}) - \mathbf{J}(\mathbf{x})^T \mathbf{y} \\ \mathbf{b} - \mathbf{g}(\mathbf{x}) - \mathbf{m} \end{bmatrix}. \tag{15.14}$$

The solution to this system is the Newton step.

15.5.3 Step Size and Slackness Parameter

Although the Newton step provides both a direction and a step size, the step size must be modified to create a useful algorithm. First, as was the case in the linear programming version, the step could lead to infeasible solutions $\mathbf{x}$ and $\mathbf{y}$. Also, Newton's method may not converge if the initial point is far from the optimal solution. We address the first difficulty by reducing the step size to maintain strict feasibility.

With nonlinear constraints, there is no easy method, like the ratio test for linear constraints, to maintain feasibility. Instead, *backtracking line search* can be used. First, we use a ratio test to maintain dual feasibility ($\mathbf{y} \geq \mathbf{0}$). Let

$$\lambda_{\max} = \min \left\{ 1, \min \left\{ \frac{y_i}{-d_i^y} : d_i^y < 0 \right\} \right\}. \tag{15.15}$$

Then each $y_i + \lambda_{\max} d_i^y \geq 0$. Next we backtrack until the constraints are satisfied:

- Choose parameters $0 < \beta < 1$ and $0 < \alpha < 1$.
- Set $\lambda = \lambda_{\max}$.
- If a constraint is violated, $g_i(\mathbf{x} + \lambda \mathbf{d}^x) > b_i$ for some i, then replace λ with $\beta\lambda$. Repeat until all constraints are satisfied.
- Replace λ with $\alpha\lambda$.

The update using this final step size is

$$(\mathbf{x}^{k+1}, \mathbf{y}^{k+1}) = (\mathbf{x}^k, \mathbf{y}^k) + \lambda(\mathbf{d}^x, \mathbf{d}^y). \tag{15.16}$$

The parameter α maintains strict feasibility. Each dual variable is reduced by no more than a factor of $1 - \alpha$. The constraint $g_i(\mathbf{x}) \leq b_i$ also remains strictly feasible.

As in Section 11.3, we want the barrier, or complementary slackness, parameter μ to approach 0 as the algorithm progresses. As $\mu \to 0$ the solution of (15.12) approaches an optimal solution of the problem. To accomplish this, we will multiply μ by γ at each iteration, where $0 < \gamma < 1$. Initially, μ should be relatively large to keep the new point away from the constraints $g_i(\mathbf{x}) \leq b_i$ and $\mathbf{y} \geq \mathbf{0}$. The method described in Section 11.3 could also be used. See Boyd and Vandenberghe (2004, Section 11.7.2) for more discussion

15.5.4 Primal-dual Interior Point Algorithm

Now we can state the overall algorithm. It is applied to problem (15.1). It requires a strictly feasible solution to the primal. Any initial $\mathbf{y} > \mathbf{0}$ may be used; it is a strictly feasible solution to the dual (15.9).

Algorithm 15.2 (Primal-dual Interior Point Algorithm)

Step 0. Initialization. Start with a strictly feasible solution **x** to the primal problem, **y** > **0**, initial μ, step size parameters $0 < \alpha < 1$ and $0 < \beta < 1$, duality gap parameter $0 < \gamma < 1$, and a stopping tolerance $\varepsilon > 0$. Compute the slack **s**.
  Step 1. Compute Newton step. Compute the step $(\mathbf{d}^x, \mathbf{d}^y)$ using (15.14).
  Step 2. Compute step size λ using (15.15) and backtracking.
  Step 3. Update solution. Update the solution $(\mathbf{x}, \mathbf{y})$ using (15.16) and **s** using (15.11). Update $\mu^{\text{new}} = \gamma\mu$.
  Step 4. Stopping condition. If $\|\mathbf{d}\| < \varepsilon$, stop. The current solution is approximately optimal. Otherwise go to Step 1.

A one-variable problem will help illustrate the basic ideas.

Example 15.9. Consider the problem

$$\max \quad 10x - e^x$$
$$\text{s.t.} \quad x \le 2.$$

The derivatives are $f'(x) = 10 - e^x$, $f''(x) = -e^x$, and $g'(x) = g''(x) = 0$. Equation (15.14) for the Newton step is

$$\begin{bmatrix} e^x & 0 \\ 0 & \dfrac{2-x}{y} \end{bmatrix} \begin{bmatrix} d^x \\ d^y \end{bmatrix} = \begin{bmatrix} 10 - e^x \\ 2 - x - \dfrac{\mu}{y} \end{bmatrix}$$

with solution

$$d^x = 10e^{-x} - 1, \quad d^y = y - \frac{\mu}{2-x}.$$

We apply the algorithm starting at $x = 1$, $y = 0.5$, and $\mu = 2$. However, instead of backtracking, we can just compute the slack in the constraint and include the ratio $-(2-x)/d^x$ in the step size formula (15.15). We use $\alpha = 0.8$, i.e., the final step size is (15.15) multiplied by 0.8, and $\gamma = 0.1$ to decrease μ. The table shows that the algorithm converges to the optimal solution $x = 2$. Note that $\lambda < 0.8$. At most iterations, we can see that y is the limiting factor on step size, not x, as y decreases by 80%, the maximum allowed by α. At iteration 4, x is the limiting factor as the slack in the constraint decreases by 80%. At every iteration d^x is larger than feasible; in this case, the specific value of μ does not affect the step taken (but a larger μ would). The quadratic fit has a maximum at $x \gg 2$, causing the Newton step to overshoot $x = 2$.

k	μ	x	y	λ
0	2	1	0.5	0.267
1	0.2	1.71	0.1	0.133
2	0.02	1.82	0.02	0.174
3	0.002	1.929	0.004	0.126
4	2×10^{-4}	1.986	0.0010	0.031
5	–	1.997	0.0006	–

The next example demonstrates the matrix notation but also illustrates that it can be challenging to pick good parameter values.

Example 15.10. Consider the problem

$$\max \quad 10 + 10x_1 - 8x_2 - 4e^{x_1} - e^{x_1 - x_2}$$
$$\text{s.t.} \quad x_2 - x_1^{0.5} \le 0$$
$$-x_2 + x_1^{1.5} \le 0.$$

The derivatives needed by the algorithm are

$$\nabla f(\mathbf{x}) = \begin{bmatrix} 10 - 4e^{x_1} - e^{x_1 - x_2} \\ -8 + e^{x_1 - x_2} \end{bmatrix}, \quad \nabla^2 f(\mathbf{x}) = \begin{bmatrix} -4e^{x_1} - e^{x_1 - x_2} & e^{x_1 - x_2} \\ e^{x_1 - x_2} & -e^{x_1 - x_2} \end{bmatrix},$$

$$\nabla g_1(\mathbf{x}) = \begin{bmatrix} -0.5x_1^{-0.5} \\ 1 \end{bmatrix}, \quad \nabla^2 g_1(\mathbf{x}) = \begin{bmatrix} 0.25x_1^{-1.5} & 0 \\ 0 & 0 \end{bmatrix},$$

$$\nabla g_2(\mathbf{x}) = \begin{bmatrix} 1.5x_1^{0.5} \\ -1 \end{bmatrix}, \quad \nabla^2 g_2(\mathbf{x}) = \begin{bmatrix} 0.75x_1^{-0.5} & 0 \\ 0 & 0 \end{bmatrix},$$

$$\mathbf{J}(\mathbf{x}) = \begin{bmatrix} -0.5x_1^{-0.5} & 1 \\ 1.5x_1^{0.5} & -1 \end{bmatrix}.$$

The progress of the algorithm is shown in the table below using the initial value $\mu = 1$ and multiplying it by the factor $\gamma = 0.1$ each iteration. The initial $\mathbf{x}$ is near the middle of the feasible region and $\mathbf{y}$ was chosen to give a reasonable first step. The Newton steps are all feasible and no backtracking is needed with the parameter $\alpha = 0.8$, so the step size is $\lambda = \alpha = 0.8$ and β is not used. The optimal solution with objective function value 5.293 occurs at $\mathbf{x} = (0.164, 0.066)$ on the boundary where $g_2(\mathbf{x}) = 0$. Progress toward this point is fairly slow.

k	μ	x	y	$-g_1(x)$	$-g_2(x)$	$f(x)$
0		(0.5, 0.6)	(5, 10)	0.107	0.246	2.700
1	1	(0.198, 0.261)	(7.37, 14.02)	0.183	0.173	4.075
2	0.1	(0.324, 0.323)	(16.05, 22.96)	0.246	0.139	4.124
3	0.01	(0.301, 0.280)	(30.37, 37.33)	0.269	0.115	4.346
4	0.001	(0.327, 0.289)	(56.29, 63.25)	0.283	0.102	4.371
5	10^{-4}	(0.324, 0.279)	(102.7, 109.6)	0.290	0.095	4.429

Practical implementations of this algorithm address several issues that we have not considered. First, an initial feasible solution is needed. A different, Phase 1 procedure is used to find one. Second, the stopping criterion is not very satisfying. The difficulty is that the total complementary slackness violation $s^T y$ does not bound the difference between the value of the current primal value $f(x)$ and its optimal value z^*. Solving a linearization of the optimality equations (15.12) does not necessarily produce a feasible primal solution, because g_i is nonlinear in the third equation. More importantly, it does not produce a dual solution that satisfies the first equation, which is also nonlinear. When the algorithm is applied to linear programs, these difficulties do not arise: the primal and dual solution at each iteration are feasible and the difference between their values is $s^T y$, giving a bound on z^*.

More sophisticated stopping criteria, involving the amounts by which these equations in (15.12) are violated, are preferable. To insure convergence of the algorithm, the backtracking method can be modified using the idea of a *merit function*. The objective function value $f(x)$ may not increase at each iteration. However, if a merit function is constructed by subtracting the infeasibility measures from $f(x)$, this merit function can be shown to improve at every iteration. Knowing that the merit function improves at every iteration, plus analyzing its rate of improvement, leads to a proof that the algorithm converges.

The third major concern is that the Hessian of f and of each g_i must be computed at each iteration. Not only are there about $mn^2/2$ second derivatives in these symmetric $n \times n$ matrices, computing each second derivative may take significant computation. Thus, using second derivatives can be inefficient. The quasi-Newton method described in Section 15.3 addresses this by using function values and first derivatives to update an estimated Hessian at each iteration.

Strong duality holds for the barrier problem because it is a convex program. As a result, assuming that it has a unique optimal solution $x(\mu)$ for every μ, there is a corresponding dual feasible point $y(\mu)$, found from (15.12). These points for different μ form a *central path*, as in the linear programming case. Dual feasible solutions to the barrier problem satisfy the first equation in (15.12) and $\mathbf{y} \geq \mathbf{0}$. These do not involve the barrier parameter, so dual feasibility is the same for the original problem. Thus, $\mathbf{x}(\mu), \mathbf{y}(\mu)$ form a primal-dual optimal pair for the barrier problem, with no duality gap, and a primal-dual feasible pair for the original problem, with objective functions that differ by the duality gap $\mathbf{s}^T\mathbf{y}$, which appears in the complementary slackness condition. Thus, on the dual path, the objective function value is an upper bound on z^*. Iterations of the algorithm are generally not on the central path, so the algorithm does not produce this bound.

There is a close connection between this algorithm and the inequality constrained quadratic program. The linear approximation used in the Newton step is a linear approximation of first derivatives. One could start with a quadratic approximation of the functions f and g_i and arrive at the same first-order optimality equations. Using a Newton step for these equations, then, is equivalent to performing one iteration of a solution method for the quadratic program that locally approximates the convex program. This quadratic program is convex because its objective function is $\frac{1}{2}\mathbf{x}^T\mathbf{Q}\mathbf{x}$, where $\mathbf{Q}$ is given by (15.13). In particular, f and all g_i are convex, meaning that their Hessian matrices are positive semidefinite, making $\mathbf{Q}$ positive semidefinite. A different quadratic program is used at each iteration as $\mathbf{Q}$ changes. This method is called *sequential quadratic programming*.

Problems

1 Consider the portfolio optimization problem

$$\max \quad 0.07x_1 + 0.04x_2 + 0.05x_3 - 0.025x_1^2 - 0.015x_2^2 - 0.01x_3^2$$
$$\text{s.t.} \quad x_1 + x_2 + x_3 = 1$$
$$x_j \geq 0,$$

where x_j is the proportion of the budget spent on investment j.
   (a) Solve without the nonnegativity constraints using the method of Section 15.4.
   (b) One investment in (a) has a negative optimal value. Eliminate this investment and resolve. Is the solution nonnegative?

2 (a) Use a solver and (15.2) to find the analytic center of the polyhedron

$$x_1 + x_2 \leq 12$$
$$x_1 \qquad \leq 5$$
$$x_2 \leq 10$$
$$x_j \geq 0$$

graphed in Figure 10.1.

(b) Multiply the first constraint by 5 and resolve. Rewrite the objective function to show that this problem must have the same optimal solution as (a).

(c) Add the redundant constraint $1.5x_1 + x_2 \leq 16$ and resolve.

3 Which of the following functions are separable?

(a) $f(x_1, x_2) = \dfrac{4x_1^2 x_2 + x_2^5}{x_2}$,

(b) $g(x_1, x_2) = 10e^{-2x_1} + 5e^{-x_2}$,

(c) $h(x_1, x_2) = 10e^{-2x_1 - x_2} + x_1^2 + 2x_2^2$.

4 Find the Hessian matrix for the functions in Exercise 15.3. Which ones are diagonal? How is this related to whether they are separable?

5 Construct a linear program to approximate the following separable convex program using linear interpolation between the points $x_1 = 0, 1, 2, 3, 4$. Do not solve.

$$\max \quad \frac{1}{1 + 2^{-x_1}} + 0.1x_2$$
$$\text{s.t.} \qquad x_1 + x_2 \leq 4$$
$$x_j \geq 0.$$

6 Construct a linear program to approximate the following separable convex program using linear interpolation between the points $x_j = 0, 1, 2$. Do not solve.

$$\max \qquad 4x_1 + 5x_2$$
$$\text{s.t.} \quad 4x_1 + x_1^3 + 4x_2 + 2x_2^2 \leq 16$$
$$x_1 + x_2 \leq 2$$
$$x_j \geq 0.$$

7 Suppose gradient search is used for the unconstrained problem

$$\min \quad f(x_1, x_2, x_3) = x_1^4 + x_1^2 - 2x_1 x_2 + 2x_2^2 + 0.5x_1^2 x_3^2 + 0.5x_3^2 - 8x_1 - 4x_3.$$

Find the search direction from
(a) $(0, 0, 0)$.
(b) $(1, 0, 1)$.

8 For Example 15.7 using gradient search,
(a) Find the search direction from $(0, 0)$.

(b) Perform two iterations of gradient search from $(0, 0)$. Which initial point, $(0, 0)$ or $(4, 0)$ (used in the example) results in the second iteration being closer to optimal?

(c) Find the search direction of the third iteration. Explain using Figure 15.2 why convergence to optimal is slower (or faster) starting from $(0, 0)$ compared with $(4, 0)$.

9 For Exercise 15.7,

(a) Find the quadratic approximation (15.3) at $(0, 0, 0)$.

(b) Find the next point using the Newton step (15.4) from the point $(0, 0, 0)$.

(c) Find the next point using the Newton step (15.4) from the point $(1, 0, 1)$.

10 For the equality constrained quadratic program

$$\min \quad 3x_1^2 + x_2^2 - 2x_1x_2 - 6x_1 - 12x_2$$
$$\text{s.t.} \quad x_1 + x_2 = 24.$$

(a) Write the first order conditions (15.8).

(b) Solve the first-order conditions to find the optimal solution.

11 (a) Is a linear program a separable program? Explain.

(b) Is the quadratic program (15.7) a separable program? Explain.

(c) Suppose Q is diagonal. Is (15.7) a separable program? Explain.

12 Suppose the Newton step (15.16) at $\mathbf{x} = (1, 1)$ and $\mathbf{y} = (2, 2)$ uses the directions $\mathbf{d}^x = (1, 3)$ and $\mathbf{d}^y = (-3, -4)$ for a problem with constraints

$$x_2 + x_1^2 - 2x_1 \leq 1,$$
$$3x_1 + x_2 \leq 6.$$

(a) Find the step size $\lambda_{\max}$ that preserves $\mathbf{y} \geq \mathbf{0}$. See (15.15).

(b) Is this step size feasible? If not, use backtracking with $\beta = 0.75$ to find a feasible step.

(c) Repeat (b) with $\mathbf{x} = (1, 0)$, using the same $\mathbf{y}$, $\mathbf{d}^x$ and $\mathbf{d}^y$ as before.

13 For Example 15.10, suppose the initial solution is $\mathbf{x} = (0.5, 0.5)$ and $\mathbf{y} = (1, 10)$ and we use $\mu = 0.1$.

(a) Find the Newton step. If we use this step, i.e. $\lambda = 1$ in (15.16), is the new solution feasible for the primal? Is it dual feasible $(\mathbf{y} > \mathbf{0})$?

(b) Compute the step size for the Newton step found in (a) using $\beta = 0.75$ for backtracking and $\alpha = 0.8$. Also compute the new solution $\mathbf{x}$ and $\mathbf{y}$ using this step size.

A

Linear Algebra and Calculus Review

A.1 Sets and Other Notation

If x is a member of a set S, we write $x \in S$. The *empty set* is denoted $\emptyset$. The set of real numbers is denoted $\mathbb{R}$ and the set of integers is denoted $\mathbb{Z}$. The set of n-tuples of real numbers is denoted $\mathbb{R}^n$; the set of n-tuples of integers is $\mathbb{Z}^n$. Likewise, $\mathbb{R}^n_+$ and $\mathbb{Z}^n_+$ are the set of n-tuples of nonnegative reals and integers. We use notation such as $\{(x, y) : x + y \le 4\}$ to mean all $(x, y) \in \mathbb{R}^2$ satisfying the inequality. The set $S \in \mathbb{R}^n$ is *bounded* if there is an M for which $\|x\| \le M$ for all $x \in S$. Here $\|x\|$ is the length of x.

The set S is a *subset* of T, denoted $S \subset T$, if every element in S is also in T. The *union* of sets S and T is denoted $S \cup T$ and is the set containing the elements that belong to S or T. The *intersection* of sets S and T is denoted $S \cap T$ and is the set containing the elements that belong to both S and T. The *difference* $S \backslash T$ is the set of all elements in S but not in T. For example,

$$\{1, 2, 3, 4\} \backslash \{3, 4, 6\} = \{1, 2\}.$$

The *cardinality* of a finite set S, denoted $|S|$, is the number of elements in the set.

The floor function $\lfloor x \rfloor$ rounds down to the nearest integer and the ceiling function $\lceil x \rceil$ rounds up.

A.2 Matrix and Vector Notation

A *matrix* is a rectangular array of numbers, called the elements of the matrix. We use boldface capital letters for matrices, and the lowercase to denote elements of the matrix, with a double subscript to denote the row and column. Thus, we write

$$\mathbf{A} = \begin{bmatrix} a_{11} & a_{12} & \cdots & a_{1n} \\ a_{21} & a_{22} & \cdots & a_{2n} \\ \vdots & \vdots & & \vdots \\ a_{m1} & a_{m2} & \cdots & a_{mn} \end{bmatrix}$$

for a matrix having m rows and n columns. The *size* of the matrix is $m \times n$. The number a_{ij} in row i and column j is called the (i,j)th element of $\mathbf{A}$. To specify a matrix by defining its size and a general element, we may write a formula such as $a_{ij} = (-1)^{i-1}b_{ij}$, which multiplies the even rows of $\mathbf{B}$ by -1. We also use the notation $\mathbf{A} = [a_{ij}]$, so the same matrix can be defined by $\mathbf{A} = [(-1)^{i-1}b_{ij}]$.

A *zero matrix* is a matrix containing all 0s, denoted $\mathbf{0}$. A *square* matrix is one with the same number of rows and columns. An *identity matrix*, denoted $\mathbf{I}$, is a square matrix whose elements $a_{ij} = 0$ for $i \neq j$ and $a_{ii} = 1$ for $i = 1, \ldots, n$. For example, the 3×3 identity matrix is

$$\mathbf{I} = \begin{bmatrix} 1 & 0 & 0 \\ 0 & 1 & 0 \\ 0 & 0 & 1 \end{bmatrix}.$$

A *row vector* is a matrix having a single row; a *column vector* is a matrix having a single column. We usually use lower-case boldface letters for vectors. A vector with n entries is called an n-vector. To economize space, column vectors are sometimes written in a row using parentheses. Thus, $\mathbf{v} = (3, 1, 8)$ is the same as

$$\mathbf{v} = \begin{bmatrix} 3 \\ 1 \\ 8 \end{bmatrix}$$

and is a column vector, while $[3\ 1\ 8]$ is a row vector. A *zero vector* is a vector containing all 0's, also denoted $\mathbf{0}$. A vector containing all 1s is denoted $\mathbf{1}$. The vector $\mathbf{e}_k$ is a vector whose kth entry is 1 and other entries are 0. For example, for 3-vectors,

$$\mathbf{e}_1 = \begin{bmatrix} 1 \\ 0 \\ 0 \end{bmatrix}, \quad \mathbf{e}_2 = \begin{bmatrix} 0 \\ 1 \\ 0 \end{bmatrix}, \quad \mathbf{e}_3 = \begin{bmatrix} 0 \\ 0 \\ 1 \end{bmatrix}.$$

The size of vectors and matrices can often be inferred from context.

It is convenient to use block notation to partition a matrix by rows or columns. For example, if

$$\mathbf{B} = \begin{bmatrix} 1 & 2 & 3 \\ 4 & 5 & 6 \end{bmatrix}, \quad \mathbf{C} = \begin{bmatrix} 10 & -1 \\ 12 & -4 \end{bmatrix}$$

then

$$[\mathbf{B}|\mathbf{C}] = \begin{bmatrix} 1 & 2 & 3 & 10 & -1 \\ 4 & 5 & 6 & 12 & -4 \end{bmatrix}.$$

In particular, we refer to the columns of an $m \times n$ matrix $\mathbf{A}$ as $\mathbf{a}_j, j = 1, \ldots, n$ and the rows as $\mathbf{A}_i, i = 1, \ldots, m$. Thus,

$$\mathbf{A} = [\mathbf{a}_1|\mathbf{a}_2|\cdots|\mathbf{a}_n] = \begin{bmatrix} \mathbf{A}_1 \\ \mathbf{A}_2 \\ \vdots \\ \mathbf{A}_m \end{bmatrix}.$$

The length of an n-vector, also called the Euclidean norm, is $\|\mathbf{v}\| = (v_1^2 + \cdots + v_n^2)^{1/2}$. The elements of $\mathbb{R}^n$ are n-vectors. Using the notion of length, a set $S \in \mathbb{R}^n$ is *bounded* if there is an M for which $\|\mathbf{x}\| \leq M$ for all $\mathbf{x} \in S$. A set $S \in \mathbb{R}^n$ is *closed* if it contains all of its limit points. More precisely, S is closed if, for any convergent sequence $\mathbf{x}_1, \mathbf{x}_2, \mathbf{x}_3, \ldots$ of points in S, $\lim_{\mathbf{x} \to \infty} \mathbf{x}_k \in S$. A closed interval $[a, b]$ is a closed and bounded set, while an open interval (a, b) is not a closed set. The positive orthant $\mathbb{R}^n = \{\mathbf{x} \in \mathbb{R}^n : \mathbf{x} \geq \mathbf{0}\}$ is closed but not bounded; a set such as $\{(x, y) : x + y < 4\}$ is not closed because it does not contain the points on the boundary $x + y = 4$.

A.3 Matrix Operations

The sum $\mathbf{A} + \mathbf{B}$ of two matrices of the same size is the matrix whose elements are the sum of the corresponding elements of $\mathbf{A}$ and $\mathbf{B}$. For example,

$$\begin{bmatrix} 10 & -1 \\ 12 & -4 \end{bmatrix} + \begin{bmatrix} 7 & 0 \\ 3 & 2 \end{bmatrix} = \begin{bmatrix} 17 & -1 \\ 15 & -2 \end{bmatrix}.$$

The product of a scalar, or number, k and a matrix $\mathbf{A}$ is the matrix $k\mathbf{A}$ obtained by multiplying each element of $\mathbf{A}$ by k. This operation is called *scalar multiplication*. For example,

$$3 \begin{bmatrix} 7 & 0 \\ 3 & 2 \end{bmatrix} = \begin{bmatrix} 21 & 0 \\ 9 & 6 \end{bmatrix}.$$

The product of an $m \times r$ matrix $\mathbf{A}$ and a $r \times n$ matrix $\mathbf{B}$ is the $m \times n$ matrix $\mathbf{C}$ with elements $c_{ij} = \sum_{k=1}^{r} a_{ik} b_{kj}$. Matrices can only be multiplied if the dimensions match in this way. For example, if

$$\mathbf{A} = \begin{bmatrix} 2 & 1 & 4 \\ 3 & 0 & 2 \end{bmatrix}, \quad \mathbf{B} = \begin{bmatrix} 1 & -2 \\ 0 & 5 \\ 2 & 1 \end{bmatrix}$$

then

$$\mathbf{AB} = \begin{bmatrix} 2 & 1 & 4 \\ 3 & 0 & 2 \end{bmatrix} \begin{bmatrix} 1 & -2 \\ 0 & 5 \\ 2 & 1 \end{bmatrix}$$

$$= \begin{bmatrix} 2(1) + 1(0) + 4(2) & 2(-2) + 1(5) + 4(1) \\ 3(1) + 0(0) + 2(2) & 3(-2) + 0(5) + 2(1) \end{bmatrix}$$

$$= \begin{bmatrix} 10 & 5 \\ 7 & -4 \end{bmatrix}.$$

We often multiply a row vector by a column vector, giving a scalar. For example,

$$\begin{bmatrix} 2 & 1 & 4 \end{bmatrix} \begin{bmatrix} 1 \\ 0 \\ 2 \end{bmatrix} = 10,$$

which is different than

$$\begin{bmatrix} 1 \\ 0 \\ 2 \end{bmatrix} \begin{bmatrix} 2 & 1 & 4 \end{bmatrix} = \begin{bmatrix} 2 & 1 & 4 \\ 0 & 0 & 0 \\ 4 & 2 & 8 \end{bmatrix}.$$

Matrix multiplication is not commutative: in general, $\mathbf{AB} \neq \mathbf{BA}$ even if their sizes are such that both are defined. The identity matrix has the property that

$$\mathbf{AI} = \mathbf{IA} = \mathbf{A}.$$

The *transpose* of an $m \times n$ matrix $\mathbf{A}$ is the $n \times m$ matrix $\mathbf{A}^T$ with elements a_{ji}. Here is an example of a matrix and its transpose:

$$\mathbf{A} = \begin{bmatrix} 2 & 1 & 4 \\ 3 & 0 & 2 \end{bmatrix}, \quad \mathbf{A}^T = \begin{bmatrix} 2 & 3 \\ 1 & 0 \\ 4 & 2 \end{bmatrix}.$$

A square matrix $\mathbf{A}$ is *symmetric* if $\mathbf{A}^T = \mathbf{A}$. A useful property is that $(\mathbf{AB})^T = \mathbf{B}^T \mathbf{A}^T$. Transposes convert row vectors into column vectors. Given two n-vectors $\mathbf{u}$ and $\mathbf{v}$, their dot product is

$$\mathbf{u} \cdot \mathbf{v} = \sum_{i=1}^{n} u_i v_i,$$

which is the same as the matrix multiplications

$$\mathbf{u} \cdot \mathbf{v} = \mathbf{u}^T \mathbf{v} = \mathbf{v}^T \mathbf{u}.$$

Two vectors are *orthogonal* if their dot product is zero.

A.4 Matrix Inverses

The *inverse* of a square matrix $\mathbf{A}$, denoted $\mathbf{A}^{-1}$, is a matrix such that $\mathbf{A}^{-1}\mathbf{A} = \mathbf{I} = \mathbf{A}\mathbf{A}^{-1}$. A matrix that has an inverse is called *invertible* or *nonsingular*. The inverse of a 2×2 matrix is

$$\mathbf{A}^{-1} = \frac{1}{a_{11}a_{22} - a_{12}a_{21}} \begin{bmatrix} a_{22} & -a_{12} \\ -a_{21} & a_{11} \end{bmatrix}.$$

An example of a matrix that does not have an inverse is

$$\begin{bmatrix} 1 & 2 \\ 1 & 2 \end{bmatrix}.$$

If we try to use the formula for the inverse earlier, it results in division by 0. The following properties apply to invertible matrices (of the same size):

$$\mathbf{I}^{-1} = \mathbf{I},$$
$$(\mathbf{A}^{-1})^{-1} = \mathbf{I},$$
$$(\mathbf{AB})^{-1} = \mathbf{B}^{-1}\mathbf{A}^{-1}.$$

We also write $\mathbf{A}^{-T} = (\mathbf{A}^{-1})^T$.

The *determinant* of a 2×2 matrix is the expression in the denominator earlier,

$$\det(\mathbf{A}) = a_{11}a_{22} - a_{12}a_{21}.$$

Determinants of larger matrices can be computed recursively from determinants of their submatrices called minors. For an $n \times n$ matrix $\mathbf{A}$, the ijth *minor*, denoted M_{ij}, is the determinant of the $(n-1) \times (n-1)$ submatrix of $\mathbf{A}$ formed by removing row i and column j. It is also called a *first minor* because only one row and column were removed. Then, for any row i of $\mathbf{A}$,

$$\det(\mathbf{A}) = \sum_{j=1}^{n} (-1)^{i+j} a_{ij} M_{ij}.$$

This is a recursive formula because it expresses the determinant of an $n \times n$ matrix in terms of determinants of $(n-1) \times (n-1)$ matrices.

A square matrix $\mathbf{A}$ is *diagonal* if $a_{ij} = 0$ for $i \neq j$. Thus, a diagonal matrix has the form

$$\mathbf{D} = \begin{bmatrix} d_1 & 0 & 0 & \cdots \\ 0 & d_2 & 0 & \cdots \\ 0 & 0 & \ddots & \\ \vdots & \vdots & & d_n \end{bmatrix},$$

also denoted by $\mathbf{D} = \text{diag}(d_1, \ldots, d_n)$. The inverse of a diagonal matrix is $\mathbf{D}^{-1} = \text{diag}(\frac{1}{d_1}, \ldots, \frac{1}{d_n})$, assuming it exists.

A.5 Systems of Linear Equations

A system of linear equations, such as

$$3x_1 + 5x_2 = 4,$$
$$2x_1 + 4x_2 = 7,$$

can be written in matrix form as

$$\begin{bmatrix} 3 & 5 \\ 2 & 4 \end{bmatrix} \begin{bmatrix} x_1 \\ x_2 \end{bmatrix} = \begin{bmatrix} 4 \\ 7 \end{bmatrix}.$$

Thus, a system of m linear equations in the variables $x_1, \ldots, x_n$ can be written as

$$\mathbf{A}\mathbf{x} = \mathbf{b},$$

where $\mathbf{A}$ is an $m \times n$ matrix containing the coefficients, $\mathbf{b}$ is a column vector containing the m right-hand sides, and $\mathbf{x} = (x_1, \ldots, x_n)$. Each row of $\mathbf{A}$ corresponds to an equation. For example, $\mathbf{A}_1 \mathbf{x} = b_1$ is the equation for row 1. If $\mathbf{A}$ is square ($m = n$) and invertible, then the unique solution is

$$\mathbf{x} = \mathbf{A}^{-1}\mathbf{b}.$$

The solution can also be found by apply a sequence of elementary row operations to the *augmented matrix* $[\mathbf{A}|\mathbf{b}]$, converting it to the form $[\mathbf{I}|\mathbf{A}^{-1}\mathbf{b}]$. This procedure, known as row reduction, can also be used to determine if the system of equations has no solution or to describe all solutions if it has multiple solutions.

There are three types of *elementary row operations* on a matrix $\mathbf{A}$:

1. Multiply a row of $\mathbf{A}$ by a nonzero constant.
2. Interchange two rows of $\mathbf{A}$.
3. Add a constant multiple of one row of $\mathbf{A}$ to another row.

For the matrix

$$C = \begin{bmatrix} 2 & 2 & 4 & 2 \\ 3 & 0 & 2 & 1 \\ 1 & 2 & 3 & 0 \end{bmatrix},$$

multiplying row 1 by $\frac{1}{2}$ results in

$$\begin{bmatrix} 1 & 1 & 2 & 1 \\ 3 & 0 & 2 & 1 \\ 1 & 2 & 3 & 0 \end{bmatrix}.$$

If we add -3 times row 1 to row 2 (and replace row 2 with this), we get

$$\begin{bmatrix} 1 & 1 & 2 & 1 \\ 0 & -3 & -4 & -2 \\ 1 & 2 & 3 & 0 \end{bmatrix}.$$

Further row operations produce the matrix

$$\begin{bmatrix} 1 & 0 & 0 & -3 \\ 0 & 1 & 0 & -6 \\ 0 & 0 & 1 & 5 \end{bmatrix}.$$

If we interpret C as the augmented matrix for the system of linear equations

$$2x_1 + 2x_2 + 4x_3 = 2,$$
$$3x_1 + 0x_2 + 2x_3 = 1,$$
$$x_1 + 2x_2 + 3x_3 = 0,$$

then the final matrix tells us that the solution is $(x_1, x_2, x_3) = (-3, -6, 5)$.

Elementary row operations can also be performed by premultiplying by an invertible matrix. For the first row operation earlier,

$$\begin{bmatrix} \frac{1}{2} & 0 & 0 \\ 0 & 1 & 0 \\ 0 & 0 & 1 \end{bmatrix} C = \begin{bmatrix} 1 & 1 & 2 & 1 \\ 3 & 0 & 2 & 1 \\ 1 & 2 & 3 & 0 \end{bmatrix}$$

and for the second

$$\begin{bmatrix} 1 & 0 & 0 \\ -3 & 1 & 0 \\ 0 & 0 & 1 \end{bmatrix} \begin{bmatrix} 1 & 1 & 2 & 1 \\ 3 & 0 & 2 & 1 \\ 1 & 2 & 3 & 0 \end{bmatrix} = \begin{bmatrix} 1 & 1 & 2 & 1 \\ 0 & -3 & -4 & -2 \\ 1 & 2 & 3 & 0 \end{bmatrix}.$$

To perform all the row operations that reduce $\mathbf{C} = [\mathbf{A}|\mathbf{b}]$ to $[\mathbf{I}|\mathbf{A}^{-1}\mathbf{b}]$, we can pre-multiply by the matrix that corresponds to each row operation, in order. Since

$$\mathbf{A}^{-1}[\mathbf{A}|\mathbf{b}] = [\mathbf{I}|\mathbf{A}^{-1}\mathbf{b}],$$

the product of these elementary row operation matrices must be $\mathbf{A}^{-1}$.

A.6 Linear Independence and Rank

We say that $\mathbb{R}^n$ has dimension n. The solution space to a system of linear equations also has a dimension. The following definitions make these precise.

A *linear combination* of the vectors $\mathbf{v}_1, \ldots, \mathbf{v}_k$ is a vector of the form $a_1\mathbf{v}_1 + \cdots + a_k\mathbf{v}_k$ for some scalars $a_1, \ldots, a_k$. A set of vectors $\mathbf{v}_1, \ldots, \mathbf{v}_k$ is *linearly dependent* if there are scalars $a_1, \ldots, a_k$, not all zero, such that $a_1\mathbf{v}_1 + \cdots + a_k\mathbf{v}_k = 0$. If no such linear combination exists, they are *linearly independent*. The set of all linear combinations is the set *spanned* by the vectors. This set is a *subspace* of $\mathbb{R}^n$. Every subspace is the set spanned by some set of vectors; if the set of vectors is linearly independent and spans the subspace, it is a *basis* for the subspace. The dimension of the subspace is the maximum number of linearly independent vectors in it, which is the number of vectors in a basis.

Now consider the solution set $S = \{\mathbf{x} \in \mathbb{R}^n : \mathbf{A}\mathbf{x} = \mathbf{b}\}$ of a system of linear equations, where $\mathbf{A}$ is $m \times n$, so that there are m equations in n unknowns. Assume the system has a solution ($S \neq \emptyset$). Then S is closely related to the null space, or kernel,

$$\text{Null}(\mathbf{A}) = \{\mathbf{x} \in \mathbb{R}^n : \mathbf{A}\mathbf{x} = 0\}.$$

Given any $\mathbf{x}_0 \in S$, shifting $\text{Null}(\mathbf{A})$ by $\mathbf{x}_0$ gives S:

$$S = \text{Null}(\mathbf{A}) + \mathbf{x}_0 = \{\mathbf{x}_0 + \mathbf{x} : \mathbf{x} \in \text{Null}(\mathbf{A})\}.$$

The rank of a matrix, $\text{rank}(\mathbf{A})$, is the maximum number of linearly independent columns; it also equals the maximum number of linearly independent rows. Since there are n columns and m rows, $\text{rank}(\mathbf{A}) \leq \min\{m, n\}$. A matrix has full rank if $\text{rank}(\mathbf{A}) = \min\{m, n\}$. Dimensions are related by

$$\dim(\text{Null}(\mathbf{A})) = n - \text{rank}(\mathbf{A}).$$

Using row rank, this equation says that the dimension of the null space (and of the solution space S) is the number of variables minus the number of linearly independent equations. For example, if $\mathbf{A}$ has one row, the solution space of that equation is a hyperplane and has dimension $n - 1$; if $n = 3$, the solution space is a plane with dimension 2. If $\mathbf{A}$ is invertible, then $\dim(\text{Null}(\mathbf{A})) = n - n = 0$. If $m < n$ and $\mathbf{A}$ has full rank m, then $\dim(\text{Null}(\mathbf{A})) = n - m$.

A.7 Quadratic Forms and Eigenvalues

The quadratic expression $x_1^2 + 3x_1x_2 - 4x_2^2$ can be written in matrix form as $\frac{1}{2}\mathbf{x}^T\mathbf{A}\mathbf{x}$, where $\mathbf{x} = (x_1, x_2)$ and

$$\mathbf{A} = \begin{bmatrix} 2 & 3 \\ 3 & -8 \end{bmatrix}.$$

Because $\mathbf{A}$ is symmetric, the off-diagonal elements each appear twice ($a_{12} = a_{21} = 3$ in this example). Multiplying by $\frac{1}{2}$ makes these entries equal to the coefficient of the corresponding term in the quadratic ($3x_1x_2$ in this example). Generalizing, any pure quadratic function of n variables can be written as $\frac{1}{2}\mathbf{x}^T\mathbf{A}\mathbf{x}$, where $\mathbf{A}$ is an $n \times n$ symmetric matrix. We call $\mathbf{x}^T\mathbf{A}\mathbf{x}$ a *quadratic form*. A separable quadratic function $a_1x_1^2 + a_2x_2^2 + \cdots + a_nx_n^2$ can be written using a diagonal matrix $\mathbf{D} = \text{diag}(a_1, \ldots, a_n)$ as $\mathbf{x}^T\mathbf{D}\mathbf{x}$.

The function $f(x) = ax^2$ is convex if $a \geq 0$. Similarly, the function $f(\mathbf{x}) = a_1x_1^2 + a_2x_2^2 + \cdots + a_nx_n^2$ is convex if $a_i \geq 0$ for $i = 1, \ldots, n$. Notice that this is equivalent to $\mathbf{x}^T\mathbf{D}\mathbf{x} \geq 0$ for all $\mathbf{x}$. Extending this property to other matrices, a symmetric matrix $\mathbf{A}$ is *positive definite* if $\mathbf{x}^T\mathbf{A}\mathbf{x} > 0$ for all nonzero vectors $\mathbf{x}$. It is *positive semidefinite* if $\mathbf{x}^T\mathbf{A}\mathbf{x} \geq 0$ for all vectors $\mathbf{x}$. For a symmetric matrix $\mathbf{A}$, the function $\mathbf{x}^T\mathbf{A}\mathbf{x}$ is convex if and only if $\mathbf{A}$ is positive semidefinite.

Concave quadratic functions can be described similarly. A symmetric matrix $\mathbf{A}$ is *negative semidefinite* if $\mathbf{x}^T\mathbf{A}\mathbf{x} \leq 0$ for all vectors $\mathbf{x}$. For a symmetric matrix $\mathbf{A}$, the function $\mathbf{x}^T\mathbf{A}\mathbf{x}$ is concave if and only if $\mathbf{A}$ is negative semidefinite.

Checking whether a matrix is positive semidefinite is requires significant computation and is related to its eigenvalues. For a square matrix $\mathbf{A}$, a scalar λ and a nonzero vector $\mathbf{x}$ satisfying $\mathbf{A}\mathbf{x} = \lambda\mathbf{x}$ are called an *eigenvalue* and *eigenvector* of $\mathbf{A}$. The eigenvalues can be found by solving the equation $\det(\mathbf{A} - \lambda\mathbf{I}) = 0$. Expanding this equation gives an nth-order polynomial that can have up to n distinct roots λ that are the eigenvalues. Now assume that $\mathbf{A}$ is symmetric. Then

(i) The eigenvalues of $\mathbf{A}$ are real.
(ii) There is a set of n orthogonal eigenvectors of $\mathbf{A}$. These form a basis for $\mathbb{R}^n$.

Let $\mathbf{u}_1, \ldots, \mathbf{u}_n$ be such a basis, which is also normalized, so that $\|\mathbf{u}_i\| = 1$ for $i = 1, \ldots, n$, and $\mathbf{Q} = [\mathbf{u}_1 | \mathbf{u}_2 | \cdots | \mathbf{u}_n]$. Since $\mathbf{u}_i^T\mathbf{u}_j = 0$, $i \neq j$, we have $\mathbf{Q}^T\mathbf{Q} = \mathbf{I}$, which makes $\mathbf{Q}^T = \mathbf{Q}^{-1}$. Such a matrix is called *orthogonal*. Let λ_i be the eigenvalue associated with $\mathbf{u}_i$, so that $\mathbf{A}\mathbf{u}_i = \lambda_i\mathbf{u}_i$. Then

$$\mathbf{Q}^{-1}\mathbf{A}\mathbf{Q} = \mathbf{Q}^T\mathbf{A}\mathbf{Q} = \text{diag}(\lambda_1, \ldots, \lambda_n).$$

Now, if we let $\mathbf{y} = \mathbf{Q}^T\mathbf{x}$, then $\mathbf{x} = \mathbf{Q}\mathbf{y}$ and

$$\mathbf{x}^T\mathbf{A}\mathbf{x} = \mathbf{y}^T(\mathbf{Q}^T\mathbf{A}\mathbf{Q})\mathbf{y} = \sum_{i=1}^{n} \lambda_i y_i^2. \tag{A.1}$$

For **A** to be positive semidefinite, (A.1) must be nonnegative for all **x**, or equivalently, for all **y**. Thus, a symmetric matrix **A** is positive semidefinite if and only if all eigenvalues of **A** are nonnegative.

A 2×2 symmetric matrix is positive semidefinite if and only if

$$\det(\mathbf{A}) = a_{11}a_{22} - a_{12}^2 \geq 0 \text{ and } a_{11} \geq 0 \tag{A.2}$$

and negative semidefinite if and only if

$$\det(\mathbf{A}) = a_{11}a_{22} - a_{12}^2 \geq 0 \text{ and } a_{11} \leq 0.$$

A *leading principal minor* of a square $n \times n$ matrix is the determinant of a submatrix made up of its first m rows and columns, $m < n$. A symmetric matrix is positive semidefinite if and only if its determinant and all of its leading principal minors are nonnegative. These minors have sizes from $(n - 1) \times (n - 1)$ to 1×1. The matrix **A** is negative semidefinite if $-\mathbf{A}$ is positive semidefinite. In a symmetric negative definite matrix, the leading principal minors alternate in sign, with the 1×1 minor a_{11} negative, the 2×2 minor positive, etc.

For example, the matrix

$$\mathbf{A} = \begin{bmatrix} 8 & -5 & 0 \\ -5 & 4 & 0 \\ 0 & 0 & 3 \end{bmatrix}$$

has

$$\det(8) = 8, \quad \det \begin{pmatrix} 8 & -5 \\ -5 & 4 \end{pmatrix} = 7, \quad \det \begin{pmatrix} 8 & -5 & 0 \\ -5 & 4 & 0 \\ 0 & 0 & 3 \end{pmatrix} = 21.$$

Because these are all positive, it is positive definite. For $-\mathbf{A}$,

$$\det(-8) = -8, \quad \det \begin{pmatrix} -8 & 5 \\ 5 & -4 \end{pmatrix} = 7, \quad \det \begin{pmatrix} -8 & 5 & 0 \\ 5 & -4 & 0 \\ 0 & 0 & -3 \end{pmatrix} = -21.$$

The alternating signs indicate that $-\mathbf{A}$ is negative definite.

A.8 Derivatives and Convexity

A function f of one variable has derivative $f'(x)$ and second derivative $f''(x)$. For a function $f(x_1, \ldots, x_n)$ of n variables, the derivative with respect to x_i, with all other variables held constant, is called the partial derivative $\frac{\partial f(\mathbf{x})}{\partial x_i}$. The gradient of f is the vector

$$\nabla f(\mathbf{x}) = \left(\frac{\partial f(\mathbf{x})}{\partial x_1}, \frac{\partial f(\mathbf{x})}{\partial x_2}, \dots, \frac{\partial f(\mathbf{x})}{\partial x_n} \right).$$

If f is differentiable at $\mathbf{x}$, then the gradient exists and is the direction of maximum increase of f at $\mathbf{x}$.

Taking the partial derivative first with respect to x_j then with respect to x_i gives the second order partial derivative $\frac{\partial^2 f(\mathbf{x})}{\partial x_i \, \partial x_j}$. The Hessian of f is the $n \times n$ matrix

$$\mathbf{H}(\mathbf{x}) = \left[\frac{\partial^2 f(\mathbf{x})}{\partial x_i \partial x_j} \right].$$

As long as these second partial derivatives are continuous, the order in which we take the derivatives does not matter, i.e.

$$\frac{\partial^2 f(\mathbf{x})}{\partial x_i \partial x_j} = \frac{\partial^2 f(\mathbf{x})}{\partial x_j \partial x_i},$$

and the Hessian is symmetric.

One important use of the Hessian is checking convexity of f. Convexity of a function is defined in Section 6.2. A differentiable function of one variable is convex if and only if $f''(x) \leq 0$ for all x. A differentiable function $f(x_1, x_2)$ is convex if and only if

$$\frac{\partial^2 f}{\partial x_1^2} \frac{\partial^2 f}{\partial x_2^2} - \left(\frac{\partial^2 f}{\partial x_1 \partial x_2} \right)^2 \geq 0 \text{ and } \frac{\partial^2 f}{\partial x_1^2} \geq 0 \tag{A.3}$$

for all (x_1, x_2) and concave if and only if

$$\frac{\partial^2 f}{\partial x_1^2} \frac{\partial^2 f}{\partial x_2^2} - \left(\frac{\partial^2 f}{\partial x_1 \partial x_2} \right)^2 \geq 0 \text{ and } \frac{\partial^2 f}{\partial x_1^2} \leq 0$$

for all (x_1, x_2). Comparing (A.2) and (A.3), we see that a differentiable function of two variables is convex if and only if the determinant of the Hessian and its leading minor are nonnegative for all $\mathbf{x}$, or equivalently, that the Hessian is positive semidefinite. This test extends to any number of variables: a differentiable function is convex if and only if its Hessian is positive semidefinite for all $\mathbf{x}$ and concave if and only if its Hessian is negative semidefinite for all $\mathbf{x}$.

We can also apply this test for convexity to the quadratic form $\mathbf{x}^T \mathbf{A} \mathbf{x}$ by setting $f(x_1, x_2) = \frac{1}{2} \mathbf{x}^T \mathbf{A} \mathbf{x}$. Then its second partial derivatives are

$$\frac{\partial^2 f}{\partial x_i \partial x_j} = a_{ij}.$$

Thus, $\mathbf{A}$ is the Hessian of f, so f is convex (concave) if and only if $\mathbf{A}$ is positive (negative) semidefinite.

Bibliography

O. Aarvik and P. Randolph. The application of linear programming to the determination of transmission fees in an electrical power network. *Interfaces*, 6 (1): 47–49, 1975.

R.K. Ahuja, T.L. Magnanti, and J.B. Orlin. *Network Flows*. Prentice Hall, Englewood Cliffs, NJ, 1993.

S.C. Albright and W.L. Winston. *Spreadsheet Modeling and Applications: Essentials of Practical Management Science*. South-Western Pub, Mason, OH, 2005.

E.D. Andersen and K.D. Andersen. The MOSEK interior point optimizer for linear programming: an implementation of the homogeneous algorithm. In H. Frenk, K. Roos, T. Terlaky, S. Zhang, editors, *High Performance Optimization*, pages 197–232. Springer, New York, 2000.

J.M. Andrews, V.F. Farias, A.I. Khojandi, and C.M. Yan. Primal–dual algorithms for order fulfillment at Urban Outfitters, Inc. *INFORMS Journal on Applied Analytics*, 49 (5): 355–370, 2019.

D.L. Applegate, R.E. Bixby, V. Chvátal, and W.J. Cook. *The Traveling Salesman Problem: A Computational Study*. Princeton University Press, Princeton, NJ, 2006.

F. Babonneau, G. Corcos, L. Drouet, and J.P. Vial. NeatWork: a tool for the design of gravity-driven water distribution systems for poor rural communities. *INFORMS Journal on Applied Analytics*, 49(2). 129–136, 2019.

C. Barnhart, P. Belobaba, and A.R. Odoni. Applications of operations research in the air transport industry. *Transportation Science*, 37 (4): 368–391, 2003.

E.M.L. Beale. Cycling in the dual simplex algorithm. *Naval Research Logistics Quarterly*, 2 (4): 269–275, 1955.

J.C. Bean, C.E. Noon, S.M. Ryan, and G.J. Salton. Selecting tenants in a shopping mall. *Interfaces*, 18 (2): 1–9, 1988.

J.E. Beasley, M. Krishnamoorthy, Y.M. Sharaiha, and D. Abramson. Scheduling aircraft landings-the static case. *Transportation Science*, 34 (2): 180–197, 2000.

R. Bellman. On a routing problem. *Quarterly of Applied Mathematics*, 16 (1): 87–90, 1958.

Linear and Convex Optimization: A Mathematical Approach, First Edition. Michael H. Veatch.
© 2021 John Wiley & Sons, Inc. Published 2021 by John Wiley & Sons, Inc.
Companion website: www.wiley.com/go/veatch/convexandlinearoptimization

R.E. Bellman and S.E. Dreyfus. *Applied Dynamic Programming*. Princeton University Press, Princeton, NJ, 2015.

A. Ben-Tal, L. El Ghaoui, and A. Nemirovski. *Robust Optimization*. Princeton University Press, Princeton, NJ, 2009.

D.P. Bertsekas. *Nonlinear Programming*. Athena Scientific, Belmont, CA, 2nd edition, 1999.

D.P. Bertsekas. *Dynamic Programming and Optimal Control*, Volume 2. Athena Scientific, Belmont, CA, 2007.

D.P. Bertsekas and J.N. Tsitsiklis. *Parallel and Distributed Computation: Numerical Methods*. Athena Scientific, Belmont, CA, 2015.

D. Bertsimas and J.N. Tsitsiklis. *Introduction to Linear Optimization*. Athena Scientific, Belmont, CA, 1997.

D. Bertsimas and R. Weismantel. *Optimization Over Integers*. Dynamic Ideas, Belmont, CA, 2005.

S. Bhulai, G. Koole, and A. Pot. Simple methods for shift scheduling in multiskill call centers. *Manufacturing & Service Operations Management*, 10 (3): 411–420, 2008.

R.E. Bixby. Implementing the simplex method: the initial basis. *ORSA Journal on Computing*, 4 (3): 267–284, 1992.

R.E. Bixby. A brief history of linear and mixed-integer programming computation. *Documenta Mathematica*, Extra Volume: ISMP: 107–121, 2012.

R.G. Bland. New finite pivoting rules for the simplex method. *Mathematics of Operations Research*, 2 (2): 103–107, 1977.

S. Boyd and L. Vandenberghe. *Convex Optimization*. Cambridge University Press, New York, 2004.

J.A. Carbajal, P. Williams, A. Popescu, and W. Chaar. Turner blazes a trail for audience targeting on television with operations research and advanced analytics. *INFORMS Journal on Applied Analytics*, 49 (1): 64–89, 2019.

Y. Chen, P. Berkhin, B. Anderson, and N.R. Devanur. Real-time bidding algorithms for performance-based display ad allocation. In *Proceedings of the 17th ACM SIGKDD International Conference on Knowledge Discovery and Data Mining*, pages 1307–1315. ACM, 2011.

D.M. Chickering and D. Heckerman. Targeted advertising on the web with inventory management. *Interfaces*, 33 (5): 71–77, 2003.

V. Chvátal. Edmonds polytopes and a hierarchy of combinatorial problems. *Discrete Mathematics*, 4 (4): 305–337, 1973.

V. Chvátal. *Linear Programming*. Macmillan, New York, 1983.

F.E. Clark. Remark on the constraint sets in linear programming. *The American Mathematical Monthly*, 68 (4): 351–352, 1961.

M.C. Cohen and G. Perakis. Optimizing promotions for multiple items in supermarkets. In S. Ray, S. Yin, editors, *Channel Strategies and Marketing Mix in a Connected World*, pages 71–97. Springer, 2020.

M.C. Cohen, N.H.Z. Leung, K. Panchamgam et al. The impact of linear optimization on promotion planning. *Operations Research*, 65 (2): 446–468, 2017.

G. Cornuéjols. Revival of the Gomory cuts in the 1990's. *Annals of Operations Research*, 149 (1): 63–66, 2007.

G.B. Dantzig. Maximization of a linear function of variables subject to linear inequalities. In T.C. Koopmans, editor, *Activity Analysis of Production and Allocation*, pages 339–347. Wiley, New York, 1951.

G.B. Dantzig. *Linear Programming and Extensions*. Princeton University Press, Princeton, NJ, 1963.

G.B. Dantzig. The diet problem. *Interfaces*, 20 (4): 43–47, 1990.

D.P. De Farias and B. Van Roy. The linear programming approach to approximate dynamic programming. *Operations Research*, 51 (6): 850–865, 2003.

D.P. De Farias and B. Van Roy. On constraint sampling in the linear programming approach to approximate dynamic programming. *Mathematics of Operations Research*, 29 (3): 462–478, 2004.

D. Den Hertog. *Interior Point Approach to Linear, Quadratic and Convex Programming: Algorithms and Complexity*. Springer Science & Business Media, Berlin, 2012.

E.W. Dijkstra. A note on two problems in connexion with graphs. *Numerische Mathematik*, 1 (1): 269–271, 1959.

L.R. Ford Jr. Network flow theory. Technical report P-923, Rand Corp, Santa Monica, CA, 1956.

J.J. Forrest and D. Goldfarb. Steepest-edge simplex algorithms for linear programming. *Mathematical Programming*, 57 (1–3): 341–374, 1992.

D. Freund, S.G. Henderson, E. O'Mahony, and D.B. Shmoys. Analytics and bikes: riding tandem with Motivate to improve mobility. *INFORMS Journal on Applied Analytics*, 49 (5): 310–323, 2019.

D. Goldfarb and J.K. Reid. A practicable steepest-edge simplex algorithm. *Mathematical Programming*, 12 (1): 361–371, 1977.

R.E. Gomory. Outline of an algorithm for integer solutions to linear programs. *Bulletin of the American Mathematical Society*, 64 (5): 275–278, 2010.

P.M.J. Harris. Pivot selection methods of the devex LP code. *Mathematical Programming*, 5 (1): 1–28, 1973.

M. Held and R.M. Karp. The traveling-salesman problem and minimum spanning trees. *Operations Research*, 18 (6): 1138–1162, 1970.

M. Held and R.M. Karp. The traveling-salesman problem and minimum spanning trees: Part II. *Mathematical Programming*, 1 (1): 6–25, 1971.

A.J. Hoffman and J.B. Kruskal. Integral boundary points of convex polyhedra. In H.W. Kuhn and A.W. Tucker, editors, *Linear Inequalities and Related Systems*, pages 223–246. Princeton University Press, Princeton, NJ, 1956.

M.P. Johnson. Community-based operations research: introduction, theory, and applications. In M.P. Johnson, editor, *Community-Based Operations Research*, pages 3–36. Springer, New York, 2012.

E.D. Kim and F. Santos. An update on the Hirsch conjecture. *Jahresbericht der Deutschen Mathematiker-Vereinigung*, 112 (2): 73–98, 2010.

V. Klee and G.J. Minty. How good is the simplex algorithm. *Inequalities*, 3 (3): 159–175, 1972.

M. Kojima, N. Megiddo, and S. Mizuno. A primal-dual infeasible-interior-point algorithm for linear programming. *Mathematical Programming*, 61 (1–3): 263–280, 1993.

H.W Kuhn. Nonlinear programming: a historical view. In R.W. Cottle and C.E. Lemke, editors, *Nonlinear Programming, SIAM-AMS Proceedings*, Volume IX, pages 1–26. American Mathematical Society, Providence, RI, 1976.

H.W. Kuhn and A.W. Tucker. Nonlinear programming. In J. Neyman, editor, *Proceedings of the Second Berkeley Symposium on Mathematical Statistics and Probability*, pages 481–492. University of California Press, Berkeley, CA, 1951.

L.M. Lancaster. The evolution of the diet model in managing food systems. *Interfaces*, 22 (5): 59–68, 1992.

E.L. Lawler, J.K. Lenstra, A.H.G. Rinnooy Kan, and D.B. Shmoys. *The Traveling Salesman Problem: A Guided Tour of Combinatorial Optimization*. Wiley, New York, 1985.

D.G. Luenberger and Y. Ye. *Linear and Nonlinear Programming*. Springer, Reading, PA, 2016.

I.J. Lustig, R.E. Marsten, and D.F. Shanno. Interior point methods for linear programming: computational state of the art. *ORSA Journal on Computing*, 6 (1): 1–14, 1994.

H. Marchand and L.A. Wolsey. Aggregation and mixed integer rounding to solve MIPs. *Operations Research*, 49 (3): 363–371, 2001.

C.H. Martin, D.C. Dent, and J.C. Eckhart. Integrated production, distribution, and inventory planning at Libbey-Owens-Ford. *Interfaces*, 23 (3): 68–78, 1993.

C.E. Miller, A.W. Tucker, and R.A. Zemlin. Integer programming formulation of traveling salesman problems. *Journal of the ACM*, 7 (4): 326–329, 1960.

J. Moline, J. Goentzel, and E. Gralla. Approaches for locating and staffing FEMA's disaster recovery centers. *Decision Sciences*, 50 (5): 917–947, 2019.

G.L. Nemhauser and L.A. Wolsey. *Integer and Combinatorial Optimization*. Wiley, New York, 1988.

C.H. Papadimitriou and K. Steiglitz. *Combinatorial Optimization*. Prentice Hall, Englewood Cliffs, NJ, 1982.

S.M. Pollock and M.D. Maltz. Operations research in the public sector: an introduction and a brief history. *Operations Research and the Public Sector*, 6: 1–22, 1994.

M.L. Puterman. *Markov Decision Processes: Discrete Stochastic Dynamic Programming*. Wiley, Hoboken, NJ, 2014.

D.J. Rader. *Deterministic Operations Research: Models and Methods in Linear Optimization*. Wiley, New York, 2010.

W.A. Robbins, and N. Tuntiwongpiboom. Linear programming a useful tool in case-mix management. *Healthcare Financial Management: Journal of the Healthcare Financial Management Association*, 40(6). 114, 1986.

A. Schrijver. *Theory of Linear and Integer Programming*. Wiley, New York, 1986.

P.J. Schweitzer and A. Seidmann. Generalized polynomial approximations in Markovian decision processes. *Journal of Mathematical Analysis and Applications*, 110 (2): 568–582, 1985.

M. Udell and O. Toole. Optimal design of efficient rooftop photovoltaic arrays. *INFORMS Journal on Applied Analytics*, 49 (4): 281–294, 2019.

M.H. Veatch. Approximate linear programming for average cost MDPs. *Mathematics of Operations Research*, 38 (3): 535–544, 2013.

L.A. Wolsey and G.L. Nemhauser. *Integer and Combinatorial Optimization*. Wiley, Hoboken, NJ, 2014.

Y. Ye, M.J. Todd, and S. Mizuno. An O(SQRT nL)-iteration homogeneous and self-dual linear programming algorithm. *Mathematics of Operations Research*, 19 (1): 53–67, 1994.

Index

Linear and Convex Optimization: A Mathematical Approach, First Edition. Michael H. Veatch.
© 2021 John Wiley & Sons, Inc. Published 2021 by John Wiley & Sons, Inc.
Companion website: www.wiley.com/go/veatch/convexandlinearoptimization